G-PROTEINS AS MEDIATORS OF CELLULAR SIGNALLING PROCESSES

WILEY SERIES ON MOLECULAR PHARMACOLOGY OF CELL REGULATION

Volume 1
G-Proteins as Mediators of Cellular Signalling Processes
  Edited by Miles D. Houslay and Graeme Milligan

Volume 2
Isoenzymes of Cyclic Nucleotide Phosphodiesterases
  Edited by Joseph Beavo and Miles D. Houslay

Further volumes in preparation

Cover illustration by kind permission of Y.-K. Ho and V. N. Hingorani

G-PROTEINS AS MEDIATORS OF CELLULAR SIGNALLING PROCESSES

Edited by
Miles D. Houslay and **Graeme Milligan**
Institute of Biochemistry, University of Glasgow, UK

A Wiley–Interscience Publication

WILEY

JOHN WILEY & SONS
Chichester • New York • Brisbane • Toronto • Singapore

Copyright © 1990 by John Wiley & Sons Ltd,
Baffins Lane, Chichester,
West Sussex PO19 1UD, England

All rights reserved.

No part of this book may be reproduced by any means, or transmitted, or translated into a machine language without the written permission of the publisher.

Other Wiley Editorial Offices

John Wiley & Sons, Inc., 605 Third Avenue,
New York, NY 10158-0012, USA

Jacaranda Wiley Ltd, G.P.O. Box 859, Brisbane,
Queensland 4001, Australia

John Wiley & Sons (Canada) Ltd, 22 Worcester Road,
Rexdale, Ontario M9W 1L1, Canada

John Wiley & Sons (SEA) Pte Ltd, 37 Jalan Pemimpin 05-04,
Block B, Union Industrial Building, Singapore 2057

Library of Congress Cataloging-in-Publication Data:

G-proteins as mediators of cellular signalling processes / edited by
  Miles D. Houslay and Graeme Milligan.
    p. cm.—(Wiley series on molecular pharmacology ; v. 1)
  'A Wiley-Interscience publication.'
  Includes bibliographical references.
  ISBN 0-471-92338-9
    1. G proteins—Physiological effect. 2. Cell interaction.
  I. Houslay, Miles D. II. Milligan, Graeme. III. Series.
  QP552.G16G23 1990   89-38932
  612'.0157'5—dc20   CIP

British Library Cataloguing in Publication Data:

G-proteins as mediators of cellular signalling processes.
  (Wiley series on molecular pharmacology, V.1)
  1. Organisms. Glycoproteins
  I. Houslay, Miles D. II. Milligan, Graeme
  574.1'9245

  ISBN 0-471-92338-9

Typeset by APS, Salisbury, Wiltshire
Printed in Great Britain by Biddles Ltd, Guildford, Surrey

CONTRIBUTORS

Richard C. Bruch, *Department of Neurobiology and Physiology, Northwestern University, Evanston*

Annette C. Dolphin, *Department of Pharmacology, St George's Hospital Medical School, London*

Peter Gierschik, *Pharmakologisches Institute der Universität Heidelberg, Heidelberg*

Goeffrey H. Gold, *Monell Chemical Senses Center, Philadelphia*

A. Hall, *Institute of Cancer Research, Royal Cancer Hospital, Chester Beatty Laboratories, London*

Vijay N. Hingorani, *Department of Biological Chemistry, University of Illinois at Chicago, Health Sciences Center, Chicago*

Yee-Kin Ho, *Department of Biological Chemistry, University of Illinois at Chicago, Health Sciences Center, Chicago*

Miles D. Houslay, *Molecular Pharmacology Group, Institute of Biochemistry, University of Glasgow, Glasgow*

Karl H. Jakobs, *Pharmakologisches Institute der Universität Heidelberg, Heidelberg*

Yoshito Kaziro, *The Institute of Medical Science, University of Tokyo, Minato-ku, Tokyo*

Alex Levitzki, *Faculty of Science, Department of Biological Chemistry, The Hebrew University of Jerusalem, Jerusalem*

Irene Litosch, *Department of Pharmacology, University of Miami School of Medicine, Miami*

Graeme Milligan, *Molecular Pharmacology Group, Departments of Biochemistry and Pharmacology, University of Glasgow, Glasgow*

Allen M. Spiegel, *Molecular Pathophysiology Branch, National Institute of Diabetes, Digestive and Kidney Disease, National Institutes of Health, Bethesda*

CONTENTS

Preface	xi
Chapter 1: Dual Control of Adenylate Cyclase	1
Levitzki, A.	
Structure of hormonally regulated adenylate cyclase	1
Universality of the system	2
Stimulation of adenylate cyclase	3
The G_s–G_i subunit dissociation mechanism for hormonal activation and inhibition	7
Mechanism of G_i action	10
Chapter 2: Structure and Identification of G-Proteins: Isolation and Purification	15
Spiegel, A. M.	
Introduction and overview	15
G-protein purification and identification: General aspects	16
G-protein purification and identification: Specific aspects	17
Future prospects	27
Chapter 3: Immunological Probes and the Identification of Guanine Nucleotide Binding Proteins	31
Milligan, G.	
Introduction	31
Early studies with polyclonal antisera	33
Anti-peptide antisera	34
Quantitation of G-protein levels using anti-peptide antisera	36
Immunological assessments of the cellular location of G-proteins	37
Immunological examination of covalent modification and G-protein function	39
Antisera as probes for functional domains of G-proteins	40
Immunological examination of G-protein levels in clinical disorders	42
Conclusions	44
Chapter 4: Molecular Biology of G-proteins	47
Kaziro, Y.	
Introduction	47
Isolation of cDNA clones for G-protein α subunits from mammalian cells	47
Isolation of human Gα genes	51
Conservation of primary structure of each Gα among mammalian species	55

Structural model and mutational analysis 57
Beta and gamma subunits of mammalian G-proteins 59
G-proteins from *Saccharomyces cerevisiae* 61
Comparison of the amino acid sequences of yeast GP1α and GP2α with those of rat brain $G_i\alpha$ and $G_o\alpha$. 61

Chapter 5: Receptor-Stimulated GTPase Activity of G-Proteins ... 67
Gierschik, P., and Jakobs, K. H.
Introduction 67
GTP hydrolysis by purified G-proteins 68
Receptor-regulated GTP hydrolysis in membranes 70
Regulation of receptor-stimulated GTP hydrolysis by Mg^{2+} and guanine nucleotides 70
Influence of cholera and pertussis toxin on receptor-stimulated GTP hydrolysis 75
Relationship between agonist–receptor binding and agonist-stimulated GTP hydrolysis 78
Conclusions 81

Chapter 6: On the Specificity of Interactions Between Receptors, Effectors and Guanine Nucleotide Binding Proteins ... 83
Milligan, G.
Introduction 83
Historical perspective 84
The multiplicity of pertussis-toxin-sensitive G-proteins 86
Reconstitution studies as an assay for G-protein function 86
The specificity of receptor–G-protein–effector interactions 89

Chapter 7: Transducin: The Signal Transducing G-Protein of Photoreceptor Cells ... 95
Hingorani, V. N., and Ho, Y.-K.
Transducin and phototransduction 95
The coupling cycle of transducin 96
The reaction dynamics of the transducin cycle 97
Organization of the transducin subunits 98
The primary sequence and the tryptic peptide map of transducin subunits 100
Identification of the functional domains of Tα 101
A three-dimensional model of Tα 101
Mapping the GTP binding site of Tα 102
Proposed coupling mechanism of Tα 106
Comparison of Tα of rod and cone photoreceptor cells 107
Functional role of T$\beta\gamma$ 108
Perspectives 109

Chapter 8: G-Protein-Mediated Signalling in Olfaction 113
Bruch R. C., and Gold, G. H.
    Introduction   113
    G-protein identification in olfactory neurones   114
    G-protein-mediated signalling events   115
    Conclusions   122

Chapter 9: G-Proteins and the Regulation of Ion Channels 125
Dolphin, A. C.
    Introduction   125
    Modulation of ion channel function by cyclic AMP   126
    Modulation of ion channel function by protein-kinase-C-dependent phosphorylation   128
    Other second messenger effects on ion channels   131
    Direct interaction of ion channels with GTP binding proteins   132
    Direct interaction of second messengers with ion channels   145
    Conclusions   145

Chapter 10: Inositol Phosphate Metabolism and G-Proteins 151
Litosch, I.
    Introduction   151
    Characteristics of the phosphoinositide response   152
    Studies in cell-free systems   157
    Properties of purified phospholipase C   161
    G-protein involvement in the regulation of phospholipase C: Studies in reconstituted systems   165
    Conclusions   166

Chapter 11: Ras and Ras-Related Guanine Nucleotide Binding Proteins . . 173
Hall, A.
    Introduction   173
    Ras proteins   173
    Ras-related proteins   185
    Discussion   191

Chapter 12: Altered Expression and Functioning of Guanine Nucleotide Binding (Regulatory) Proteins During Growth, Transformation, Differentiation and in Pathological States 197
Houslay, M. D.
    Introduction   197
    Pseudohypoparathyroidism   206
    Chronic ethanol administration   207
    Growth-hormone-secreting pituitary adenomas   211
    S49 lymphoma cell line   212
    Adrenalectomy   214

Hypothyroidism 216
Diabetes and insulin-resistant states 216
Heart failure 223
Growth, differentiation and ageing 225
Conclusions 226

Index 231

PREFACE

Co-ordinating responses in multicellular organisms demands an efficient network for performing signal transduction between remote areas. This is often achieved using specific hormones/neurotransmitters/growth factors/cytokines. Such receptors can fall into two brand classes. The first of these generates a signal exhibiting ion-channel or tyrosyl kinase activity. The second, however, stimulates another plasma membrane protein to exert its actions. Receptors in this second class appear to achieve this effect by employing specific members of a growing family of guanine nucleotide binding proteins to act as 'go betweens' which couple the receptor to an appropriate 'signal generator'. It now appears that the guanine nucleotide binding proteins or, as they are more commonly known, G-proteins provide a pervasive family of proteins which can bind and hydrolyse GTP and play pivotal roles in normal signal transduction processes and in oncogenes. Studies on the identification, analysis and functioning of this rapidly expanding family have taken advantage of a spectrum of techniques which range from fundamental biochemical analysis to those involving cellular and molecular biology. In this volume a number of acknowledged experts in the field chart the progress of our understanding from the discovery of the importance of a GTP binding protein in the regulation of adenylate cyclase to the present day. We now find specific G-proteins regulating visual transduction, olfaction, the myriad of events surrounding cell growth control and ionic events at the cell membrane. Our increasing knowledge of the 3-D structure of 'single' G-proteins provides the basis for relating structure to function and this is emphasised in recent discoveries of changes in G-protein function in transformed states and the use of mutagenesis to probe functional domains in G-proteins.

1
DUAL CONTROL OF ADENYLATE CYCLASE

Alexander Levitzki

Department of Biological Chemistry, Hebrew University of Jerusalem, 91904 Jerusalem, Israel

STRUCTURE OF HORMONALLY REGULATED ADENYLATE CYCLASE

The hormonally regulated adenylate cyclase is a multicomponent system composed of five functional units (Figure 1.1):

(1) The stimulatory receptor R_s which binds the stimulatory ligand. We shall discuss in some detail the β-adrenoceptor which is a member of this superfamily of receptors [1,2]. This receptor, like all known receptors which interact with G-proteins, is a transmembrane glycoprotein with seven putative transmembrane spanning sequences. It is the only member of this large family of receptors which has been purified, reconstituted with other pure components of the adenylate cyclase system [3,4] and been cloned.

(2) The G_s protein which binds GTP and is tightly associated with the catalytic unit. G_s [5] is composed of three subunits: α_s, which possesses the GTP binding site and is the target for ADP-ribosylation by NAD^+ catalysed by cholera toxin; the β subunit and the γ subunit, which are tightly associated with each other ([6] for review). The heterotrimeric G_s protein is localized in the inner leaflet of the plasma membrane and is much less hydrophobic than either the β-adrenoreceptor or the catalytic unit of adenylate cyclase, C.

(3) The catalytic unit C is a hydrophobic protein with a single transmembrane spanning domain carrying its catalytic site at the cytoplasmic side of the membrane.

(4) An inhibitory receptor, R_i, homologous in structure to R_s [7–9].

(5) The inhibitory GTP binding protein G_i. As in G_s, this protein is a heterotrimer composed of three subunits: α_i, β and γ, where the $\beta\gamma$ complex is highly similar or identical to the one found in G_s ([6] for review).

The subunit α_i is homologous to α_s and is a substrate for ADP-ribosylation catalysed by the toxin of *Bordetella pertussis* (pertussis toxin) but not by cholera toxin [10]. The homology is, as expected, high in the sequences involved in the binding of GTP. At the

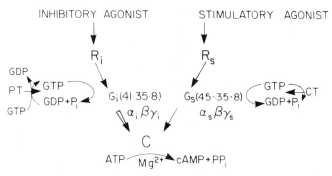

Figure 1.1 Dual regulation of adenylate cyclase. The catalyst C responds to the stimulatory G-protein G_s and to the inhibitory G-protein G_i. G_s is tightly associated with C whereas G_i is loosely bound to the system. It is not yet clear whether G_i interacts directly with C or not. The stimulatory signal is initiated by the stimulatory receptor R_s and the inhibitory signal by the inhibitory receptor R_i. Cholera-toxin-catalysed ADP-ribosylation of G_s inhibits the GTPase activity of G_s whereas pertussis-toxin-catalysed ADP-ribosylation of G_i inhibits GDP to GTP exchange on G_i. In both cases the transmission of the message to C is modified. Inhibition of the GTPase G_s locks the G_sC complex in its active state. Inhibition of the GDP to GTP exchange on G_i locks G_i in its inactive conformation thus nullifying inhibitory action on C

carboxy terminal regions which participate in G-protein-to-receptor interactions, the homology is lower. It is also likely that the other regions of divergence in the sequence are involved in the interactions with the effector. (For a comprehensive discussion of the subject, see Chapters 4 and 6.) Since α_s and α_i are likely to interact at different domains of the catalyst C (page 11), it is to be expected that the effector domain of α will be different for each G-protein. So far it has been generally accepted that it is the α subunit that transmits the signal to the effector. Recently, however, it was pointed out that there may be differences between the $\beta\gamma$ subunits of different G-proteins [6] (see also Chapter 4) and therefore each class of receptors may recognize a specific set of $\alpha\beta\gamma$ structure. When more will come to be known on the diversity of $\beta\gamma$s it will also become clearer whether those subunits play a role in the specificity of the signal produced by the G-protein involved. As far as G_s and G_i are concerned, however, it seems that the $\beta\gamma$s are identical and functionally interchangeable.

UNIVERSALITY OF THE SYSTEM

It has been shown that stimulatory receptors from one cell type can hybridize with a heterologous adenylate cyclase system when transplanted into another cell type. For example, β-adrenoceptors from turkey erythrocytes and glucagon receptors from rat liver can be transferred into Friend erythroleukaemia cells and activate their adenylate cyclase [11,12]. Kinetic experiments also demonstrated that two different receptor types in the same cell couple to a single pool of adenylate cyclase, namely A2 adenosine receptors and β_1-adrenoceptors, both in turkey erythrocytes [13,14] and rat brain [15]. These experiments suggested already in the late 1970s that receptors that differ markedly in their pharmacological specificity possess common structural domains that participate in the interaction between the stimulatory receptor and G_s. Recent cloning and sequencing studies on β_2- and β_1-adrenoceptors support this hypothesis. It was found [2] that certain sequences are involved with ligand binding while others are involved in interactions with the G-protein. This feature seems to be generally true for receptors which interact with G-proteins. Chimaeras of β_2- and α_2-adrenoceptors have been constructed, such that the pharmacological specificity is that of the cyclase inhibitory α_2-adrenoceptor whereas the biochemical activity is of the β-type, namely the receptor activates adenylate cyclase [16]. This experiment is an elegant proof for the concept of structural separation between the ligand binding domain and the G-protein interaction domain on the receptor. Most strikingly, the mechanism of interaction between these two domains seems to be independent of the specificity of the domains. This is probably why an α_2-adrenergic ligand (clonidine), which is normally associated with adenylate cyclase inhibition, can activate the enzyme when it interacts with the $\alpha_2-\beta_2$ adrenoreceptor chimaera where the G_s/β-adrenoceptor is preserved [16].

G_s is universal since it couples to a variety of C units from different sources. For example, restoration of a functional β-receptor-dependent adenylate cyclase in G_s-deficient S49cyc$^-$ (AC$^-$)-lymphoma cells can be achieved by G_s from various sources: wild type S49 lymphoma cells, rabbit liver, turkey erythrocytes and human erythrocytes ([17] for review). Recently we have shown that the catalytic unit of adenylate cyclase from two different species: bovine brain and rabbit heart couple equipotently with G_s from either turkey erythrocytes or rabbit liver [3]. This result suggests that C is also a highly conserved molecule. Studies using monoclonal antibodies against C reveal that bovine brain C and rabbit myocardial C are similar but not identical [18]. A few variants of G_i are also known from cloning studies but the functional differences are not yet understood [6] (see Chapter 4). It is not clear at this point how many G_i-type proteins are truly involved with the inhibition of adenylate cyclase and whether they participate in other signal transdution pathways. The universality of the components has allowed more flexibility in reconstitution experiments in which components from various species have been co-reconstituted to produce functional interactions [3,4].

STIMULATION OF ADENYLATE CYCLASE

General features

Hormone binding to R_s and GTP binding to G_s at the α_s subunit are both required to induce activation of C to the cAMP-producing state. The active state of the system decays concomitantly with the hydrolysis of GTP to GDP and P_i at the G_s regulatory site. Replenishment of G_s with GTP and the continued presence of hormone at the receptor allow the system to regain its active cAMP-producing state. This type of 'on–off' cycle accounts well for the properties of hormone-dependent adenylate cyclase ([1] for review):

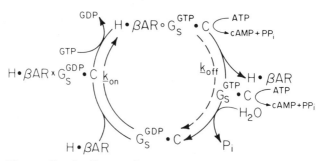

Figure 1.2 The on–off cycle of hormonally stimulated adenylate cyclase. The hormone-bound β-adrenoceptor interacts briefly with the G_s(GDP)C complex. During the brief encounter and 'opening' of the α_s subunit (see below) GDP is exchanged with GTP to form the activated cyclase G_s(GTP)C. The agonist-bound receptor dissociates from the complex and moves on to activate another enzyme molecule. The lifetime of GTP at the site is between 2 to 3 seconds before it hydrolyses to GDP. The cycle repeats itself as long as GTP and the hormone are present and as long as desensitization of the system ($t_{1/2} \sim 1$ min) does not begin. Desensitization eliminates or significantly reduces the ability of the βAR to couple to G_s

(1) Hormone-dependent adenylate cyclases always possess hormone-dependent GTPase activity. This activity has indeed been demonstrated directly in glucagon-dependent cyclase, pancreozymin-dependent cyclase, and prostaglandin E_1-dependent cyclase ([19] for review) as well as in β-adrenoceptor G_s reconstituted systems [2,10].

(2) Non-hydrolysable GTP analogues such as Gpp(NH)p and GTPγS, generate a permanently active enzyme due to the complete blockade of the 'turn-off' GTPase reaction ([10] for review).

(3) Inhibition of the GTPase reaction by cholera-toxin-catalysed ADP-ribosylation of α_s activates the cyclase due to the slow-down of the 'off' GTPase reaction ([10,17,19] for reviews).

(4) From independent measurements of the rate of enzyme activation by hormone and guanyl nucleotide (the 'on' reaction) and the decay of the cAMP producing state to its basal state (the 'off' reaction), one can compute the fraction of the total pool of cyclase which is in the active state in the presence of GTP. These measurements [20] demonstrate that the suggested two-state model [21] is sufficient to account for the 'on–off' cycle. This has not been experimentally verified for the β-adrenoceptor-dependent adenylate cyclase (Figure 1.2).

Catalytic role of the receptor and 'collision coupling'

A series of experiments demonstrated, rather surprisingly at the time, that the role of the receptor is catalytic ([1] for review), i.e. one receptor can activate numerous adenylate cyclase molecules. When the number of β-adrenoceptors on turkey erthrocyte membranes was progressively reduced using a specific β-adrenoreceptor-directed affinity label, the rate of cyclase activation was proportionally reduced but the maximal specific activity attainable was unchanged when hormone and Gpp[NH]p are used. The rate constant of cyclase activation, k_{on}, was found to be linearly dependent on receptor concentration (equation 1.1).

$$\text{Activity} = \text{Activity}_{max} \times \{1 - \exp(-k_{on})t\}$$
$$= \text{Activity}_{max} \times \{1 - \exp(-k[R]_T t\} \quad (1.1)$$

Where k_{on} is the measured first-order constant, k the intrinsic first-order constant and $[R]_T$ the receptor concentration in the system.

This discovery led to the formulation of the 'collison coupling' model ([1] for review) (Figure 1.2) where the steps of adenylate cyclase activation are described in the following steps:

$$\begin{aligned}
HR_s + G_s(GDP)C &\rightleftharpoons HR_sG_s(GDP)C \\
HR_sG_s(GDP)C + GTP &\rightleftharpoons HR_sG_s(GTP)C + GDP \\
HR_sG_s(GTP)C &\rightleftharpoons HR_sG'_s(GTP)C' \quad (1.2)\\
HR_sG'_s(GTP)C' &\rightleftharpoons HR_s + G'_s(GTP)C' \\
G'_s(GTP)C' + H_2O &\rightleftharpoons G_s(GDP)C + Pi
\end{aligned}$$

where the intermediate $HR_sG_s(GTP)C$, does not accumulate.

This formulation predicts: (a) strict first-order kinetics of adenylate cyclase activation by hormone and Gpp[NH]p or GTP [22]; (b) linear dependence of the rate constant of activation k_{on} on receptor concentration [3,22,23]; and (c) a Michaelian (non-cooperative) dependence of k_{on} on the hormone concentration [22]. From the hormone dependence curve of k_{on} one can compute receptor–agonist dissociation constants. The numerical values for these constants should be identical to those obtained by direct binding measurements [24].

These predictions were indeed verified experimentally. A large series of agonists and partial agonists were tested as activators of adenylate cyclase in the turkey erythrocyte system. It was found that the better the agonist, the higher the k_{on}, whereas the GTPase turn-off rate is identical for all agonists [20]. From this finding it seems that agonists differ from each other in the fraction of active enzyme that they can induce.

G_s shuttle mechanisms

A number of investigators suggested alternative mechanisms for activation of adenylate cyclase by R_s and G_s. These models postulate that the G_s unit, being the 'transducer' element, physically shuttles between the receptor and the catalytic unit. The basic idea is that the inactive form G_s (GDP) is idle and separate from C within the membrane bilayer. In its active form, G'_s(GTP) associates with the catalyst C to form the active species of the enzyme G_s('GTP)C'.

The suggestion that G_s physically shuttles [25,26] between the β-receptor and the catalytic unit stems from the following observations: (a) activation of G_s by the β-adrenoceptor can be achieved in the absence of the catalytic unit (see above); (b) G_s associates with the β-adrenoceptor in the presence of β-agonists but in the absence of GTP it is the GDP-bound state of G_s that displays high affinity towards the agonist bound R_s [26,27] or its analogues; and (c) GTP and its analogues reduce β-adrenoceptor affinity towards agonists [26,27]. The essence of this proposed mechanism is that G_s(GDP) first interacts with the β-adrenoceptor, GDP is exchanged with GTP and G_s(GTP) then physically separates from the receptor, migrates away from the receptor and associates with C to generate the activate cyclase G'_s(GTP)C'. Upon hydrolysis of GTP to GDP plus P_i, G_s(GDP) separates from C and re-associates with the receptor.

Rigorous mathematical analysis showed [28] that 'shuttle' models predict non-first-order complex kinetics of adenylate cyclase activation as well as complex dependence of the rate of activation on receptor concentration. Furthermore, the shuttle model predicts a dependence of the rate of activation on the concentration of G_s and C. Kinetic experiments on native membranes [29] and on systems reconstituted from purified components ([1] for review) establish that activation of the C unit is first order while the rate of adenylate cyclase activation depends linearly on receptor concentration. These findings exclude the G_s shuttle model as well as some modified versions that can be envisioned [28] because none of them can account for the experimentally observed kinetic features. These kinetic experiments actually mean that G_s is associated with C at all times, namely during the *entire* course of the C to C' transition. Biochemical evidence also supports the notation that G_s and C are tightly associated. In the presence of phospholipids, a stoichiometric complex C_s–C can be purified 250-fold both in the GDP basal state and in the Gpp[NH]p preactivated state. These separation experiments demonstrate that G_s(GDP) is not separate from C as assumed by the G_s-shuttle model but is associated with C [5].

Conformational transitions

During the interaction between HR_s and G_s, the nucleotide binding site assumes an 'open' conformation. As a result of this opening, GDP falls off the site and is replaced by incoming GTP. The open conformation can be demonstrated in kinetic experiments and the properties of the open state measured. It was found, for example, that the G-protein in its 'open' but nucleotide-free state has a longer lifetime than in the presence of a nucleotide such as Gpp[NH]p [30]. The conversion of the 'open' metastable conformation to the 'closed' conformation, which is unable to exchange the nucleotide, is facilitated by the bound nucleotide. GTP, Gpp[NH]p and GDPβS facilitate the 'closure' reaction whereas the metastable 'open' state can be trapped for a long time by GMP [30].

The 'open' conformation of G_s is also manifested in its increased affinity for the agonist-bound β-adrenoceptor. Thus, in the presence of β-agonist and in the absence of GTP one can separate a complex between HR_s and G_s (or G_sC) [27]. Once GDP is replaced by GTP, G_s decays to its 'closed' conformation which also displays lower affinity towards HR_s and thus the complex between these two entities dissociates. An antagonist is unable to bring about the conformational change in G_s and thus the affinity of R_s to G_s is not modulated by antagonists [26,27]. The availability of pure G_s, pure R_s

and pure C calls for direct physicochemical measurements of the conformational transitions in each of the proteins during hormone activation. These measurements can then be correlated with the different conformational states of the system which were defined by kinetic mechanistic experiments [27,30].

THE G_s–G_i SUBUNIT DISSOCIATION MECHANISM FOR HORMONAL ACTIVATION AND INHIBITION

General considerations

Exposure of G_s to the non-hydrolysable GTP analogue GTPγS and high Mg^{2+} in Lubrol PX induces the irreversible dissociation of the protein:

$$G_s(GDP) + GTP\gamma S \longrightarrow \alpha_s(GTP\gamma S) + \beta\gamma + GDP \qquad (1.3)$$

This finding led to a model [31] which suggests that hormonal stimulation in the presence of the natural ligand GTP also leads to the reversible dissociation of G_s to $\alpha'_s(GTP)$ which then associates and activates the catalytic unit C to form $\alpha'_s(GTP)C'$. Upon hydrolysis of GTP to GDP, $\alpha_s(GDP)$ dissociates from C and re-associates with $\beta\gamma$ (see Figure 1.3). This model then suggests that $\alpha'_s(GTP)$ binds strongly to C but weakly to $\beta\gamma$ whereas the situation is reversed for $\alpha_s(GDP)$. According to this scheme the $\beta\gamma$ subunits compete with C for α_s. Hence, elevation of the level of the $\beta\gamma$ subunits causes α_s to be less accessible to C and therefore induces inhibition. This scheme therefore immediately suggests a mechanism for adenylate cyclase inhibition: an agonist bound to an inhibitory receptor (R_i), facilitates the dissociation of G_i to $\alpha_i(GTP)$ and $\beta\gamma$, the latter scavenging α_s and thus inhibiting C (Figure 1.3). This scheme does not exclude the direct inhibition of C by $\alpha_i(GTP)$, as an additional effect. The involvement of $\alpha_i(GTP)$ as an inhibiting species is not essential in this model and is believed [31] to play a minor role, if any. This scheme

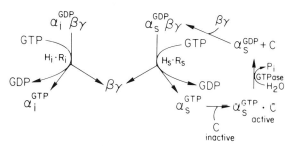

Figure 1.3 The G_s–G_i dissociation model for adenylate cyclase activation and inhibition

accounts for the failure to demonstrate a G_i to C interaction in reconstituted systems [32], and provides a mechanism by which G_i confers inhibition of C without direct physical interaction. This mechanism explains why, in systems composed of G_s, C and G_i in detergent solutions, the inhibitory effect of G_i is largely due to the presence of $\beta\gamma$. Furthermore, the $\beta\gamma$-inhibitor effect depends on the presence of G_s [33]. This effect, however, is actually due to the experimental set-up of these experiments: G_s was activated by GTPγS and its dissociated α_s(GTPγS) subunit became the species activating C. When G_i was added to the G_s/GTPγS mixture, the $\beta\gamma$ subunits released from G_i(GTPγS) scavenged the α_s(GTPγS) subunits generated from G_s(GTPγS), thus reducing the amount of free α_s(GTPγS) available for interaction with C. GTPγS was used rather than GTP, and the G_i/G_s mixture contained detergent (0.1% Lubrol PX). It remains to be clarified how these experimental conditions relate to the physiological conditions in vivo, and whether indeed in the membrane $\beta\gamma$ subunits dissociate during activation (see below).

This scheme (Figure 1.3) implies that the α_s subunit shuttles between G_s and the catalytic unit C and that G_s and C are physically separate entities within the membrane bilayer. Any mechanism, however, which implies G_sC dissociation is not compatible with the kinetic data in native membranes and in reconstituted systems as previously explained (page 6). Furthermore, direct biochemical evidence demonstrated that G_s, or at least α_s, is permanently associated with C (see [5] and page 6). A modified G_s dissociation model [34,35] (Figure 1.4), however, can accommodate the 'collision coupling' model. The

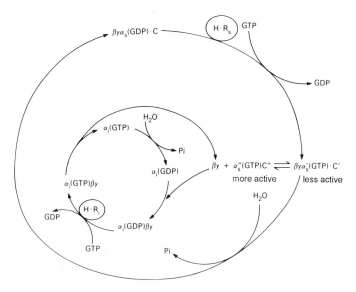

Figure 1.4 The G_s–G_i partial dissociation mechanism for adenylate cyclase regulation; α_s''(GTP)C" is more active than $\beta\gamma\alpha_s'$(GTP) · C'

modified G_s dissociation scheme suggesting that both species $G_s'(GTP)C'$ and $\alpha_s''(GTP)C''$ represent active cyclase molecules and that the $G_s(GTP)C$ species is less active. An important constraint on the system is that nucleotide exchange at α_s requires the presence of $\beta\gamma$ subunits, namely the associated species $G_s(GTP)$ $\{\gamma\beta\alpha_s(GTP)\}$ must exist. This has been directly demonstrated when the Gpp[NH]p preactivated form of G_s was reconstituted with purified β1-adrenergic receptor and GTP. In this experiment [23], it was found that an exchange of Gpp[NH]p to GTP takes place only in the presence ($\sim$ ten-fold molar exess) of $\beta\gamma$ subunits [23]. (This excess is probably required to shift the equilibrium to favour the associated state of G_s such that Gpp[NH]p to GTP exchange can take place.)

These results also mean that if $\beta\gamma$ dissociates from $G_s(GTP)$, they must be able also to re-associate with $\alpha_s(GTP)$. Interestingly, the activation by GTPγS is irreversible [36] probably because $\beta\gamma$ has no measurable affinity to $\alpha_s(GTP\gamma S)$. Direct measurements demonstrate that the tendency of pure heterotrimeric G-protein to dissociate depends on the nature of the guanyl nucleotide bound. Dissociation was found to be complete for GTPγS, whereas GTP did not induce any dissociation [37]. This finding supports the view that in the presence of GTP, limited or no subunit dissociation occurs during the activation of adenylate cyclase. A recent study claims that activation of G_s in intact S49 membranes by β-agonists and GTP results in subunit dissociation. This was demonstrated by extracting the membranes with sodium cholate *subsequent* to activation and then probing the extract with anti-α_s antibodies [38]. Clearly, this protocol is improper since the high concentration of sodium cholate undoubtedly promoted α_s dissociation from $\beta\gamma$. It is therefore still not clear to what extent G_s undergoes dissociation subsequent to activation with GTP. So far, no experiments aimed at monitoring the oligomeric state of G_s in the native membrane have been performed. It is not immediately apparent how one should actually design such an experiment without perturbing the subunit interactions within the G_s oligomer.

Beta subunits are tightly associated with activated cyclase

Recently, we purified Gpp[NH]p preactivated adenylate cyclase from turkey erythrocytes [39] and from bovine brain [40] at low detergent (Lubrol PX) to protein ratios and found that the β subunit co-purifies stoichiometrically with the enzyme and with its associated α_s subunit: α_s was monitored using [^{35}S]GTPγS binding and the β subunit using anti-β antibodies. These results demonstrate for the first time that, under mild conditions that do not force the dissociation of G_s, activated α_s and $\beta\gamma$ subunits may remain associated with each other through extensive purification. Bovine brain $G_s(Gpp[NH]p)C$ does not lose its β subunit when it is affinity purified on a forskolin affinity column at low salt. When the forskolin column is washed first with high salt, subsequent to the absorption of the enzyme, but prior to its elution with forskolin, the purified enzyme is devoid of the β subunit and represents the $\alpha_s(Gpp[NI I]p)C$ species [39]. These results further demonstrate that the α_s to $\beta\gamma$ interaction is sensitive to high salt. A high salt wash of the forskolin column to which cyclase has been absorbed is a necessary step for the purification of the naked catalyst of adenylate cyclase. It is interesting to note that the β subunit also leaks out of the Gpp[NH]p activated complex during extensive centrifugation (16 h) in sucrose gradients (5% to 20%) [39,40]. It is more than probable that, in the membrane, dissociation of α_s from $\beta\gamma$ is even less likely. Certainly the environment of the membrane bilayer and intracellular matrix will provide very different conditions to those experienced in vitro in our experimental protocol.

The turkey erythrocyte Gpp[NH]p preactivated enzyme which is purified on Ultragel AcA34, MonoQ and phenylsuperose, under mild conditions also retains its β subunits. Although the experimental conditions used employ high monovalent salt concentrations, no loss of β subunit from the G_sC complex is observed [39]. This result further supports the hypothesis that it is highly unlikely that $\beta\gamma$ dissociation is part of the mechanism of adenylate cyclase activation. This conclusion has far reaching implications since it precludes the dissociation mechanism and its modified version as the basis for understanding adenylate cyclase inhibition by G_i. This conclusion is not in contrast with the claims that $\beta\gamma$ subunits may themselves be messengers to other effector systems (see below).

Kinetic considerations

Since the GTPase off-rate is 15 to 20 times faster than the on-rate of activation, there is only 5% to 7% of G_s which is active at any given time in the presence of hormone and GTP [21]. This means that one should detect only a small fraction of α_s as free subunits, unless there is an environmental condition that pulls the equilibrium towards dissociation. Such an environmental condition could be a detergent such as sodium cholate used to examine the release of α_s from membranes of S49 cells subsequent to their activation by a β-agonist and GTP [38].

MECHANISM OF G_i ACTION

The most commonly accepted model for the action of G_i is the dissociation mechanism (Figures 1.3 and 1.4). This model assumes that G_i, G_s and the catalytic unit of adenylate cyclase C are all separate components. Upon interaction of G_s with an agonist H-bound stimulator receptor R_s, GDP is exchanged with GTP, leading to the dissociation of G_s to the GTP bound α_s and $\beta\gamma$. α_s(GTP) then seeks C, associates with it and thus leads to its activation. Upon hydrolysis of GTP to GDP at the α_s subunit, the latter dissociates from C to regenerate free α_s(GDP), which is subsequently reassociated with $\beta\gamma$ to regenerate G_s(GDP) ($=\alpha_s$(GDP)$\beta\gamma$). In fact, it is assumed in this model that α_s and $\beta\gamma$ are in equilibrium at all times [33], where in the presence of GTP the dissociated state is favoured and in the presence of GDP the associated state is favoured:

$$\alpha_s + \beta\gamma \underset{\text{GTP}}{\overset{\text{GDP}}{\rightleftarrows}} \alpha_s\beta\gamma \qquad (1.4)$$

This feature is the key to understanding the model of G_i action according to the dissociation model. Upon activation of G_i by an adenylate cyclase inhibitory receptor R_i, it undergoes dissociation to GTP-bound α_i and $\beta\gamma$. Since $\beta\gamma$ subunits are identical in both G-proteins (G_s and G_i), their release can immediately influence the equilibrium depicted in equation 1.4. Since the associated form of G_s is inactive towards C according to the dissociation mechanism, elevated levels of $\beta\gamma$ lead to inhibition of cAMP production, since they bind free α_s. In addition, α_i(GTP) can exert a direct effect on C. As explained on page 9, the dissociation mechanism as such [10,33] cannot coexist with a number of other experimental findings.

In order to overcome these serious drawbacks of the original dissociation model [31,33], we have suggested [17,34,35] a modification. In this modification, α_s remains associated with C *at all times* whereas $\beta\gamma$ are allowed to dissociate upon exchange of GDP by GTP at the α_s subunit (see equation 1.5 and Figure 1.4):

$$C\alpha_s\beta\gamma \underset{\text{GDP}}{\overset{\text{GTP}}{\rightleftarrows}} C\alpha_s + \beta\gamma \qquad (1.5)$$

As in the original dissociation model the associated state is favoured when α_s is bound with GDP and the dissociated state is favoured when α_s is bound with GTP. The species C-α_s(GTP)$\beta\gamma$ must be allowed to account for the facilitation by $\beta\gamma$ of Gpp[NH]p to GTP exchange in the Gpp[NH]p-activated adenylate cyclase.

This modified dissociation mechanism, like the original one, also accounts for the inhibitory action of G_i in the absence of a direct G_i to C interaction. As in the original dissociation mechanism, the active adenylate cyclase species is α_s(GTP)C whereas the associated species $\gamma\beta\alpha_s$(GTP)C is less active and the $\beta\alpha_s$(GDP)C is inactive or possesses low basal activity. The basic difference between this model and the original dissociation model is that G_s and C are not separate units but are associated as a complex which *can* lose its $\beta\gamma$ subunits but not the α_s subunit. This minimal feature is essential for the model not to contradict experimental results showing that: (a) the kinetics of adenylate cyclase activation is first order; (b) the rate constant of activation is independent of the concentration of G_s or C [29] and depends only on receptor concentration; and (c) the rate of G_s or of G_sC activation is linear with receptor concentration [3,22,25,41].

As discussed on page 9, we have observed that under mild experimental conditions, $\beta\gamma$ remains associated with even the activated form of the enzyme, namely $\gamma\beta\alpha_s$(GTP)C. Therefore, one must consider mechanisms for adenylate cyclase inhibition which do not involve $\beta\gamma$ dissociation. The most plausible model is the frequently mentioned possibility that G_i specifically interacts through α_i with C [42,43,44], namely the α_i subunit, and not $\beta\gamma$, carrying the regulatory message. This argument does not negate the possibility that $\beta\gamma$ carries other regulatory messages not related to signalling of adenylate cyclase [45].

Such interaction has not been demonstrated in vitro when pure C and G_i were co-reconstituted into phospholipid vesicles [32]. These experiments, however, were conducted in the absence of G_s. G_i to C interactions are clearly seen in membranes of S49cyc$^-$ cells which possess C and G_i, but lack G_s [33,34]. It is therefore likely that the phospholipid environment used in the G_i to C reconstitution [32] were incomplete and therefore did not support the G_i to C interaction. Since the fidelity of reconstituted cyclase systems has not yet achieved the levels of native systems (see discussion, [46]), it is indeed plausible that G_i to C interactions were not expressed by the purified components in the lipid composition used. It will therefore be of interest to examine the role of lipids in the G_i to C interaction. G_i interacts weakly with the adenylate cyclase systems as compared to G_s and therefore its interactions may be more sensitive to the nature of the lipid environment. The findings that certain reagents such as low heparin [47] uncouple the cyclase inhibitory action of G_i without affecting the stimulation by G_s support the notion that G_i interacts with C with weaker forces than G_s. In purification experiments, G_i, G_o, etc., readily separate from the complex G_sC, whereas separation between G_s and C requires high salt and/or ionic detergent such as sodium cholate [48]. Since in intact membranes which possess both G_s and G_i one can demonstrate the inhibitory effects of G_i, it is reasonable to assume that G_i physically interacts with the adenylate cyclase system. If this is indeed so, one should be able to monitor a G_i to R_s cross-talk, even if G_i and R_s are not interacting directly. This hypothesis led us to explore the ability of G_i to affect the affinity of β-adrenoceptors towards an agonist in the native S49 membranes which possess both G_i and G_s. In these experiments, we measured the affinity of the receptors to (−)isoproterenol subsequent to treatment with pertussis toxin. We found that subsequent to treatment, the affinity of the receptors was reduced 3 to 3.5-fold compared to untreated membranes [49] (Figure 1.5). This effect is not seen in S49-cyc$^-$ membranes which lack G_s and of course not in turkey erythrocyte membranes which lack functional G_i [39]. These results suggest that G_i interacts better with C once it is complexed with G_s (Figure 1.6). More experiments are needed in order to examine closely the question of G_i to C interactions. Not only $α_i$ may interact directly with C, there are also experiments which show that isolated $βγ$ can inhibit C and G_sC [50,51,52] but in these experiments a relatively large excess of $βγ$ over C was used. The $βγ$ subunits

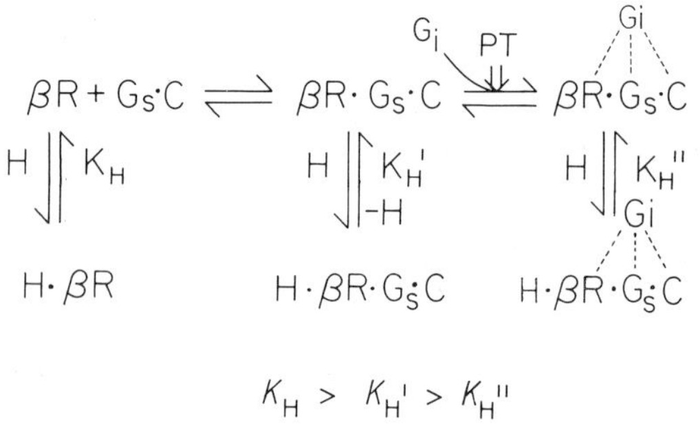

$$βR + G_s·C \rightleftharpoons βR·G_s·C \xrightarrow[PT]{G_i} βR·G_s·C$$

with vertical equilibria:
- $H \updownarrow K_H$ giving $H·βR$
- $H \updownarrow K_H'$ (−H) giving $H·βR·G_s·C$
- $H \updownarrow K_H''$ (with G_i) giving $H·βR·G_s·C$

$$K_H > K_H' > K_H''$$

Figure 1.5 The physical interaction between G_i and the adenylate cyclase system

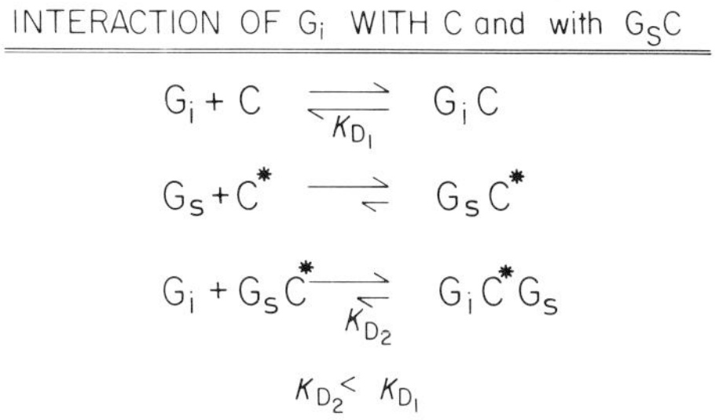

INTERACTION OF G_i WITH C and with G_sC

$$G_i + C \underset{K_{D_1}}{\rightleftharpoons} G_iC$$

$$G_s + C^* \rightleftharpoons G_sC^*$$

$$G_i + G_sC^* \underset{K_{D_2}}{\rightleftharpoons} G_iC^*G_s$$

$$K_{D_2} < K_{D_1}$$

Figure 1.6 Interaction of G_i and C is improved by G_s

in a reconstituted system inhibit the basal activity of C with little effect on the hormone-stimulated activity [52].

It is actually highly unlikely that free $\beta\gamma$ subunits play a direct regulatory role because they seem to be functionally interchangeable between different G-proteins. Thus, any time a G-protein is activated, adenylate cyclase becomes affected, a rather displeasing biological non-specificity.

The physiological meaning of these results remains to be established. It also remains for future experiments to establish whether $\beta\gamma$ subunits can exert direct effects on adenylate cyclase or phospholipase C, phospholipase A_2 and K^+ channels (for review see [43]) and, if so, whether they are of biological significance.

REFERENCES

1. Levitzki, A. (1988) *Science*, **241**, 800–806
2. Lefkowitz, R. Y., and Caron, M. G. (1988) *J. Biol. Chem.*, **263**, 4993–4996.
3. Feder, D., Hekman, M., Klein, H. W., Levitzki, A., Helmreich, E. J. M., and Pfeuffer, T. (1986) *EMBO J.*, **5**, 1509–1514.
4. May, D. C., Ross, E. M., and Gilman, A. G. (1985) *J. Biol. Chem.*, **260**, 15829–15833.
5. Arad, H., Rosenbusch, J., and Levitzki, A. (1984) *Proc. Natl. Acad. Sci. USA*, **81**, 6579–6583.
6. Cassey, P. J., and Gilman, A. G. (1988) *J. Biol. Chem.*, **263**, 2577–2580.
7. Kubo, T., Fakuda, K., Mikami, A., Maeda, A., Takakashi, H., Mishima, M., Haga, T., Haga, K., Ichigama, A., Kangawa, D., Kojima, M., Matsuo, H., Thirose, T., and Numa S. (1986) *Nature (London)*, **323**, 411–416.
8. Bonner, T. I., Buckley, J. N. J., Young, A. C., and Brann, M. R. (1987) *Science*, **237**, 527–532.
9. Kobilka, B. K., Matsui, H., Kobilka, T. S., Yang-Feng, T. L., Franke, U., Caron, M. G., Lefkowitz, R. J., and Regan, J. W. (1987) *Science*, **238**, 650–656.
10. Gilman, A. G., (1987) *Ann. Rev. Biochem.*, **56**, 615–649.
11. Schramm, M., Orly, J., Eimerl, S., and Korner, M. (1977) *Nature (London)*, **268**, 310–313.
12. Schramm, M. (1979) *Proc. Natl. Acad. Sci. USA*, **76**, 1174–1178.
13. Sevilla, N., Tolkovsky, A. M., and Levitzki, A. (1977) *FEBS Lett.*, **81**, 339–341.
14. Tolkovsky, A. M., and Levitzki, A. (1978) *Biochemistry*, **17**, 3811–3817.
15. Braun, S., and Levitzki, A. (1979) *Mol. Pharmacol.*, **16**, 737–748.
16. Kobilka, B. K., Kobilka, T. S., Daniel, L., Regan, J. W., Caron, M. G., and Lefkowitz, R. J. (1988) *Science*, **240**, 1310–1316.
17. Levitzki, A. (1986) *Physiol Rev.*, **66**, 819–842.
18. Mollner, S., and Pfeuffer, T. (1988) *Eur. J. Biochem.*, **17**, 265–271.

19. Levitzki, A. (1982) *Trends Pharmacol. Sci. (TIPS)*, **3**, 203–208.
20. Arad, M., and Levitzki, A. (1979) *Mol. Pharmacol.*, **16**, 748–756.
21. Levitzki, A. (1977) *Biochem. Biophys. Res. Commun.*, **74**, 1154–1159.
22. Tolkovsky, A. M., and Levitzki, A. (1978) *Biochemistry*, **17**, 3795–3810.
23. Hekman, M., Feder, D., Keenan, A. K., Gal, A., Klein, H. W., Pfeuffer, T., Levitzki, A., and Helmreich, E. J. M. (1984) *EMBO J.*, **3**, 3339–3345.
24. Levitzki, A., Sevilla, N., Atlas, D., Steer, M. L. (1975) *J. Mol. Biol.*, **37**, 35–53.
25. Citri, Y., and Schramm, M., (1980) *Nature (London)*, **287**, 297–300.
26. De Lean, A., Stadel, J. M., and Lefkowitz, Z. J. (1980) *J. Biol. Chem.*, **255**, 7108–7117.
27. Limbird, L. E., Gill, D. M., Stadel, J. M., Mickey, A. R., and Lefkowitz, R. J. (1980) *J. Biol. Chem.*, **255**, 1854–1861.
28. Tolkovsky, A. M., and Levitzki, A. (1981) *J. Cyclic. Natl. Res.*, **1**, 139–150.
29. Tolkovsky, A. M., Braun, S., and Levitzki, A. (1982) *Proc. Natl. Acad. Sci. USA*, **79**, 213–217.
30. Braun, S., Tolkovsky, A. M., and Levitzki, A. (1982) *J. Cyclic. Nucl. Res.*, **8**, (3), 133–147.
31. Gilman, A. G. (1984) *Cell*, **36**, 577–579.
32. Smigel, M. D. (1986) *J. Biol. Chem.*, **261**, 1976–1982.
33. Katada, T., Bokoch, G., Smigel, M. D., Ui, M., and Gilman, A. G. (1984) *J. Biol. Chem.*, **259**, 3586–3595.
34. Levitzki, A. (1984) *J. Rec. Res.*, **4**, 399–409.
35. Levitzki, A. (1987) *FEBS Lett.*, **211**, 113–118.
36. Cassel, D., and Selinger, Z. (1977) *Biochem. Biophys. Res. Commun.*, **77**, 868–873.
37. Huff, R. M., and Neer, E. J. (1986) *J. Biol. Chem.*, **261**, 1105–1110.
38. Rasnas, L. A., and Insel, P. A., (1988) *J. Biol. Chem.*, **263**, 17239–17242.
39. Bar-Sinai, A., Minich, M., Shorr, R. G. L., and Levitzki, A. (1990) *J. Biol. Chem.*, in press.
40. Marback, I., Bar-Sinai, A., Minich, M., Seamon, K., Shorr, R. G. L., and Levitzki, A. (submitted).
41. Pedersen, S. E., and Ross, E. M. (1982) *Proc. Natl. Acad. Sci. USA*, **79**, 7228–7232.
42. Hildebrandt, J. D., Codina, J., Risinger, R., and Birnbaumer, L. (1984) *J. Biol. Chem.*, **259**, 2039–2042.
43. Bourne, H. R. (1989) *Nature (London)*, **337**, 504–505.
44. Hildebrandt, J. D., and Birnbaumer, L. (1983) *J. Biol. Chem.*, **258**, 13141–13147.
45. Toro, M. J., Montoya, E., and Birnbaumer, L. (1987) *Mol. Endocrinol.*, **1**, 669–676.
46. Levitzki, A. (1985) *Biochim. Biophys. Acta*, **822** 127–153.
47. Willuweit, B., and Aktories, K., (1988) *Biochem. J.*, **249**, 857–863.
48. Ross, E. M. (1981) *J. Biol. Chem.*, **256**, 1949–1953.
49. Marbach, I., Shiloach, J., and Levitzki, A. (1988) *Eur. J. Biochem.*, **172**, 239–246.
50. Katada, T., Kusakabe, K., Oinuma, M., and Ui, M. (1987) *J. Biol. Chem.*, **262**, 11897–11900.
51. Enomoto, K., and Asakawa, T. (1986) *FEBS Lett.*, **202**, 63–68.
52. Im, M.-J., Holzhofer, A., Keenan, A. K., Imekman, M., Helmreich, E. J. M., and Pfeuffer, T. (1987) *J. Receptor Res.*, **I**(1–4), 17–42.

2
STRUCTURE AND IDENTIFICATION OF G-PROTEINS: ISOLATION AND PURIFICATION

Allen M. Spiegel

Molecular Pathophysiology Branch, National Institute of Diabetes, Digestive and Kidney Disease, Building 10, Room 8D-17, National Institutes of Health, Bethesda, MD, 20892, USA

INTRODUCTION AND OVERVIEW

The discovery by Rodbell and co-workers of the critical role of guanine nucleotides in mediating hormonal stimulation of adenylyl cyclase [1] led to the isolation and purification of G_s, the G-protein coupling receptors to stimulation of adenylyl cyclase, and eventually, to the isolation of many other members of what has emerged as a large family of heterotrimeric, signal-transducing G-proteins [2–4]. G_s was initially resolved from other components of the receptor–adenylyl cyclase complex by GTP affinity chromatography [5,6]. Studies with the mutant S49 mouse lymphoma cell line, cyc$^-$, provided further evidence that receptors, adenylyl cyclase, and G_s are separate entities [7]. The mutant cell line, deficient in G_s activity, provided a powerful functional assay for the purification of G_s [8]. Purified G_s was shown initially to consist of α subunits of either 45 kDa or 52 kDa (termed 42 kDa and 47 kDa, respectively, by others) and a 35 kDa β subunit [8]. The α subunits corresponded to those previously identified as the guanine nucleotide binding entity, as well as the substrate for cholera-toxin-catalysed ADP-ribosylation [9]. Subsequently, a low molecular weight γ subunit was also found to be associated with G_s and other G-proteins [10], as had already been shown for the retinal G-protein, transducin or G_t [11].

Indeed, parallel work on the mechanism of visual transduction in retinal photoreceptor rod cells led to the isolation of a heterotrimeric G-protein, G_t, that couples the photon receptor, rhodopsin, to cGMP phosphodiesterase [12]. Ease of preparation of rod outer segment membranes, relative abundance of G_t in these membranes and extractibility of G_t in aqueous buffers facilitated its purification in quantities large enough for detailed biochemical studies [13], generation of specific antisera [14] and amino acid sequencing of each of the three subunits [15–19].

The initial work of the Rodbell group also indicated a role for guanine nucleotides in agonist-mediated inhibition of adenylyl cyclase [1]. The discovery by Ui and co-workers that adenylyl cyclase inhibition could be blocked by pertussis-toxin-catalysed ADP-ribosylation of an approximately 41 kDa protein [20] focused attention on purification of a distinct G_i protein responsible for mediating adenylyl cyclase inhibition. Using pertussis-toxin-catalysed ADP-ribosylation as an assay, putative G_i proteins were purified from rabbit liver [21] and human erythrocyte [22]. Definitive functional identification of a specific G-protein as G_i, however, has not yet been achieved. As with G_s and G_t, G_i was found to be a heterotrimer with a unique α subunit but an apparently functionally equivalent $\beta\gamma$ subunit.

Subsequently, several additional heterotrimeric G-proteins have been identified and purified. Measurement of high affinity, specific guanine nucleotide binding in brain suggested that this might be a particularly rich source of GTP binding proteins [23]. This was confirmed by purification of a novel G-protein, G_o, comprising approximately 1–2% of brain membrane protein [23–25]. Immunochemical studies of neutrophils, a cell type with an abundance of pertussis toxin substrate, suggested the existence of novel forms of G-protein pertussis toxin substrates [26]. Likewise, cloning of complementary DNAs (cDNAs) encoding putative G-protein α subunits [2–4] indicated considerable heterogeneity in the G-protein family. Additional functions regulated by G-proteins, e.g. activity of certain ion channels and of other enzymes of second messenger metabolism such as phospholipase C, have been identified [27]. In several instances, novel G-proteins corresponding to a particular cDNA have been purified [28,29] and identified by using peptide-specific antisera [30,31] or by direct amino acid sequencing [32]. Functional assignment of particular G-proteins has, however, lagged behind structural identification.

Proteins of 21 kDa associated with the cytoplasmic surface of the plasma membrane, products of the H-, K-, and N-*ras* genes, are structurally related to the G-protein α subunits [15]. Recently, it has become clear that there is a very large family of low (20–30 kDa) molecular weight ras-related GTP binding proteins [33]. A detailed discussion of the purification and identification of membranes of this family is beyond the scope of this chapter (see chapter 11 for details), but I will briefly review this subject, particularly as it relates to purification of members of the heterotrimeric G-protein family.

G-PROTEIN PURIFICATION AND IDENTIFICATION: GENERAL ASPECTS

The starting material for purification of G-proteins (with the exception of G_t, see below) has been crude or purified plasma membrane preparations. By a variety of criteria, little if any G_s, G_i, or G_o is found in the cytosol, but the possibility that these or other novel G-proteins might be found in subcellular compartments other than the plasma membrane has not been excluded.

Solubilization of G-proteins from plasma membranes is the first step in purification. Initially, non-ionic detergents such as Lubrol PX were employed [5–7]. An ionic detergent, sodium cholate, proved more efficient in extracting G-proteins from membranes, and, presumably because of a lower critical micellar concentration, facilitated subsequent G-protein purification [8]. By adjusting ionic strength in the extraction buffer (see below), significant purification can be achieved even in this first step. Subsequent purification strategies have generally involved sequential chromatography by ion exchange (most often DEAE–Sephacel), gel permeation (Ultrogel AcA34), and one or more of the following steps: hydrophobic columns (e.g. heptylamine–Sepharose), hydroxyapatite columns, additional ion exchange steps (e.g. mono-Q column). Purification of resolved α and $\beta\gamma$ subunits may be achieved by treating extracts with G-protein activators such as 'AMF' ($AlCl_3$, $MgCl_2$ and NaF) before further purification, or by resolving the dissociated subunits (e.g. by gel permeation chromatography) after G-protein purification.

Several types of assay may be employed in monitoring G-protein purification. Specific functional assays could involve ability to couple to receptors and/or effectors. Only G_s has been purified in this way, based on its ability to reconstitute agonist-stimulated adenylyl cyclase activity in the mutant cyc[−] cell line [8]. High affinity, specific guanine nucleotide (e.g. GTPγS) binding is a more general assay, first successfully employed in purification of brain G_o [23]. G-protein pertussis toxin substrates may be purified by monitoring toxin-catalysed ADP-ribosylation with [28], or without [21,22], added $\beta\gamma$ subunits. In principle, specific antibodies can also be used to monitor purification.

Direct amino acid sequencing and comparison with predicted sequences of cloned cDNAs is the definitive method for identification of purified G-proteins. Because there is a high degree of sequence homology, many peptides may need to be sequenced to provide definitive identification. Antibodies, particularly those raised against unique, defined synthetic peptides [30,31], and whose specificity has been rigorously defined, can also be very useful in identifying a purified G-protein. Since several G-proteins serve as pertussis toxin substrates, this method is of limited utility in identifying individual subtypes. Likewise, the similar mobility of many G-proteins on one-dimensional SDS–PAGE limits the information gained with this method. Modifications such as lowering the concentration of Bis-acrylamide in gel recipes does afford greater resolution in the 39–41 kDa range, and may allow distinction between subtypes such as G_{i1} (41 kDa) and G_{i2} (40 kDa). Proteolytic peptide mapping, and 2-D gel electrophoresis can also help discriminate G-protein subtypes, but these techniques rarely allow definitive identification. Functional assays, e.g. GTPase kinetics, receptor and effector reconstitution, can provide data relevant to G-protein identification. Since the degree of specificity for certain G-protein interactions is not yet known, such functional assays do not necessarily provide definitive identification of a particular G-protein.

G-PROTEIN PURIFICATION AND IDENTIFICATION: SPECIFIC ASPECTS

Transducin (G_t)

Light absorption by retinal photoreceptor rod outer segments (ROS) activates a GTPase activity [34]. This reflects photon activation of rhodopsin, and facilitation of guanine nucleotide exchange on G_t by activated rhodopsin. If guanine nucleotides are withheld, G_t binds tightly to light-activated rhodopsin on ROS membranes [11]. Unlike other G-proteins (see below), which require detergents to remove them from plasma membranes, G_t is soluble in aqueous buffers. In particular, the $\beta\gamma$ complex binds to ROS membranes in an ionic-strength-dependent manner [11]. These properties facilitate purification of milligram quantities of G_t [12–14].

Sequential extraction of light-activated ROS membranes with aqueous isotonic and hypotonic buffers removes most peripheral membrane proteins, including the '48 k protein (arrestin)' and cGMP phosphodiesterase. Subsequent extraction with hypotonic buffer containing guanine nucleotide yields >90% pure G_t (Figure 2.1). Purification to homogeneity can be achieved by hexylagarose chromatography of G_t eluted from ROS membranes [13]; α and $\beta\gamma$ subunits of G_t can be readily resolved (Figure 2.2) by Blue Sepharose chromatography [35–37]. This method is effective irrespective of whether GTP or GTPγS has been used to elute G_t (in the former case G_t exists as the holoprotein,

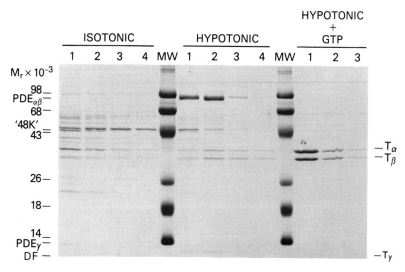

Figure 2.1 Elution of G_t from ROS membranes. Retinas were dissected under ambient light from cows' eyes freshly obtained from the slaughterhouse. Vortexing retinas in buffer containing 20% sucrose shears off ROS. ROS membranes are purified by sucrose density gradient centrifugation, and may be stored frozen in liquid nitrogen or immediately extracted. ROS membranes are sequentially extracted six times in isotonic buffer (10 mM Tris/HCl, pH 7.5, 0.1, mM EDTA, 100 mM NaCl, 5 mM $MgCl_2$), six times in hypotonic buffer (same buffer but omit NaCl and $MgCl_2$), and three times in hypotonic buffer plus 0.1 mM GTP (or non-hydrolysable GTP analogue). Equal volumes of isotonic extracts 1–4, hypotonic extracts 1–4, and hypotonic + GTP extracts 1–3 were loaded on a 15% polyacrylamide gel, and proteins separated by SDS–gel electrophoresis. The figure shows a representative gel stained with Coomassie Blue. The size of molecular weight markers (MW) is indicated along the left. The principle protein extracted under isotonic conditions is the '48K' protein (also termed arrestin). Hypotonic buffer releases cGMP phosphodiesterase α and β subunits ($PDE_{\alpha,\beta}$, poorly resolved) as well as the γ subunit (PDE_γ). The majority of G_t subunits (T_α, T_β, T_γ) are not released until GTP is added to the hypotonic buffer

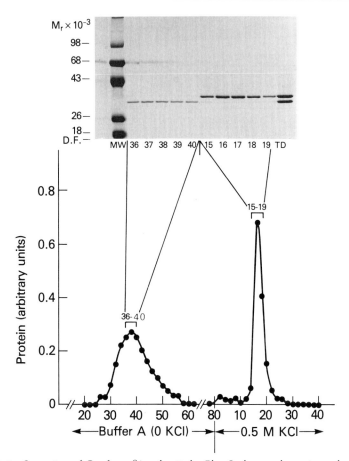

Figure 2.2 Separation of $G_{t,\alpha}$ from β/γ subunits by Blue Sepharose chromatography. Approximately 10 mg of G_t extracted from ROS membranes was exchanged into buffer A (10 mM Tris/HCl, pH 7.5, 6 mM $MgCl_2$, 1 mM dithiothreitol, 25% glycerol) with a Sephadex PD-10 column. The G_t-containing fractions were applied to a Blue Sepharose GL-6B column equilibrated with buffer A. The column was eluted with 500 ml of buffer A, and then with 500 ml of buffer A containing 0.5 M KCl: 5 ml fractions were collected and analysed for protein concentration (shown in arbitrary units in bottom graph). 40 μl samples from selected fractions (36–40 from buffer A elution and 15–19 from buffer A + KCl elution) were analysed by SDS–gel electrophoresis on a 10% polyacrylamide gel. The upper part of the figure shows a Coomassie-Blue-stained gel of these fractions runs alongside molecular weight markers (MW) and a sample of G_t holoprotein (lane marked TD). The positions of $G_t\alpha$ and β subunits are indicated. The γ subunit is not resolved from the dye front (DF) on a 10% gel

in the latter case as dissociated α and $\beta\gamma$ subunits). An ion exchange mechanism, critically dependent upon Mg^{2+} concentration [37], is the probable basis for separation of G_t subunits by Blue Sepharose. The efficacy of this method for resolution of the subunits of other G-proteins has not been reported. Chromatography with ω-amino octylagarose has also been successfully employed to resolve α and $\beta\gamma$ subunits of Gpp(NH)p-eluted G_t [13].

G_t purified from ROS membranes has been extensively characterized, both structurally and functionally. Partial amino acid sequences determined for purified bovine $G_t\alpha$ [15,38], β [16,17] and γ [18,19] subunits have been shown to correspond to the amino acid sequences predicted for the corresponding cDNAs cloned from bovine retinal libraries. In contrast, a cDNA cloned from a bovine retinal library encoding a putative α subunit closely related to that of G_t predicts an amino acid sequence that has yet to be confirmed through sequencing of a purified protein [39]. Immunochemical evidence suggests that this cDNA encodes a form of $G_t\alpha$ specifically expressed in cone photoreceptor cells [40,41].

G_s

G_s is defined as the G-protein that couples receptors for agonists that stimulate cAMP production to the effector, adenylyl cyclase. The protein is ubiquitously expressed, and thus, in principle, could be purified from any source. The initial purification was from rabbit liver. Using a 1% sodium cholate, low ionic strength buffer to extract G_s from crude liver plasma membranes, and four chromatographic steps (DEAE–Sephacel, Ultrogel AcA34, Heptylamine–Sepharose and a second DEAE–Sephacel), Sternweis et al. [42], were able to achieve a 1300-fold purification with a 15% yield. Because purified G_s is relatively unstable in cholate, the detergent was exchanged for Lubrol in the second DEAE–Sephacel step. AMF was employed throughout purification to activate and stabilize the protein. Although this would be expected to dissociate α and $\beta\gamma$ subunits (see below), the final preparation consisted of 45 kDA and 52 kDa proteins corresponding to the two forms of $G_s\alpha$ and a 35 kDa protein corresponding to the β subunit [42]. Presumptively, this reflects the large excess of β over α_s subunits. A prominent, approximately 41 kDa contaminant, in retrospect, represents co-purifying α_i (see below). Similar methods were used to purify G_s from turkey [43] and human [44] erythrocytes. The major difference is that high ionic strength (0.6–0.8 M NaCl) was required in the cholate-containing buffer to extract G_s from erythrocyte membranes. Also, only the 45 kDa form of α_s was obtained from erythrocytes. As stated earlier, purified G_s was identified initially by functional assay. Indeed, the cDNA encoding α_s was identified, not by comparison to protein-derived amino acid sequence, but rather using a combination of specific peptide antibodies and mRNA hybridization to cyc$^-$ cells which lack α_s mRNA [45]. Purified G_s, after activation by AMF or GTPγS, was resolved into α [46] and $\beta\gamma$ [47] subunits, and each was shown to possess distinct activity, i.e. ability to activate cyclase and ability to block holoG$_s$ dissociation, respectively.

G_i, G_o and other pertussis toxin substrates

G_i was initially purified from rabbit liver [21] and human erythrocytes [22] by methods essentially identical to those employed for purification of G_s. G_i co-migrates with G_s

during the initial chromatographic steps, but can be separated from G_s by Heptylamine-Sepharose [21] or hydroxyapatite [22] column chromatography. Pertussis-toxin-catalysed ADP-ribosylation was used to identify G_i. The properties of the purified, resolved rabbit liver G_i subunits were studied with respect to adenylyl cyclase inhibition. While activated α_i subunits are capable of inhibiting cyclase activity in mutant cyc$^-$ membranes, in wild type membranes and in membranes from other types of cells, $\beta\gamma$ subunits proved more effective in inhibiting cyclase activity [48]. This has led to the suggestion that inhibition of adenylyl cyclase is, in fact, mediated by $\beta\gamma$ rather than α subunits [3], and leaves open the question of which G_i subtype is the true functional G_i. In fact, the identity of purified liver G_i has not yet been established but immunochemical studies (Spiegel, unpublished observations) indicate that it consists of, at a minimum, G_{i2} and G_{i3}.

Codina et al. [49,50] have described in detail a protocol for purification of G_s and G_i from human erythrocytes without using activators such as AMF. An approximate 5000-fold purification with a yield of 10% was achieved: 30% ethylene glycol was found to stabilize G-protein activity. More recently, erythrocyte G_i has been found to activate certain membrane potassium channels, and hence has been termed G_K. Direct amino acid sequencing of the human erythrocyte G_K has identified it as G_{i3} [51], but immunochemical studies (Spiegel, unpublished observations) indicate that in addition to G_{i3}, human erythrocyte G_i contains G_{i2}. Thus, although G_{i3} may well be capable of functioning as G_K, the identity of endogenous G_K is still an open question.

Specific, high affinity guanine nucleotide binding, using [^{35}S]GTPγS and a rapid filtration method [23], showed that brain contains a greater amount of binding activity than other tissues and cells (e.g. ten-fold > liver). Purification of bovine [23–25] and rat [52] brain pertussis toxin substrates led to the identification of two G-proteins with α subunits of 41 kDA and 39 kDa, respectively. The α_{39} is particularly abundant, about 1% of brain plasma membrane protein, in large measure accounting for the high guanine nucleotide binding activity found in brain. By a variety of criteria, including differences in GTPase activity, peptide maps and antibody reactivity, the brain α_{41} and α_{39} were shown to be distinct proteins. Direct amino acid sequencing of the purified proteins confirmed this and identified α_{41} and α_{39} as G_{i1} and G_o, respectively [53,54]. G-proteins purified from human erythrocyte, bovine brain, and bovine ROS show similarities and differences in migration on 1-D SDS–PAGE (Figure 2.3).

In one of the initial reports [24] on purification of brain pertussis toxin substrates, an additional α_{40} band was noted on Coomassie-Blue-stained gels of the purified proteins. Using a purification protocol involving DEAE–Sephacel, Ultrogel AcA34, DEAE–toyopearl 650S, and mono-Q chromatography, Katada et al. succeeded in resolving the porcine brain α_{40}-containing G-protein from G_{i1} and G_o [29]. Similar protocols were used to resolve these proteins from bovine brain [55,56]. Using peptide antibodies, α_{40} was tentatively identified as G_{i2} [55,56]; this was confirmed by direct amino acid sequencing [32].

Analysis of brain G-protein-containing fractions from mono-Q chromatography with specific peptide antibodies identified an additional α subunit [55]. This protein is indistinguishable from G_o with a panel of antisera, but precedes G_o in elution from a

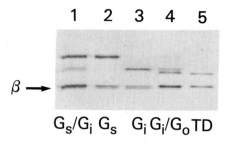

Figure 2.3 Comparison of purified G-proteins on SDS–polyacrylamide gel electrophoresis. G_s and G_i purified from human erythrocyte membranes (first three lanes), G-proteins purified from bovine brain membranes (G_i/G_o), and G_t purified from bovine ROS membranes (TD) were resolved on a 10% gel and stained with Coomassie Blue. Only the portion of the gel containing α and β subunits is shown. The β subunit of G_t is exclusively the 36 kDa form, and that of brain G-proteins, predominantly the 36 kDa form. Erythrocyte-derived G-proteins contain larger amounts of the 35 kDa form of the β subunit. The predominant α subunits in the brain G-protein preparation are G_{i1} (41 kDa) and G_o (36 kDa). The latter co-migrates with $G_t\alpha$. Erythrocyte G_i consists primarily of G_{i3} (see text) which co-migrates with G_{i1}. Erythrocyte $G_s\alpha$ is predominantly the form (~ 43 kDa) encoded by the 'short' form of $G_s\alpha$ mRNA

mono-Q column (Figure 2.4). On 1-D SDS–PAGE, moreover, the mobility of this novel α subunit is intermediate between that of G_{i2} and that of G_o (Figure 2.5). 2-D gel electrophoresis (Figure 2.6) indicates that the pI of the novel form of G_o is distinctly more basic than that of 'conventional' G_o. The basis for these differences between the two forms of G_o (e,g. post-translational modifications, primary sequence) awaits clarification. 2-D gel electrophoresis, moreover, indicates considerable charge heterogeneity even for purified G-proteins that appear homogeneous on 1-D SDS–PAGE (compare Figures 2.5 and 2.6). Again, this raises the possibility of post-translational modifications that lead to forms with subtle charge differences.

Immunochemical evidence had suggested that the major pertussis toxin substrate of human neutrophils differs from other pertussis toxin substrates including G_t, G_{i1} and G_o [26]. The major pertussis toxin substrate of bovine neutrophils was purified from a cholate extract of a purified plasma membrane fraction [28]. Treatment with AMF, which leads to subunit dissociation (see Figure 2.7), was found necessary for successful purification. Toxin-catalysed ADP-ribosylation of fractions, with added exogenous $\beta\gamma$ subunits, was used to monitor purification of α subunits (Figure 2.7). The predominant protein purified in this way migrates as an α_{40} on SDS–PAGE, and has been identified using peptide-specific antibodies as G_{i2} [30]. A similar protein has been purified from HL-60 cells [57,58] and rabbit neutrophil [59] membranes. An α_{41} has also been purified from HL-60 cells [58], and identified with peptide antibodies as G_{i3} [31]. G_{i2}, which may be expressed ubiquitously [60], has also been identified by direct sequencing of a protein purified from human platelets [61], and has been tentatively identified as an α_{40} purified from bovine lung [62]. In both platelet [61] and lung [62], additional α_{41} proteins were purified, but were not identified. A novel α_{43} pertussis toxin substrate has been identified in preparations of G-proteins purified from human erythrocytes [63]. It appears to cross-

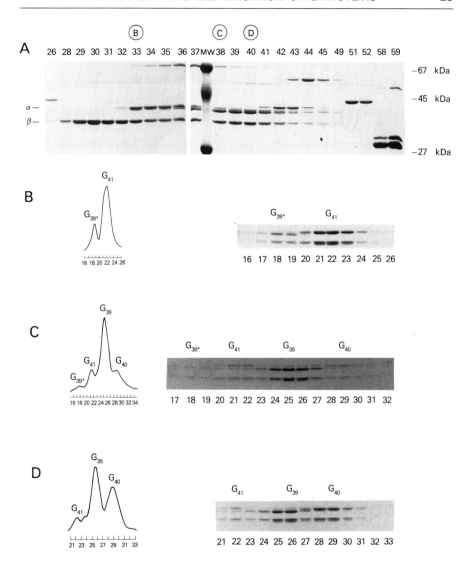

Figure 2.4 Purification of bovine brain G-proteins by (A) TSK–toyopearl and (B–D) mono-Q chromatography. G-proteins were purified from a cholate extract of bovine brain membranes by monitoring high affinity guanine nucleotide binding. Following fractionation by DEAE–Sephacel and AcA34 Ultrogel chromatography, the major guanine nucleotide binding peak was applied to a DEAE–toyopearl 650(S) column. The latter was eluted with a gradient of increasing NaCl. Fractions were analysed by SDS–gel electrophoresis and Coomassie Blue staining as shown in (A). Selected fractions (labelled B, C and D) were subjected to further chromatography on a mono-Q column. Panels B–D show the tracing of OD_{280} absorption for the G-protein-containing fractions on the left, and the Coomassie-Blue-stained gel of the corresponding fractions on the right. (Reproduced with permission of the American Chemical Society from Goldsmith et al. (1988) Biochemistry, **27**)

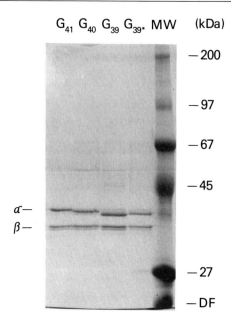

Figure 2.5 SDS–polyacrylamide gel electrophoresis and Coomassie Blue staining of purified bovine brain G-proteins. Aliquots of proteins purified by mono-Q chromatography (as in Figure 2.4) were separated on a 10% gel. The positions of the α subunits, of the β doublet (predominantly the 36 kDa form), and of molecular size markers (MW) are indicated. Presumptive γ subunits run with the dye front (DF). (Reproduced with permission of the American Chemical Society from Goldsmith et al. (1988) Biochemistry, **27**)

react with a peptide antiserum raised against a conserved GTP binding sequence, but its structure has not otherwise been defined.

Heterogeneity of purified $\beta\gamma$ subunits has also been found. Whereas only a 36 kDa β subunit is associated with G_t, variable amounts of 36 kDa and 35 kDa forms of the β subunit are associated with other purified G-proteins (see Figure 2.3). In brain, the 36 kDa form predominates; in liver and blood-derived cells roughly equal amounts of 35 kDa and 36 kDa forms are present. In placenta, the 35 kDa form predominates, and has been purified from this source [64]. Sequencing of purified 36 kDa β subunit matches the predicted sequence of a cloned cDNA [16,17]. Specific peptide antibodies identify the 35 kDa form of β as the product of a distinct cDNA [65]. As yet, only the sequence of the G_t-γ subunit has been defined [18,19]. Preliminary evidence suggests that γ subunits associated with other G-proteins, not only differ from that of G_t, but are themselves heterogeneous [23,64].

Low molecular weight GTP-binding proteins

Three distinct *ras* genes (H-, K-, and N-*ras*) encode approximately 21 kDa proteins that share several features in common with G-protein α subunits; *ras* proteins specifically bind guanine nucleotides, possess GTPase activity and are tethered to the cytoplasmic surface of the plasma membrane. Unlike G-protein α subunits, ras proteins bind guanine nucleotides even after denaturation and transfer from SDS–polyacrylamide gels to nitrocellulose paper. Results using this technique, as well as the results of cDNA cloning, now indicate that there exists a large family of ras-related, lower molecular weight GTP binding proteins [33].

In no case as yet has the specific function of a member of this family been identified, but the high degree of evolutionary conservation of amino acid sequence suggests a critical role in cell function. Several members of this family have been purified, and

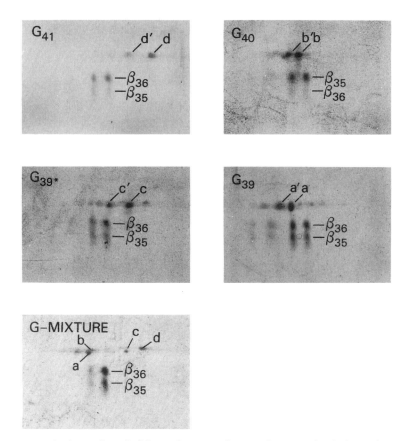

Figure 2.6 Analysis of purified brain G-proteins by two-dimensional gel electrophoresis. A mixture of purified brain G-proteins (G-MIXTURE) and individual purified G-proteins (as in Figure 2.5) were subjected to two-dimensional gel electrophoresis and silver stained. The positions of the two forms of β subunit are indicated. Individual α subunits are labelled as follows: a = α_{39}, b = α_{40}, c = $\alpha_{39'}$, and d = α_{41}. a', b', c', and d' indicate more acidic forms of the major spots. The approximate pI range is from 6.5 to 5.1 (right to left). (Reproduced with permission of the American Chemical Society from Goldsmith et al. (1988) Biochemistry, 27)

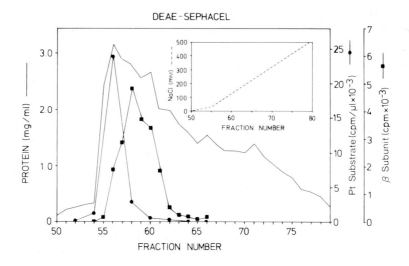
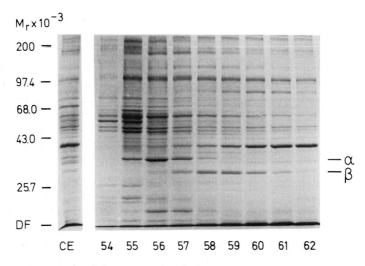

Figure 2.7 DEAE–Sephacel chromatography of a bovine neutrophil membrane cholate extract. AMF (20 μM $AlCl_3$, 10 mM $MgCl_2$, 10 mM NaF) was added to the cholate extract which was then applied to a DEAE–Sephacel column equilibrated with buffer containing AMF, 20 mM Tris/HCl, pH 7.5, 1 mM dithiothreitol, 1 mM EDTA and 1% (w/v) sodium cholate. The column was washed with the same buffer and then eluted with a linear gradient of NaCl (see inset, top of figure). Fractions were assayed (top of figure) for protein concentration, pertussis toxin (PT) substrate (in an assay buffer supplemented with $G_t\beta\gamma$), and β subunit (by quantitative immunoblot). The lower part of the figure shows an analysis of the cholate extract (CE) and selected fractions from DEAE–Sephacel chromatography by SDS–gel electrophoresis and Coomassie Blue staining. The positions of α and β subunits are indicated. (Reproduced with permission of the Federation of European Biochemical Societies from P. Gierschik et al. (1987) Eur. J. Biochem., **165**, 185–194)

identified by a variety of techniques, most rigorously, by direct amino acid sequencing. As with G-protein α subunits, low molecular weight GTP binding proteins have generally been purified from cholate extracts of plasma membranes. Using chromatographic methods identical to those used in purification of G-proteins from rabbit liver [42], two low molecular weight GTP binding proteins termed 'ARF' and 'G_p' have been purified from rabbit liver [66] and human placenta [67], respectively. Both proteins have also been purified from bovine brain [68,69]. In both cases, the lower molecular weight GTP binding proteins were resolved from the larger G-proteins by gel permeation chromatography. ARF was initially identified as a protein cofactor necessary for cholera-toxin-catalysed ADP-ribosylation of G_s, and this assay forms the basis for its purification. The protein, however, is itself a GTP binding protein. G_p (misleadingly named for source of purification, placenta) was identified solely as a low molecular weight GTP binding protein, rather than on the basis of a specific function. Immunochemical studies show it to be distinct from ARF and from *ras* proteins [67], but its structure has not been defined. G_p was reported to co-purify with $\beta\gamma$ complexes indistinguishable from those in heterotrimeric G-proteins [67]. This is the only example to data of a low molecular weight GTP binding protein associated with $\beta\gamma$ subunits. Rigorous evidence that G_p functionally associates with $\beta\gamma$ subunits has not yet been presented.

Using a different series of chromatographic steps, another group has succeeded in purifying as many as six distinct low molecular weight GTP binding proteins from a cholate extract of bovine brain membrane [70]. Initial chromatography on Ultrogel AcA-44 permits separation of a first peak of heterotrimeric G-proteins and a second peak of low molecular weight GTP binding proteins. The latter are fractionated through additional steps on ϕ-Sepharose, hydroxyapatite, and mono-Q columns. c-K-*ras* [71], the *rho* gene product [72], and a novel 24 kDa GTP-binding protein [73] have been purified in this way, and identified by amino acid sequencing. The ARF protein is also separated as a distinct peak by this method, but the relationship between any of these proteins and G_p has not been defined. Several of the low molecular weight GTP binding proteins, including the *rho* gene product, apparently serve as substrates for ADP-ribosylation catalysed by certain botulinum toxins [72]. A botulinum toxin substrate of 22 kDa, and other low molecular weight GTP binding proteins of 24 kDa and 26 kDa have also been purified from human neutrophil membranes [74].

Some of the low molecular weight GTP binding proteins may also exist in soluble form. Thus ARF has been purified from brain cytosol [75,76], as have several other low molecular weight GTP binding proteins including the novel 24 kDa protein [77]. Whether the soluble forms of these proteins differ from the membrane-bound forms, e.g. in post-translational modification, has not yet been established.

FUTURE PROSPECTS

It seems likely that additional heterotrimeric G-proteins (and certainly low molecular weight GTP binding proteins) remain to be purified. In some instances, newly cloned cDNAs imply the existence of a novel G-protein yet to be purified. It is possible that subcellular compartments other than the plasma membrane harbour novel G-proteins, since as discussed above, most purification efforts have focused on plasma membrane extracts. Expression of G-protein subunits using recombinant DNA technology, e.g. in *Escherichia coli*, offers a powerful approach to identification of the functions of individual, defined G-proteins, but at the same time, purification of G-proteins from such sources may require new strategies. Possible novel approaches include immunoaffinity chromatography using specific peptide and/or monoclonal antibodies. Finally, co-purification of G-protein–receptor and G-protein–effector complexes, if feasible, could provide important insights into the specificity of G-protein receptor–effector coupling.

REFERENCES

1. Rodbell, M. (1980) *Nature (London)*, **284**, 17–21.
2. Spiegel, A. (1987) *Mol. Cell. Endocrinol.*, **49**, 1–16.
3. Gilman, A. (1987) *Ann. Rev. Biochem.*, **56**, 615–649.
4. Lochrie, M. I., and Simon, M. I., (1988) *Biochemistry*, **27**, 4957–4965.
5. Pfeuffer, T., and Helmreich, E. J. M. (1975) *J. Biol. Chem.*, **250**, 867–876.
6. Spiegel, A. M., Downs, R. W., and Aurbach, G. D. (1979) *J. Cyclic Nucleotide Res.*, **5**, 3–17.
7. Ross, E. M., and Gilman, A. G. (1977) *J. Biol. Chem.*, **252**, 6966–6969.
8. Northup, J. K., Sternweis, P. C., Smigel, M. D., Schleifer, L. S., Ross, E. M., and Gilman, A. G. (1980) *Proc. Natl. Acad. Sci. USA*, **77**, 6516–6520.
9. Cassel, D., and Pfeuffer, T. (1978) *Proc. Natl. Acad. Sci. USA*, **75**, 2669–2673.
10. Hildebrandt, J. D., Codina, J., Risinger, R., and Birnbaumer, L. (1984) *J. Biol. Chem.*, **259**, 2039–2042.
11. Kuhn, H. (1980) *Nature (London)*, **283**, 587–589.
12. Baehr, W., Morita, E. A., Swanson, R. J., and Applebury, M. L. (1982) *J. Biol. Chem.*, **257**, 6452–6460.
13. Fung, B. K.-K. (1983) *J. Biol. Chem.*, **258**, 10495–10502.
14. Gierschik, P., Simons, C., Woodard, C., Somers, R., and Spiegel, A. (1984) *FEBS Lett.*, **172**, 321–325.
15. Hurley, J. B., Simon, M. I., Teplow, D. B., Robishaw, J. D., and Gilman, A. G. (1984) *Science*, **226**, 860–862.
16. Sugimoto, K., Nukada, T., Tanabe, T., Takahashi, H., Noda, M., Minamino, N., Kangawa, K., Matsuo, H., Hirose, T., Inayama, S., and Numa, S. (1985) *FEBS Lett.*, **191**, 235–240.
17. Fong, H. K. W., Hurley, J. B., Hopkins, R. S., Miake-Lye, R., Johnson, M. S., Doolittle, R. F., and Simon, M. I. (1986) *Proc. Natl. Acad. Sci. USA*, **83**, 2162–2166.
18. Hurley, J. B., Fong, H. K. W., Teplow, D. B., Dreyer, W. J., and Simon, M. I. (1984) *Proc. Natl. Acad. Sci. USA*, **81**, 6948–6952.
19. Ovchinnikov, Y. A., Lipkin, V. M., Shuvaeva, T. M., Bogachuk, A. P., and Shemyakin, V. V. (1985) *FEBS Lett.*, **179**, 107–110.
20. Ui, M. (1984) *Trends Pharmacol. Sci.*, **5**, 277–279.
21. Bokoch, G. M., Katada, T., Northup, J. K., Hewlett, E. L., and Gilman, A. G. (1983) *J. Biol. Chem.*, **258**, 2072–2075.
22. Codina, J., Hildebrandt, J., Iyengar, R., Birnbaumer, L., Sekura, R. D., and Manclark, C. R. (1983) *Proc. Natl. Acad. Sci. USA*, **80**, 4276–4280.
23. Sternweis, P. C., and Robishaw, J. D. (1984) *J. Biol. Chem.*, **259**, 13806–13813.

24. Neer, E. J., Lok, J. M., and Wolf, L. G. (1984) *J. Biol. Chem.*, **259**, 14222–14229.
25. Milligan, G., and Klee, W. A. (1985) *J. Biol. Chem.*, **260**, 2057–2063.
26. Gierschik, P., Falloon, J., Milligan, G., Pines, M., Gallin, J. I., and Spiegel, A. (1986) *J. Biol. Chem.*, **261**, 8058–8062.
27. Birnbaumer, L., Codina, J., Mattera, R., Yatani, A., Schere, N., Toro, M.-J., and Brown, A. M. (1987) *Kidney International*, **32**, S14–S37.
28. Gierschik, P., Sidiropoulos, D., Spiegel, A., and Jakobs, K. H. (1987) *Eur. J. Biochem.*, **165**, 185–194.
29. Katada, T., Oinuma, M., Kusakabe, K., and Ui, M. (1987) *FEBS Lett.*, **213**, 353–358.
30. Goldsmith, P., Gierschik, P., Milligan, G., Unson, C. G., Vinitsky, R., Malech, H., and Spiegel, A. (1987) *J. Biol. Chem.*, **262**, 14683–14688.
31. Goldsmith, P., Rossiter, K., Carter, A., Simonds, W., Unson, C. G., Vinitsky, R., and Spiegel, A. (1988) *J. Biol. Chem.*, **263**, 6476–6479.
32. Itoh, H., Katada, T., Ui, M., Kawasaki, H., Suzuki, K., and Kaziro, Y. (1988) *FEBS Lett.*, **230**, 85–89.
33. Chardin, P. (1988) *Biochimie*, **70**, 865–868.
34. Wheeler, G. L., and Bitensky, M. W. (1977) *Proc. Natl. Acad. Sci. USA*, **74**, 4238–4242.
35. Shinozawa, T., Uchida, S., Martin, E., Cafiso, D., Hubbel, W., and Bitensky, M. (1980) *Proc. Natl. Acad. Sci. USA*, **77**, 1408–1411.
36. Pines, M., Gierschik, P., Milligan, G., Klee, W., and Spiegel, A. (1985) *Proc. Natl. Acad. Sci. USA*, **82**, 4095–4099.
37. Kleuss, C., Palast, M., Brendel, S., Rosenthal, W., and Schultz, G. (1987) *J. Chromatog.*, **407**, 281–289.
38. Tanabe, T., Nukada, T., Nishikawa, Y., Sugimoto, K., Suzuki, H., Takahashi, H., Noda, M., Haga, T., Ichiyama, A., Minamino, N., Matsuo, H., and Numa, S. (1985) *Nature (London)*, **315**, 242–245.
39. Lochrie, M. A., Hurley, J. B., and Simon, M. I. (1985) *Science*, **228**, 96–99.
40. Grunwald, G. B., Gierschik, P., Nirenberg, M., and Spiegel, A. (1986) *Science*, **231**, 856–859.
41. Lerea, C., Somers, D. E., Hurley, J. B., Klock, I. B., and Bunt-Milam, A. H. (1986) *Science*, **234**, 77–80.
42. Sternweis, P. C., Northup, J. K., Smigel, M. D., and Gilman, A. G. (1981) *J. Biol. Chem.*, **256**, 11517–11526.
43. Hanski, E., Sternweis, P. C., Northup, J. K., Dromerick, A. W., and Gilman, A. G. (1981) *J. Biol. Chem.*, **256**, 12911–12919.
44. Hanski, E., and Gilman, A. G. (1982) *J. Cyclic. Nucleotide. Res.*, **8**, 323–336.
45. Harris, B. A., Robishaw, J. D., Mumby, S. M., and Gilman, A. G. (1985) *Science*, **229**, 1274–1277.
46. Northup, J. K., Sternweis, P. C., and Gilman, A. G. (1983) *J. Biol. Chem.*, **258**, 11361–11368.
47. Northup, J. K., Smigel, M. D., Sternweis, P. C., and Gilman, A. G. (1983) *J. Biol. Chem.*, **258**, 11369–11376.
48. Katada, T., Bokoch, G. M., Smigel, M. D., Ui, M., and Gilman, A. G. (1984) *J. Biol. Chem.*, **259**, 3586–3595.
49. Codina, J., Hildebrandt, J. D., Sekura, R. D., Birnbaumer, M., Bryan, J., Manclark, C. R., Iyengar, R., and Birnbaumer, L. (1984) *J. Biol. Chem.*, **259**, 5871–5886.
50. Codina, J., Rosenthal, W., Hildebrandt, J. D., Sekura, R. D., and Birnbaumer, L. (1984) *J. Receptor Res.*, **4**, 411–442.
51. Codina, J., Olate, J., Abramowitz, J., Mattera, R., Cook, R. G., and Birnbaumer, J. (1988) *J. Biol. Chem.*, **263**, 6746–6750.
52. Katada, T., Oinuma, M., and Ui, M. (1986) *J. Biol. Chem.*, **261**, 8182–8191.
53. Nukada, T., Tanabe, T., Takahashi, H., Noda, M., Haga, K., Haga, T., Ichiyama, A., Kangawa, K., Hiranaga, K., Matsuo, H., and Numa, S. (1986) *FEBS Lett.*, **197**, 305–310.

54. Itoh, H., Kozasa, T., Nagata, S., Nakamura, S., Katada, T., Ui, M., Iwai, S., Ohtsuka, E., Kawasaki, H., Suzuki, K., and Kaziro, Y. (1986) *Proc. Natl. Acad. Sci. USA*, **83**, 3776–3780.
55. Goldsmith, P., Backlund, P. S. Jr, Rossiter, K., Carter, A., Milligan, G., Unson, C. G., and Spiegel, A. (1988) *Biochemistry*, **27**, 7085–7090.
56. Mumby, S., Pang, I.-H., Gilman, A. G., and Sternweis, P. C. (1988) *J. Biol. Chem.*, **263**, 2020–2026.
57. Oinuma, M., Katada, T., and Ui, M. (1987) *J. Biol. Chem.*, **262**, 8347–8353.
58. Uhing, R. J., Polakis, P. G., and Snyderman, R. (1987) *J. Biol. Chem.*, **262**, 15575–15579.
59. Dickey, B. F., Pyun, H. Y., Williamson, K. C., and Navarro, J. (1987) *FEBS Lett.*, **219**, 289–292.
60. Brann, M., Collins, R., and Spiegel, A. (1987) *FEBS Lett.*, **222**, 191–198.
61. Nagata, K., Katada, T., Tohkin, M., Itoh, H., Kaziro, Y., Ui, M., and Nozawa, Y. (1988) *FEBS Lett.*, **237**, 113–117.
62. Morishita, R., Kato, K., and Asano, T. (1988) *Eur. J. Biochem.*, **174**, 87–94.
63. Iyengar, R., Rich, K. A., Herberg, J. T., Grenet, D., Mumby, S., and Codina, J. (1987) *J. Biol. Chem.*, **262**, 9239–9245.
64. Evans, T., Fawzi, A., Fraser, E. D., Brown, M. L., and Northup, J. K. (1987) *J. Biol. Chem.*, **262**, 176–181.
65. Gao, B., Mumby, S., and Gilman, A. G. (1987) *J. Biol. Chem.*, **262**, 17254–17257.
66. Kahn, R. A., and Gilman, A. G. (1984) *J. Biol. Chem.*, **259**, 6228–6234.
67. Evans, T., Brown, M. L., Fraser, E. D., and Northup, J. K. (1986) *J. Biol. Chem.*, **261**, 7051–7059.
68. Kahn, R. A., and Gilman, A. G. (1986) *J. Biol. Chem.*, **261**, 7906–7911.
69. Waldo, G. L., Evans, T., Fraser, E. D., Northup, J. K., Martin, M. W., and Harden, T. K. (1987) *Biochem. J.*, **246**, 431–439.
70. Kikuchi, A., Yamshita, T., Kawata, M., Yamamoto, K., Ikeda, K., Tanimoto, T., and Takai, Y. (1988) *J. Biol. Chem.*, **263**, 2897–2904.
71. Yamashita, T., Yamamoto, K., Kikuchi, A., Kawata, M., Kondo, J., Hishida, T., Ternishi, Y., Shiku, H., and Takai, Y. (1988) *J. Biol. Chem.*, **263**, 17181–17188.
72. Yamamoto, K., Kondo, J., Hishida, T., Teranishi, Y., and Takai, Y. (1988) *J. Biol. Chem.*, **263**, 9926–9932.
73. Kikuchi, A., Yamamoto, K., Fujita, T., and Takai, Y. (1988) *J. Biol. Chem.*, **263**, 16303–16308.
74. Bokoch, G. M., Parkos, C. A., and Mumby, S. M. (1988) *J. Biol. Chem.*, **263**, 16744–16749.
75. Kahn, R. A., Goddard, C., and Newkirk, M. (1988) *J. Biol. Chem.*, **263**, 8282–8287.
76. Tsai, S.-C., Noda, M., Adamik, R., Chang, P. P., Chen, H.-C., Moss, J., and Vaughan, M. (1988) *J. Biol. Chem.* **263**, 1768–1772.
77. Yamamoto, K., Kim, S., Kikuchi, A., Takai, Y. (1988) *Biochem. Biophys. Res. Comm.*, **155**, 1284–1292.

3
IMMUNOLOGICAL PROBES AND THE IDENTIFICATION OF GUANINE NUCLEOTIDE BINDING PROTEINS

Graeme Milligan

Molecular Pharmacology Group, Departments of Biochemistry and Pharmacology, University of Glasgow, Glasgow, G12 8QQ, Scotland

INTRODUCTION

Given the extreme homology of the primary sequences of the α subunits of the known signal transducing guanine nucleotide binding proteins (G-proteins) (see Chapter 4) and that, except for the forms of $G_s\alpha$, all of these polypeptides are close to 40 kDa in molecular mass, then it is apparent that means to identify selectively each G-protein are required. The traditional strategy has been based upon the ability of particular bacterial exotoxins to transfer catalytically the [^{32}P]ADP-ribose moiety of [^{32}P]NAD to the α subunits of various G-proteins. While cholera toxin is able to catalyse ADP-ribosylation of the forms of $G_s\alpha$ on an arginine residue, this is not a highly specific reaction and autoradiographs of membrane preparations following treatment with cholera toxin and [^{32}P]NAD frequently demonstrate the incorporation of radioactivity into many polypeptides. There is no reason to assume that these additional polypeptides represent other G-proteins, although certain other G-proteins do appear to be able to act as substrates for cholera-toxin-catalysed ADP-ribosylation under appropriate artificial conditions [1–4]. One polypeptide which is frequently labelled in this reaction is the A subunit of cholera toxin itself. Further, a series of common proteins including bovine serum albumin and lysozyme can act as suitable substrates for the action of cholera toxin when present in high concentrations. While pertussis toxin catalyses an analogous NAD-dependent ADP-ribosylation to that of cholera toxin, the acceptor amino acid is a cysteine residue and the substrate specificity is much greater than for cholera toxin. Although treatment with pertussis toxin produces attenuation of receptor-mediated control of the inhibition of adenylate cyclase, and hence a substrate for this toxin is defined as the α subunit of 'G_i' it has become apparent that a considerable number of pertussis-toxin-sensitive G-proteins exist, including, G_i1, G_i2, G_i3, TD1, TD2 and G_o [5,6]. While TD1 (rod transducin) and

TD2 (cone transducin) appear to be limited in distribution solely to photoreceptor-containing tissues, the pattern of the distribution of the other forms is complex and, as such, many, if not all, of these forms are likely to be expressed in a single cell type.

It is generally possible to resolve the pertussis-toxin-sensitive G-proteins in SDS–PAGE slabs sufficiently well to note that the apparent mobility of the various forms is such that $G_o > G_i2 > G_i3 > G_i1$ (Figure 3.1). However, the differences in mobility are small, and as various covalent modifications, such as pertussis-toxin-catalysed ADP-ribosylation [7], protein kinase C-mediated phosphorylation [8] and alkylation [9] all produce noticeable alterations in the apparent electrophoretic mobility of these G-proteins, then means for more specific identification of these polypeptides are a necessity. In the last few years, immunological probes have started to provide this specificity.

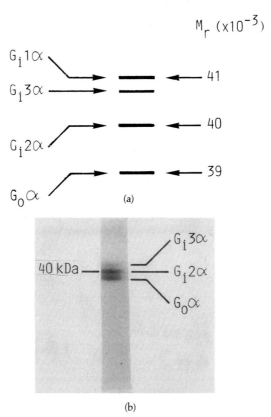

Figure 3.1 Relative mobilities of the pertussis-toxin-sensitive G-proteins in SDS–PAGE. (a) Diagrammatic representation of the relative apparent mobilities in SDS–PAGE of the pertussis-toxin-sensitive G-proteins (G_i1, G_i2, G_i3, G_o) which are found outwith photoreceptor containing tissues. (b) SDS–PAGE of membranes from neuroblastoma × glioma hybrid, NG108-15, cells following treatment with [^{32}P]NAD and thiol-activated pertussis toxin. These cells do not express detectable levels of G_i1. The identities of the three polypeptides were assessed by concurrent immunoblotting with selective anti-peptide antisera

While all of the 'classical' G-proteins are heterotrimers of distinct α, β and γ subunits, immunological studies have largely concentrated on producing reagents able to discriminate between the various α subunits. This is essentially a reflection that the α subunit is generally viewed as the polypeptide which defines the individual G-protein and as such will determine the function of the overall heterotrimeric complex. As such, this chapter will concentrate on immunological studies pertaining to the diversity and function of the α subunits. However, it is important to reflect that the functions and diversity of the β and particularly the γ subunits are ill-defined and are likely to be a fruitful area for study. Immunological probes have indeed confirmed the presence of two distinct gene products ($\beta 36$ and $\beta 35$) [10,11] corresponding to β subunits and while differences in distribution have been noted, no evidence has suggested differences in their functional properties. Immunological evidence also suggests that there are likely to be multiple variants of the γ subunit [12]. It is too early to speculate, however, on the significance of these observations.

EARLY STUDIES WITH POLYCLONAL ANTISERA

As with the understanding of many other aspects of the roles of G-proteins in cellular signalling processes it was the ready availability of relatively large amounts of transducin, isolated following elution from vertebrate rod outer segments, which allowed the generation of the first polyclonal anti-G-protein α antisera [13]. Further, it was the cross-reactivity of a polyclonal antiserum (CW6) generated against bovine rod transducin with a 41 kDa pertussis-toxin-sensitive polypeptide isolated from bovine brain [14] which allowed studies to be performed on G_i. This antiserum, while identifying brain G_i (now referred to as $G_i 1$), did not identify a 39 kDa pertussis-toxin-sensitive G-protein (G_o) which essentially co-purified with brain $G_i 1$ [14,15], providing tentative evidence that G_o was unlikely to represent a proteolytic cleavage product of $G_i 1$. Production of polyclonal antisera against mixtures of $G_i 1$ and G_o (using close to 1:1 ratios of these two proteins) succeeded in generating a number of antisera able to identify $G_o \alpha$ (e.g. RV3) but none against $G_i \alpha$ [16]. These data thus confirmed that G_o could not be derived directly from $G_i 1$ by proteolytic cleavage.

It appears to be extremely difficult to generate selective polyclonal antisera against $G_i 1$ by using this protein as antigen, as only a few reports have commented upon obtaining a useful reagent by this means [17,18]. The availability of antisera able to discriminate selectively between G_o and $G_i 1$ thus allowed both mapping of the distribution of these proteins [19–22] and the recognition that individual cell types were able both to co-express these proteins and to regulate independently their levels [23,24]. However, the inability of either of these antisera to identify the major pertussis-toxin-sensitive G-protein(s) of each of human neutrophils [25] and rat glioma cells [26], observations which could not be attributed simply to species variation, indicated that further complexity must exist in the range of pertussis-toxin-sensitive G-proteins expressed. The fact that a number of later bleeds of the anti-$G_i 1$ antiserum also displayed weak cross-reactivity with

G_o [14] suggested that these two polypeptides were likley to have considerable sequence homology and the subsequent isolation of cDNA clones corresponding to a range of individual pertussis-toxin-sensitive G-proteins has confirmed this concept (see [6] and Chapter 4 for review). Thus, it is important to be able to validate the specificity/selectivity of anti-G-protein antisera. Given the extreme similarities of the different gene products it still remains virtually impossible to produce a homogeneous preparation of a single G-protein using protein purification techniques [27] (see also Chapter 2). On the basis of these facts it would be surprising if a number of polyclonal antisera displayed complete specificity for the G-protein that they were theoretically generated against. Considerations such as this may explain the considerable debate relating to whether G_i1 and G_i2 or G_i2 and G_o represent the two major forms of pertussis-toxin-sensitive G-proteins expressed in rat adipocytes [28–30] (the former of these statements is the correct one (Figure 3.2)). Despite, these caveats, a considerable range of polyclonal antisera have been generated against 'purified' G-protein preparations and have provided a wealth of information [31–34].

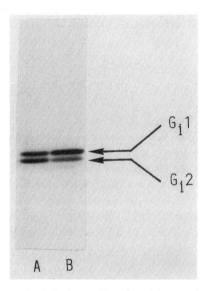

Figure 3.2 The two major pertussis-toxin-sensitive G-proteins in rat adipocyte plasma membranes are G_i1 and G_i2. Rat white adipocyte plasma membranes were resolved by SDS–PAGE and immunoblotted with an antiserum (SG1) which identifies the α subunits of G_i1 and G_i2 equally but does not cross-react with the α subunits of either G_i3 or G_o.

ANTI-PEPTIDE ANTISERA

Information obtained from cDNAs corresponding to the various G-proteins has led to a number of groups generating anti-peptide antisera which are likely to represent selective reagents for the individual G-proteins [7,35–38].

Epitope mapping of the polyclonal anti-rod transducin antiserum (CW6) which was able to cross-react with $G_i1\alpha$ demonstrated that essentially all of the immunoreactivity was directed against a single region located within the C-terminal 5 kDa peptide of transducin α [14]. As the predicted primary sequence of rod transducin was then available, via cDNA cloning, but the equivalent information for G_i1 was not, we generated anti-peptide antisera against a synthetic peptide which is equivalent to the extreme C-terminal decapeptide of rod transducin α, with the hope that the reagents so produced would also cross-react with G_i1. This was indeed the case [7,39], and with hindsight it is not surprising that this should be true as information from the cDNA corresponding to G_i1 predicts the alteration of but a single amino acid within the region used as antigen. These antisera did not cross-react with G_o which differs in amino acid sequence in five of the ten positions of the synthetic peptide. More surprisingly at the time, these antisera also identified the major pertussis-toxin-sensitive G-proteins of both neutrophils [39] and rat glioma cells [7], which had not been recognized by the polyclonal anti-G_i1 or G_o antisera described above [25,26], hence demonstrating both that G_i1 and the novel pertussis-toxin-sensitive G-protein present in these cells must have marked similarities, presumably, although not neccessarily, in their extreme C-terminal regions and if this were so that the epitope identified by the polyclonal antiserum (CW6) was not at the extreme C-terminus of transducin but further upstream.

With the subsequent cloning of cDNAs for a pertussis-toxin-sensitive G-protein from both rat glioma cells [40] and leukocytic tissues [41] and the realization that the presumptive G-protein so encoded (G_i2) was highly similar (88% overall identity to G_i1) and indeed was completely identical to G_i1 in the extreme C-terminal region used to generate the antiserum, the cross-reactivity of the antipeptide antiserum between G_i1 and G_i2 was explained. This, however, necessitated the generation of further anti-peptide antisera able to discriminate between the two forms. This has now been achieved [7,27]. The recent identification of a further cDNA (G_i3) [5,42] in which the predicted protein sequence is even more highly related to G_i1 (94% identity) than is G_i2 has required a re-examination of the antisera previously produced to assess if they display cross-reactivity with G_i3. It appears that certain of the C-terminal antisera which, as noted above cannot distinguish between G_i1 and G_i2, show essentially no cross-reactivity with G_i3 (unpublished), even though the C-terminal ten amino acids of G_i3 are highly similar to those of G_i1 and G_i2, differing in only two of the ten positions. Further, the two alterations represent extremely conservative substitutions (Figure 3.3). Furthermore, antisera raised against a synthetic peptide representing the C-terminal decapeptide of G_i3 appear not to cross-react with G_i1 and G_i2; however, they do cross-react with G_o (Figure 3.4). It must be considered likely that this is a reflection of the antigenic importance of the aromatic alcohol of the C-terminal tyrosine residue which G_i3 and G_o, but not G_i1 and G_i2, have as a common feature.

G_i1	K	N	N	L	K	D	C	G	L	F
G_i2	K	N	N	L	K	D	C	G	L	F
G_i3	K	N	N	L	K	E	C	G	L	Y
G_o	A	N	N	L	R	G	C	G	L	Y

Figure 3.3 The C-terminal decapeptides of the 'G_i-like' G-proteins

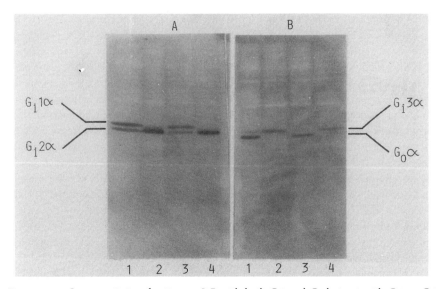

Figure 3.4 Cross-reactivity of antiserum I3B with both G_i3 and G_o but not with G_i1 or G_i2. Membranes of rat glioma C6 BU1 cells (two, 100 μg; four, 50 μg) which express high levels of G_i2 and substantial quantities of G_i3, and membranes of rat cerebral cortex (one, 100 μg; three, 50 μg) which have high levels of G_i1 and G_o and lower levels of G_i2 and G_i3 were resolved on SDS–PAGE using conditions able to resolve the various pertussis-toxin-sensitive G-proteins. Samples were immunoblotted with either antiserum SG1 (panel A) (see legend to Figure 3.2) or antiserum I3B (directed against the C-terminal decapeptide of $G_i3\alpha$) (panel B). Antiserum I3B identified a 41 kDa polypeptide in C6 BU1 cells ($G_i3\alpha$) but in cerebral cortex, while low levels of G_i3 were noted (panel B, lane 1), the antiserum also identified a 39 kDa polypeptide ($G_o\alpha$) which was not detected in C6 BU1 membranes. The cross-reactivity of this antiserum between G_i3 and G_o, but not with G_i1 or G_i2, strongly suggests that the C-terminal amino acid, (phenylalanine for G_i1 and G_i2, tyrosine for G_i3 and G_o) plays a key role in epitope identification for antisera directed against the extreme C-terminal region of the G-proteins

Possible cross-reactivity of anti-G-protein antisera with each newly identified G-protein must be considered carefully. While the most recent addition to the G-protein family (G_z) [43] (also called G_x) [44] is less homologous with the other identified G-proteins there is no reason to assume that all of the members of this class of proteins have now been identified and cloned. The specificity of antisera of defined sequence is likely to be especially difficult to validate in the case of G-proteins such as G_z which are currently known only as a presumptive coding sequence and where the protein itself remains to be isolated.

QUANTITATION OF G-PROTEIN LEVELS USING ANTI-PEPTIDE ANTISERA

While the use of selective antisera has provided a considerable increase in our knowledge of both the distribution and function (see later) of the individual G-proteins, relatively little progress has been made in using these probes in a quantitative manner. Largely this is a reflection of the lack of availability of the relevant purified proteins for use as standards. It remains a major undertaking to purify homogeneous preparations [27] (see Chapter 2) of the individual G-proteins, a task not suited, for example, to the use of such antisera as diagnostic probes in clinical laboratories (see Chapter 12). The ability to produce recombinant protein derived from relevant cDNA clones is likely to alleviate this problem [45–48]. Firstly, this approach should indeed provide homogeneous preparations of the different G-proteins and, secondly, although some difficulties in the expression of functional G-proteins by this strategy have been noted [45], the activity or otherwise of the protein is not relevant in these assays. Standard curves can thus be generated utilizing such recombinant protein on western blots to allow quantitation. It would be more convenient to adapt such procedures for ELISA systems to allow either recombinant protein, or alternatively the synthetic peptide used to generate the antiserum, to act as standard. This indeed has been reported, at least using an anti-peptide antiserum able to identify $G_s\alpha$ [49]. However, potential difficulties with these systems should be noted:

(1) The antiserum may not react equally well with the synthetic peptide and the G-protein under investigation when it is located either in cell membranes or in a detergent extract of the membranes.
(2) Many of the anti-peptide antisera so far produced demonstrate cross-reactivity with a range of other proteins besides the G-protein in question (it is noticeable that many papers employing such reagents display only the 'area of interest' of a western blot). These cross-reactivities may vary in different tissues and cells dependent upon the levels of expression of these other polypeptides. As such it must be assumed that these polypeptides would contribute to the signal recorded.
(3) As distinct short (44 kDa) and long (46 kDa) forms of $G_s\alpha$ are generated in a single cell via differential splicing of pre-mRNA transcribed from a single gene then many anti-$G_s\alpha$ antisera will identify both forms equally and hence provide a composite signal.

IMMUNOLOGICAL ASSESSMENTS OF THE CELLULAR LOCATION OF G-PROTEINS

As the function of G-proteins is to allow transfer of information between hormone-activated receptor and effector systems and that each of these entities is either a transmembrane spanning protein or is in integral association with the plasma membrane, then it should be anticipated that the G-proteins should also be in intimate contact with the plasma membrane. No evidence has yet been produced to indicate that any of the G-protein subunits is a transmembrane protein and thus it is likely that G-proteins are in contact with the cytoplasmic face of the plasma membrane. With the exception of transducin, which can be removed from membranes of rod outer segments simply by altering the ionic strength of the buffer (see Chapters 2 and 7), the G-proteins appear to require detergent treatment to produce a soluble preparation. The availability of selective anti-G-protein antisera has allowed a direct examination of the cellular localization of G-protein subunits and has produced some surprising answers.

Examination of the location of G-protein α subunits by either immunoblotting subcellular fractions from various cells [50–52] or by the use of immunocytochemistry [53] has indicated a more varied location of G-proteins than might be anticipated from their known functions. As well as being associated with the plasma membrane, G-proteins have been detected in various microsomal fractions and indeed also in the cytoplasm. While it could be suggested that the G-protein detected in the cytoplasm might reflect protein moving towards membrane fractions following synthesis, experiments using isolated membrane fractions have demonstrated that the α subunits of all of the G-proteins can be released slowly from the membrane following activation of the G-protein with poorly hydrolysed analogues of GTP [54–56] (Figure 3.5). Analogues of either ATP or GDP cannot substitute in this process. Furthermore, persistent activation of $G_s\alpha$ in whole cells following cholera-toxin-catalysed ADP-ribosylation produces a release and 'down-regulation' of $G_s\alpha$ in all cells in which this process has been examined [57] (Figure 3.6).

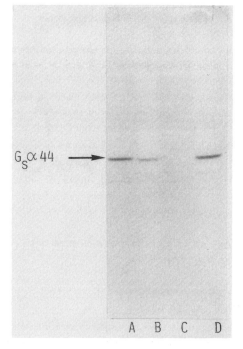

Figure 3.5 Release of $G_s\alpha$ from membranes of rat glioma C6 cells in response to activation by Gpp(NH)p. Membranes of rat glioma C6 BU1 (100 μg) cells were incubated for 60 minutes at 37 °C with Gpp(NH)p (100 μM)(A, B) or GDPβS (100 μM) (C, D). The samples were subsequently resolved into membrane-associated (B, D) and supernatant fractions (A, C) by centrifugation in an airfuge. The individual fractions were resolved on SDS–PAGE and immunoblotted with antiserum CS1 (an antipeptide antiserum which identifies the C-terminal decapeptide of forms of $G_s\alpha$)

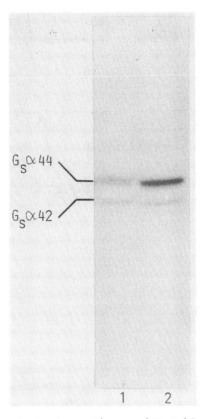

Figure 3.6 Cholera toxin pretreatment causes 'down-regulation' of $G_s\alpha$. L6 skeletal myoblasts in tissue culture were either treated (1) or not treated (2) with holomeric cholera toxin (100 ng/ml, 24 hours) and membranes then prepared. These membranes (100 μg) were resolved on SDS–PAGE and immunoblotted with antiserum CS1 (see Figure 3.5). Markedly reduced levels of $G_s\alpha$ immunoreactivity were noted in membranes of cholera-toxin-treated cells. The remaining immunoreactivity migrated more slowly through the gel, a consequence of ADP-ribosylation (see [7])

IMMUNOLOGICAL EXAMINATION OF COVALENT MODIFICATION AND G-PROTEIN FUNCTION

The availability of selective anti-G-protein antisera has also allowed examination of the presence and roles of post-translational and other covalent modifications in altering G-protein function. Immunoprecipitation of G_i and $G_s\alpha$ subunits from astrocytoma cells prelabelled with [^{3}H]myristic acid demonstrated the presence of amide-linked myristate on G_i but not on G_s [58]. Subsequent studies with purified $G_o\alpha$ have also indicated that this polypeptide is myristoylated [58,59]. It had previously been known that these proteins were resistent to Edman degradation [59] and such N-terminal modifications are consistent with these observations.

The most common mechanism whereby covalent modification can alter the functionality of proteins involves reversible phosphorylation/dephosphorylation reactions. It had originally been noted that treatment of platelets with tumour-promoting phorbol esters prevented α_2-adrenergic-mediated inhibition of adenylate cyclase [60]. While these data were compatible with the idea that a protein-kinase-C-mediated phosphorylation of G_i could functionally inactivate this protein, particularly as it was concurrently noted that protein kinase C would catalyse phosphorylation of purified G_i [61], it has required the availability of anti-$G_i\alpha$ antisera capable of immunoprecipitation to demonstrate this directly. Treatment of rat hepatocytes with either TPA or with an analogue of glucagon (TH-glucagon) which is able to stimulate the production of diacylglycerol but not of cAMP, led to the functional inactivation of G_i as assessed by the inability of low concentrations of Gpp(NH)p to inhibit forskolin-amplified adenylate cyclase activity in membranes prepared from the treated cells [8]. Immunoprecipitation of phosphorylated G_i was noted from the treated cells [8]. Interestingly, immunoprecipitation of G_i from untreated hepatocytes demonstrated a lower, basal incorporation of [^{32}P] into $G_i\alpha$ [8]. This observation has been replicated in independent experiments [62]. The basis for the basal phosphorylation of $G_i\alpha$ in hepatocytes is unclear and it remains to be ascertained if incorporation of phosphate under the two conditions is at the same site. Basal phosphorylation of G_i, however, does not appear to be a universal phenomenon ([63] and unpublished).

Phosphorylation of purified G_i by both cAMP-dependent protein kinase [64] and the tyrosyl kinase activity of the insulin receptor [65,66] has also been noted but it remains to be demonstrated if these events occur in whole cells. However, δ opioid receptor-mediated inhibition of adenylate cyclase is attenuated in membranes of neuroblastoma × glioma hybrid NG108-15 cells following exposure of the cells to dibutyryl cAMP [38], suggesting that phosphorylation and functional inactivation of G_i by cAMP-dependent protein kinase may occur.

ANTISERA AS PROBES FOR FUNCTIONAL DOMAINS OF G-PROTEINS

Given that the role of a G-protein is to allow selective communication between agonist-activated receptors and either enzymes which synthesize second messengers or certain ion channels, then particular domains of the G-proteins must define selective contact sites between receptors and the G-protein and between the G-protein and effector system. Considerable information has recently become available, via site-directed mutagenesis, deletion mutagenesis and the construction of chimaeric species, on which regions of G-protein-linked receptors are likely to be important for the interaction of receptors with G-proteins. Similar studies are under way using clones corresponding to the various G-proteins [67]. However, the availability of site-specific antisera against various G-proteins has allowed alternative strategies to be employed to address the same question. Pertussis-toxin-catalysed ADP-ribosylation prevents productive contacts between receptors and G-proteins which are substrates for this toxin. The site of this modification is only four amino acids from the C-terminus of the α subunit of each of these G-proteins. On this basis it would appear that the C-terminus of G-proteins is likely to be a key area of contact between G-proteins and receptors [68]. The interaction of a hormone-stimulated receptor with a G-protein can most appropriately and directly be examined by measuring receptor stimulation of the rate of binding and subsequent hydrolysis of GTP by the relevant G-protein [69]. Such stimulation of high affinity GTPase activity can usually be measured adequately when the G-protein in question is a substrate for pertussis toxin [70,71]. We have assessed the interaction of a δ opioid receptor with a pertussis-toxin-sensitive G-protein(s) in neuroblastoma $\times$ glioma, NG108-15 cells by incubating membranes derived from these cells with antisera directed against the extreme C-terminus of either forms of G_i (AS7) or G_o (OC1) [72,73]. Opioid peptide stimulation of high affinity GTPase activity was obliterated by using the anti-G_i antiserum (in this cell, this antiserum identifies only G_i2) but was not attenuated by pre-incubation with the anti-G_o antiserum, even though both of these antisera are able to recognize and immunoprecipitate the appropriate G-protein from a mixture of G_i and G_o isolated from bovine brain [8].

A considerable series of experiments have recently been performed using similar approaches. The vertebrate visual transduction system has been particularly amenable to such an approach. Cerione et al. [74] pre-incubated purified transducin α subunit with the same antiserum (AS7) as previously used by McKenzie et al. [72,73] and then incorporated this complex into phospholipid vesicles along with rhodopsin. Following treatment with the antiserum, there was a diminished ability of rhodopsin to stimulate the GTPase activity of transducin. In contrast, pre-incubation of transducin with non-specific rabbit IgG did not hinder the interaction of rhodopsin and transducin. Maximal inhibition of the rhodopsin-stimulated GTPase activity was achieved at calculated molar ratios of transducin to specific IgG of approximately 1:1. Further, the intrinsic basal GTPase activity of transducin α was not reduced by incubation with the antibodies, indicating that only the interaction of rhodopsin and transducin was being hindered. Rather than using site-specific anti-peptide antisera, Hamm and co-workers [75,76] have used a monoclonal antibody, which they have epitope mapped to a region close to the C-terminus of transducin [76], to interfere in a similar fashion with the light-dependent activation of transducin by rhodopsin. The monoclonal antibody used by Hamm and co-workers cross-reacts with a series of other G-proteins and as such they have also used this means to attempt to interfere with the interaction of muscarinic receptors with G_k. The activation of a set of K^+ channels in isolated membrane patches of guinea pig atrial cells by muscarinic receptor agonists is attenuated by pretreatment with pertussis toxin [77]. Application of the monoclonal antibody to the solution bathing an inside out patch of atrial membrane prevented agonist modulation of K^+ currents. By contrast, the antibody was unable to prevent the effect on K^+ current when it was added to membrane patches in which G_k has been persistently activated by the addition of GTPγS, indicating that the antibody was producing its effect by preventing activation of the G-protein and not at a subsequent stage [78]. Thus, from a series of experimental models, it appears that antisera able to interact with the C-terminus of the α subunits of a range of G-proteins are able to prevent receptor activation of the G-protein.

IMMUNOLOGICAL EXAMINATION OF G-PROTEIN LEVELS IN CLINICAL DISORDERS

It has been recognized for some time that the genetic lesion in patients with pseudohypoparathyroidism type 1a is the heterozygous non-expression of the α subunit of G_s [79]. More recently it has become apparent that a number of disorders display alterations in tissue sensitivity to hormones which function at receptors which interact with the G-protein signalling system. A particular case in point are syndromes in which one facet of the disorder is a tissue resistance to insulin. In both diabetic states [80] and in the normaglycaemic but insulin-resistant obese Zucker rat [81], a complete lack of function of G_i is noted (see Chapter 12 for details). In plasma membranes of hepatocytes isolated from streptozotocin-induced diabetic rats, markedly lower levels of immunoreactive G_i were recorded in comparison to control animals [80]. This observation may, however, be tissue specific. While no functional G_i can be measured in adipocyte membranes derived from the same diabetic animals, equivalent levels of immunoreactive G_i are present in adipocyte membranes of both control and diabetic animals (Strassheim, Milligan and Houslay, in preparation), suggesting that the alteration in G-protein function may be due to a covalent modification.

Hypothyroidism is associated in rat models with an enhanced sensitivity of adipocytes to hormones which function via the inhibition of adenylate cyclase and a reduction in sensitivity to hormones which function via a stimulation of this enzyme. Plasma membranes of adipocytes isolated from hypothyroid animals show increased levels of $G_i\alpha$ in comparison to those from euthyroid controls [29,82]. Further, they display elevated levels of the G-protein β subunit [82] (Figure 3.7). In contrast to these studies, a small study has recently reported that there are no significant alterations in levels of the α subunits of either G_s or G_i in fat cells isolated from either normal or hypothyroid human subjects [83]. The reason for this discrepancy is unclear but could relate to the greater inherent genetic variability of the human subjects employed and the relatively small sample size.

An undefined mutation in $G_s\alpha$ is likely to be responsible for the high secretory activity of a number of growth-hormone-producing pituitary adenomas [84]. Further, reduced levels of each of mRNA, polypeptide and function of $G_s\alpha$ has been noted in neuroblastoma × glioma hybrid cells which were exposed to high levels of ethanol [85]. This observation may be of significance in relation to known defects in the stimulatory arm of adenylate cyclase in tissues of alcoholics. As alterations of G-protein concentration and function appear to be associated with the failing heart [86], it is likely that studies in a range of cardio-vascular models will provide further evidence for alterations in the efficiency of G-protein-linked signalling systems in disease processes.

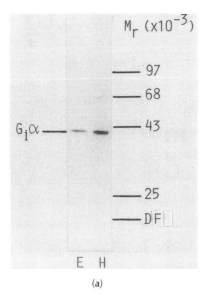

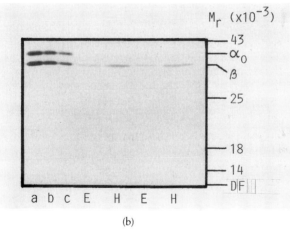

Figure 3.7 Hypothyroid and euthyroid adipocytes. Levels of $G_i\alpha$ and β subunit. (a) Elevated levels of $G_i\alpha$ in membranes of hypothyroid rats. Membranes were isolated from adipocytes of either control euthyroid (E) or hypothyroid (H) rats. After SDS–PAGE the samples were immunoblotted for G_1. Membranes from the hypothyroid rats displayed elevated G_i immunoreactivity (see [82] for fuller details). (b) Elevated levels of β subunit in membranes of hypothyroid rats. Membranes from adipocytes of either control, euthyroid (E) or hypothyroid (H) rats were resolved as in Figure 3.7(a) and immunoblotted with antiserum RV3, a polyclonal antiserum generated against partially purified bovine brain G_o [16]. Lanes a, b, and c contain varying concentrations of bovine brain G_o. Antiserum RV3 contains populations of antibodies directed against either $G_o\alpha$ or β subunit. Hypothyroid adipocytes are shown to contain higher levels of the β subunit than are membranes from adipocytes of euthyroid animals. No reactivity corresponding to $G_o\alpha$ was noted in adipocyte membranes (see also Figure 3.2 and related text)

CONCLUSIONS

The availability of a series of selective immunological probes able to identify the α subunits of individual G-proteins has quickly led to a considerable increase in our knowledge of the diversity, distribution and function of this family of homologous proteins. This process is likely to continue in parallel with information generated by the construction and expression of chimaeric G-proteins and the expression and reconstitution of defined G-proteins via either cDNAs or genomic clones [87,88].

REFERENCES

1. Owens, J. R., Frame, L. T., Ui, M., and Cooper, D. M. F. (1985) *J. Biol. Chem.*, **260**, 15946–15952.
2. Gierschik, P., and Jakobs, K. H. (1987) *FEBS Lett.*, **224**, 219–224.
3. Milligan, G., and McKenzie, F. R. (1988) *Biochem. J.*, **252**, 369–373.
4. Milligan, G. (1989) *Cell. Signalling*, **1**, 65–74.
5. Jones, D. T., and Reed, R. R. (1987) *J. Biol. Chem.*, **262**, 14241–14249.
6. Lochrie, M. I., and Simon, M. I. (1988) *Biochemistry*, **27**, 4957–4965.
7. Goldsmith, P., Gierschik, P., Milligan, G., Unson, C. G., Vinitsky, R., Malech, H., and Spiegel, A. (1987) *J. Biol. Chem.*, **262**, 14683–14688.
8. Pyne, N. J., Murphy, G. J., Milligan, G., and Houslay, M. D. (1989) *FEBS Lett.*, **243**, 77–82.
9. Sternweis, P. C., and Robishaw, J. D. (1984) *J. Biol. Chem.*, **259**, 13806–13813.
10. Gao, B., Mumby, S., and Gilman, A. G. (1987) *J. Biol. Chem.*, **262**, 17254–17257.
11. Roof, D. J., Applebury, M. L., and Sternweis, P. C. (1985) *J. Biol. Chem.*, **260**, 16242–16249.
12. Evans, T., Fawzi, A., Fraser, E. D., Brown, M. L., and Northup, J. K. (1987) *J. Biol. Chem.*, **262**, 176–181.
13. Gierschik, P., Codina, J., Simons, C., Birnbaumer, L., and Spiegel, A. (1985) *Proc. Natl. Acad. Sci. USA*, **82**, 727–731.
14. Pines, M., Gierschik, P., Milligan, G., Klee, W., and Spiegel, A. (1985) *Proc. Natl. Acad. Sci. USA*, **82**, 4095–4099.
15. Milligan, G., and Klee, W. A. (1985) *J. Biol. Chem.*, **260**, 2057–2063.
16. Gierschik, P., Milligan, G., Pines, M., Goldsmith, P., Codina, J., Klee, W., and Spiegel, A. (1986) *Proc. Natl. Acad. Sci. USA*, **83**, 2258–2262.
17. Katada, T., Oinuma, M., Kasakabe, K., and Ui, M. (1987) *FEBS Lett.*, **213**, 353–358.
18. Nagata, K.-I., Katada, T., Tohkin, M., Itoh, Y., Kaziro, Y., Ui, M., and Nozawa, Y. (1988) *FEBS Lett.*, **237**, 113–117.
19. Milligan, G., Streaty, R. A., Gierschik, P., Spiegel, A., and Klee, W. A. (1987) *J. Biol. Chem.*, **262**, 8626–8630.
20. Lutje, C. W., Gierschik, P., Milligan, G., Unson, C. G., Spiegel, A., and Nathanson, N. M. (1987) *Biochemistry*, **26**, 4876–4884.
21. Homburger, V., Brabet, P., Audigier, Y., Pantaloni, C., Bockaert, J., and Rouot, B. (1987) *Mol. Pharmacol.*, **31**, 313–319.
22. Huff, R. M., Axton, J. M., and Neer, E. J. (1985) *J. Biol. Chem.*, **260**, 10864–10871.
23. Gierschik, P., Morrow, B., Milligan, G., Rubin, C., and Spiegel, A. (1986) *FEBS Lett.*, **199**, 103–106.
24. Watkins, D. C., Northup, J. K., and Malbon, C. C. (1987) *J. Biol. Chem.*, **262**, 10651–10657.
25. Gierschik, P., Falloon, J., Milligan, G., Pines, M., Gallin, J., and Spiegel, A. (1986) *J. Biol. Chem.*, **261**, 8058–8062.

26. Milligan, G., Gierschik, P., Spiegel, A., and Klee, W. (1986) *FEBS Lett.*, **195**, 225–230.
27. Goldsmith, P., Backlund, P. S. Jr, Rossiter, K., Carter, A., Milligan, G., Unson, C. G., and Spiegel, A. (1988) *Biochemistry*, **27**, 7085–7090.
28. Rapiejko, P. J., Northup, J. K., Evans, T., Brown, J. E., and Malbon, C. C. (1986) *Biochem. J.*, **240**, 35–40.
29. Ros, M., Northup, J. K., and Malbon, C. C. (1988) *J. Biol. Chem.*, **263**, 4362–4368.
30. Hinsch, K.-D., Rosenthal, W., Spicher, K., Binder, T., Gausepohl, H., Frank, R., Schultz, G., and Joost, H. G. (1988) *FEBS Lett.*, **238**, 191–196.
31. Lutje, C. W., and Nathanson, N. M. (1988) *J. Neurochem.*, **50**, 1775–1782.
32. Worley, P. F., Baraban, J. M., Van Dop, C., Neer, E. J., and Snyder, S. H. (1986) *Proc. Natl. Acad. Sci. USA*, **83**, 4561–4565.
33. Milligan, G., Gierschik, P., and Spiegel, A. (1987) *Biochem. Soc. Trans.*, **15**, 42–45.
34. Toutant, M., Aunis, D., Bockaert, J., Homburger, V., and Rouot, B. (1987) *FEBS Lett.*, **215**, 339–344.
35. Mumby, S. M., Kahn, R. A., Manning, D. R., and Gilman, A. G. (1986) *Proc. Natl. Acad. Sci. USA*, **83**, 265–269.
36. Lerea, C. L., Somers, D. E., Hurley, J. B., Klock, I. B., and Bunt-Milan, A. H. (1986) *Science*, **234**, 77–80.
37. Chang, K.-J., Pugh, W., Blanchard, S. G., McDermed, J., and Tam, J. P. (1988) *Proc. Natl. Acad. Sci. USA*, **85**, 4929–4933.
38. Mullaney, I., Magee, A. I., Unson, C. G., and Milligan G. (1988) *Biochem. J.*, **256**, 649–656.
39. Falloon, J., Malech, H., Milligan, G., Unson, C., Kahn, R., Goldsmith, P., and Spiegel, A. (1986) *FEBS Lett.*, **209**, 352–356.
40. Itoh, H., Kozasa, T., Nagata, S., Nakamura, S., Katada, T., Ui, M., Iwai, S., Ohtsuka, E., Kawasaki, H., Suzuki, K. and Kaziro, Y. (1986) *Proc. Natl. Acad. Sci. USA*, **83**, 3776–3780.
41. Didsbury, J. R., Ho, Y.-S. and Snyderman, R. (1987) *FEBS Lett*, **211**, 160–164.
42. Suki, W. N., Abramowitz, J., Mattera, R., Codina, J., and Birnbaumer, L. (1987) *FEBS Lett.*, **220**, 187–192.
43. Fong, K. H. W., Yoshimoto, K. K., Eversole-Cire, P., and Simon, M. I. (1988) *Proc. Natl. Acad. Sci. USA*, **85**, 3066–3070.
44. Matsuoka, M., Itoh, H., Kozasa, T., and Koziro, Y. (1988) *Proc. Natl. Acad. Sci. USA*, **85**, 5384–5388.
45. Graziano, M. P., Casey, P. J., and Gilman, A. G. (1987) *J. Biol. Chem.*, **262**, 11375–11381.
46. Olate, J., Mattera, R., Codina, J., and Birnbaumer, L. (1988) *J. Biol. Chem.*, **263**, 10394–10400.
47. Yatani, A., Mattera, R., Codina, J., Graf, R., Okabe, K., Padrell, E., Iyengar, R., Brown, A. M., and Birnbaumer, L. (1988) *Nature (London)*, **336**, 680–682.
48. Goldsmith, P., Rossiter, K., Carter, A., Simonds, W., Unson, C. G., Vinitsky, R., and Spiegel, A. M. (1988) *J. Biol. Chem.*, **263**, 6476–6479.
49. Ransnas, L. A., and Insel, P. A. (1988) *J. Biol. Chem.*, **263**, 9482–9485.
50. Rotrosen, D. R., Gallin, J. I., Spiegel, A., and Malech, H. L. (1988) *J. Biol. Chem.*, **263**, 10958–10964.
51. Bokoch, G. M., Bickford, K., and Bohl, B. P. (1988) *J. Cell. Biol.*, **106**, 1927–1936.
52. Ali, N., Milligan, G., and Evans, W. H. (1989) *Biochem. J.*, **261**, 905–912.
53. Gabrion, J., Brabet, P., Dao, B. N. T., Homburger, Dumuis, A., Sebben, M., Rouot, B., and Bockaert, J. (1989) *Cell. Signalling*, **1**, 107–123.
54. McArdle, H., Mullaney, I., Magee, A. I., Unson, C., and Milligan, G. (1988) *Biochem. Biophys. Res. Comm.*, **152**, 243–251.
55. Milligan, G., Mullaney, I., Unson, C. G., Marshal, L., Spiegel, A. M., and McArdle, H. (1988) *Biochem. J.*, **254**, 391–396.
56. Milligan, G., and Unson, C. G. (1989) *Biochem. J.*, **260**, 837–841.
57. Milligan, G., Unson, C. G., and Wakelam, M. J. O. (1989) *Biochem. J.*, **262**, 643–649.

58. Buss, J. E., Mumby, S. M., Casey, P. J., Gilman, A. G., and Sefton, B. M. (1987) *Proc. Natl. Acad. Sci. USA*, **84**, 7493–7497.
59. Schultz, A. M., Tsai, S.-C., Kung, H.-F., Oroszlan, S., Moss, J., and Vaughan, M. (1987) *Biochem. Biophys. Res. Comm.*, **146**, 1234–1239.
60. Jakobs, K. H., Bauer, S., and Watanabe, Y. (1985) *Eur. J. Biochem.*, **151**, 425–430.
61. Katada, T., Gilman, A. G., Watanabe, Y., Bauer, S., and Jakobs, K. H. (1985) *Eur. J. Biochem.*, **151**, 431–437.
62. Rothenberg, P. L., and Kahn, C. R. (1988) *J. Biol. Chem.*, **263**, 15546–15552.
63. Crouch, M. F., and Lapetina, E. G. (1988) *J. Biol. Chem.*, **263**, 3363–3371.
64. Watanabe, Y., Imaizumi, T., Misaki, N., Iwakura, K., and Yoshida, H. (1988) *FEBS Lett.*, **236**, 372–377.
65. O'Brien, R. M., Houslay, M. D., Milligan, G., and Siddle, K. (1987) *FEBS Lett.*, **212**, 281–288.
66. Zick, Y., Sagi-Eisenberg, R., Pines, M., Gierschik, P., and Spiegel, A. (1986) *Proc. Natl. Acad. Sci. USA*, **83**, 9294–9297.
67. Weiss, E. R., Kelleher, D. J., Woon, C. W., Soparkar, S., Osawa, S., Heasley, L. E., and Johnson, G. L. (1988) *FASEB J.*. **2**, 2841–2848.
68. Masters, S. B., Stroud, R. M., and Bourne, H. R. (1986) *Protein. Eng.*, **1**, 47–54.
69. Milligan, G. (1988) *Biochem. J.*, **255**, 1–13.
70. Koski, G., and Klee, W. A. (1981) *Proc. Natl. Acad. Sci. USA*, **78**, 4184–4189.
71. Aktories, K., and Jakobs, K. H. (1981) *FEBS Lett.*, **130**, 235–238.
72. McKenzie, F. R., Kelly, E. C. H., Unson, C. G., Spiegel, A. M., and Milligan, G. (1988) *Biochem. J.*, **249**, 653–659.
73. McKenzie, F. R., Mullaney, I., Unson, C. G., Spiegel, A. M., and Milligan, G. (1988) *Biochem. Soc. Trans.*, **16**, 434–437.
74. Cerione, R. A., Kroll, S., Rajaram, R., Unson, C., Goldsmith, P., and Spiegel, A. M. (1988) *J. Biol. Chem.*, **263**, 9345–9352.
75. Hamm, H. E., Deretic, D., Hofmann, K. P., Schleicher, A., and Kohl, B. (1987) *J. Biol. Chem.*, **262**, 10831–10838.
76. Deretic, D., and Hamm, H. E. (1987) *J. Biol. Chem.*, **262**, 10839–10847.
77. Yatani, A., Codina, J., Brown, A. M., and Birnbaumer, L. (1987) *Science*,, **235**, 207–211.
78. Yatani, A., Hamm, H., Codina, J., Mazzoni, M. R., Birnbaumer, L., and Brown, A. M. (1988) *Science*, **241**, 828–831.
79. Carter, A., Bardin, C., Collins, R., Simons, C., Bray, P., and Spiegel, A. (1987) *Proc. Natl. Acad. Sci. USA*, **84**, 7266–7269.
80. Gawler, D. J., Milligan, G., Unson, C. G., Spiegel, A. M., and Houslay, M. D. (1987) *Nature (London)*, **327**, 229–232.
81. Houslay, M. D., Gawler, D. J., Milligan, G., and Wilson, A. (1989) *Cell. Signalling*, **1**, 9–22.
82. Milligan, G., Spiegel, A. M., Unson, C. G., and Saggerson. E. D. (1987) *Biochem. J.*, **247**, 223–227.
83. Ohisalo, J. J., and Milligan, G. (1989) *Biochem. J.*, **260**, 843–847.
84. Vallar, L., Spada, A., and Giannattasio, G. (1987) *Nature (London)*, **330**, 566–568.
85. Mochly-Rosen, D., Chang, F.-H., Cheever, L., Kim, M., Diamond, I., and Gordon, A. S. (1989) *Nature (London)*, **333**, 848–850.
86. Neumann, J., Schmitz, W., Scholz, H., von Meyerinck, L., Doring, V., and Kalmar, P. (1988) *Lancet*, **8617**, 936–937.
87. Weinstein, L. S., Spiegel, A. M., and Carter, A. D. (1988) *FEBS Lett.*, **232**, 333–340.
88. Itoh, H., Toyama, R., Kozasa, T., Tsukamoto, T., Matsuoka, M., and Kaziro, Y. (1988) *J. Biol. Chem.*, **263**, 6656–6664.

MOLECULAR BIOLO(
G-PR(

Yoshito Kaziro

Institute of Medical Science, University of Tokyo, 4-6-1, Shirokanedai, Minato-ku, Tokyo 108, JAPAN

INTRODUCTION

Signal transducing GTP binding proteins are classified largely into two groups, i.e. heterotrimeric GTP binding proteins which are referred to as G-proteins, and low molecular weight monomeric GTP binding proteins (LMG) including ras, rap, rho and rab proteins. The basic mechanism of the reaction catalysed by these proteins appears to be analogous to that proposed for translational elongation factors [37]. The GTP bound form is an active conformation which turns on the transmission of signals, and the hydrolysis of bound GTP to GDP is required to shift the conformation to an inactive form, i.e. to shut off the signal transduction.

G-proteins are involved in a variety of transmembrane signalling systems as transducers [1,2]. Two G-proteins, G_s and G_i, are involved in hormonal stimulation and inhibition, respectively, of adenylate cyclase, whereas G_o (other G protein), which is present predominantly in brain tissues, may be involved in neuronal responses. Two transducins, G_t1 and G_t2, which are present in retinal rods and cones, respectively, regulate cGMP phosphodiesterase activity and mediate visual signal transduction. There is evidence suggesting the presence of additional G-proteins, which may be involved in the activation of phospholipase C and phospholipase A_2, as well as the gating of K^+ and Ca^{2+} channels. The functions of individual G-proteins will be reviewed in other chapters of this book.

This chapter reviews briefly the structure of cDNAs for various G-protein α subunits (Gαs) from mammalian cells, the organization of human genes for Gαs, the structure–function relationship of Gα proteins, and the occurrence of other Gα genes in lower eukaryotes including *Saccharomyces cerevisiae*.

ISOLATION OF cDNA CLONES FOR G-PROTEIN α SUBUNITS FROM MAMMALIAN CELLS

Recently, much effort has been focused on the cloning of cDNAs coding for various G-protein α subunits. Table 4.1 lists the cDNA as well as genomic clones isolated from mammalian cells. These studies have revealed that there are at least three subtypes of $G_i\alpha$, designated as $G_i1\alpha$, $G_i2\alpha$, and $G_i3\alpha$, whose structures are closely related but distinct [3].

The presence of multiple $G_i\alpha$ subtypes had been suggested from the molecular heterogeneity, immunological distinction and functional differences of the pertussis toxin substrates in mammalian cells [5,34]. By comparing the deduced amino acid sequences of each $G_i\alpha$ subtype with the partial amino acid sequences of purified proteins, it was found that the 41 kDa pertussis toxin substrate mainly expressed in brain tissues corresponds to

Table 4.1 Molecular cloning of G_s, G_i, G_o, G_x and G_t α subunit genes and cDNAs

DNA library	Classification and nomenclature								References
	$G_s\alpha$	$G_i1\alpha$	$G_i2\alpha$	$G_i3\alpha$	$G_o\alpha$	$G_x\alpha$	$G_t1\alpha$	$G_t2\alpha$	
Genomic library									
Human	$G_s\alpha$								[4]
Human		$G_i1\alpha$	$G_i2\alpha$	$G_i3\alpha$					[5]
Human			$G_i2\alpha$						[6]
Human						$G_x\alpha$			[7]
cDNA library									
Human brain	$G_s\alpha$	α_i-1							[8,9]
Human brain					$G_o\alpha$				[10]
Human T-cells			α_i2	α_i3					[11]
Human monocytes (U-937)			$G_i\alpha$						[12]
Human granulocytes (HL-60)						$G_x\alpha$			[13]
Human liver	$G_s\alpha$								[14]
Human liver				α_i-3					[15]
Human liver				α_i-3					[16]
Human retina						$G_z\alpha$			[17]
Bovine brain	$G_s\alpha$								[18]
Bovine adrenal gland	$G_s\alpha$								[19]
Bovine cerebral cortex	$G_s\alpha$	$G_i\alpha$							[20,21]
Bovine pituitary gland		α_i	α_h						[22]
Bovine retina					$G_o\alpha$				[23]
Bovine cerebellum					G39				[24]
Bovine retina							$T\alpha$		[25]
Bovine retina							$T\alpha$		[26]
Bovine retina							$G_t\alpha$		[27]
Bovine retina								$T\alpha$	[28]
Rat brain						$G_x\alpha$			[7]
Rat glioma cells (C6)	$G_s\alpha$		$G_i2\alpha$	$G_i3\alpha$	$G_o\alpha$				[5,29]
Rat olfactory epithelium	$G_{\alpha s}$	$G\alpha_i1$	$G\alpha_i2$	$G\alpha_i3$	$G_o\alpha$				[30]
Mouse macrophages (PU-5)	α_s		α_i						[31]
Mouse lymphoma cells (S49)	$G_s\alpha$								[32]
Mouse lymphoma cells (S49)	$G_s\alpha$								[33]

$G_i1\alpha$, whereas 40 kDa and 41 kDa species expressed in most tissues correspond to $G_i2\alpha$ and $G_i3\alpha$, respectively.

Transducins (G_t) also have two subtypes, $G_t1\alpha$ and $G_t2\alpha$, which are expressed in rods and cones, respectively [35]. On the other hand, four different $G_s\alpha$ cDNAs are generated from a $G_s\alpha$ ($G_s1\alpha$) gene by alternative splicing. More recently, Jones and Reed [36] reported the occurrence of another subtype of $G_s\alpha$ coded by a distinct gene ($Golf\alpha$ or $G_s2\alpha$) which is expressed specifically in olfactory cells.

We have recently isolated a new $G\alpha$ clone (designated as $G_x\alpha$) which is apparently insensitive to pertussis toxin [7]. The Cys residue at the fourth position from the C-terminus which is common to all pertussis-toxin-sensitive $G\alpha$s is replaced by Ile in $G_x\alpha$. Human $G_z\alpha$ cDNA isolated independently by Fong et al. [17] from retina, may be the counterpart of $G_x\alpha$. The molecular weights, number of amino acid residues, and other properties of $G\alpha$s are listed in Table 4.2, and the amino acid sequences of rat $G_s\alpha$, $G_i2\alpha$, $G_i3\alpha$, $G_i1\alpha$, $G_o\alpha$, and $G_x\alpha$ deduced from the nucleotide sequences are shown in Figure 4.1.

Figure 4.2 shows a schematic representation of the structure of *Escherichia coli* EF-Tu, G-protein α subunits ($G\alpha$), yeast RAS2 protein, and mammalian H-ras p21 protein. A remarkable homology was found in two regions, designated as P and G sites of all GTP binding proteins. In EF-Tu, earlier biochemical studies indicated that the region around Cys-137 of EF-Tu (G site) is responsible for interaction with guanine nucleotide [37], and later the four residues Asn–Lys–Cys–Asp are found by X-ray analysis to be situated close to the guanine ring [38]. On the other hand, it has been shown that the mutation of amino acid residue 12 of p21 from Gly to Val decreases GTPase activity and increases transforming activity. The sequence homologous to this region (P site) was found in all GTP binding proteins. The consensus sequence, Gly–(Xaa)$_4$–Gly–Lys (GXXXXGK), was located close to phosphoryl residues of bound guanine nucleotide [38–40]. G' site is a unique sequence which is highly conserved in all G-proteins but is less remarkable in

Table 4.2 Properties of $G\alpha$ subtypes from mammalian cells

Species	M_r	Amino acids	Size	Expression	Sensitivity to CTX or PTX	Function
$G_s1\alpha$	45 663	394	52 & 45 K	all tissues	CTX	adenylate cyclase Ca^{2+} channel
$G_s2\alpha$	44 322	381	45 K	olfactory cells	CTX	adenylate cyclase Ca^{2+} channel
$G_i1\alpha$	40 345	354	41 K	neuronal cells	PTX	(PLC)
$G_i2\alpha$	40 499	355	40 K	all tissues	PTX	(adenylate cyclase)
$G_i3\alpha$	40 522	354	41 K	all tissues	PTX	(K^+ channel)
$G_o\alpha$	40 068	354	39 K	neuronal cells	PTX	Ca^{2+} channel
$G_x\alpha$	40 879	355	41 K	neuronal cells	—	?
$G_t1\alpha$	39 971	350	40 K	retinal rods	CTX, PTX	cGMP PDase
$G_t2\alpha$	40 143	354	40.5 K	cones	CTX, PTX	cGMP PDase

```
Gsα : MGCLGNSKTEDQRNEEKAQREANKKIEKQLQKDKQVYRATHRLLLLGAGESGKSTIVKQM  60
Gi2α: MGCTVSA-------EDKAAAERSKMIDKNLREDGEKAAREVKLLLLGAGESGKSTIVKQM  53
Gi3α: MGCTLSA-------EDKAAVERSKMIDRNLREDGEKAAKEVKLLLLGAGESGKSTIVKQM  53
Gi1α: MGCTLSA-------EDKAAVERSKMIDRNLREDGEKAAREVKLLLLGAGESGKSTIVKQM  53
Goα : MGCTLSA-------EERAALERSKAIEKNLKEDGISAAKDVKLLLLGAGESGKSTIVKQM  53
Gxα : MGCRQSS-------EEKEAARRSRRIDRHLRSESQRQRREIKLLLLGTSNSGKSTIVKQM  53

Gsα : RILHVNGFNGEGGEEDPQAARSNSDGEKATKVQDIKNNLKEAIETIVAAMSNLVPPVELA 120
Gi2α: KIIHEDGYSEE---------------ECRQYRAVVYSNTIQSIMAIVKAMGNLQIDFADP  98
Gi3α: KIIHEDGYSED---------------ECKQYKVVVYSNTIQSIIAIIRAMGRLKIDFGEA  98
Gi1α: KIIHEAGYSEE---------------ECKQYKAVVYSNTIQSIIAIIRAMGRLKIDFGDA  98
Goα : KIIHEDGFSGE---------------DVKQYKPVVYSNTIQSLAAIVRAMDTLGVEYGDK  98
Gxα : KIIHSGGFNLE---------------ACKEYKPLIIYNAIDSLTRIIRALAALKIDFHNP  98

Gsα : NPENQFRVDYILSVMNVPNFDFPPEFYEHAKALWEDEGVRACYERSNEYQLIDCAQYFLD 180
Gi2α: QRADDARQLFALSCAAEEQGMLPEDLSGVIRRLWADHGVQACFGRSREYQLNDSAAYYLN 158
Gi3α: ARADDARQLFVLAGSAEE-GVMTSELAGVIKRLWRDGGVQACFSRSREYQLNDSASYYLN 157
Gi1α: ARADDARQLFVLAGAAEE-GFMTAELAGVIKRLWKDSGVQACFNRSREYQLNDSAAYYLN 157
Goα : ERKADSKMVCDVVSRMEDTEPFSAELLSAMMRLWGDSGIQECFNRSREYQLNDSAKYYLD 158
Gxα : DRAYDAVQLFALTGPAESKGEITPELLGVMRRLWADPGAQACFGRSSEYHLEDNAAYYLN 158

Gsα : KIDVIKQADYVPSDQDLLRCRVLTSGIFETKFQVDKVNFHMFDVGGQRDERRKWIQCFND 240
Gi2α: DLERIAQSDYIPTQQDVLRTRVKTTGIVETHFTFKDLHFKMFDVGGQRSERKKWIHCFEG 218
Gi3α: DLDRISQTNYIPTQQDVLRTRVKTTGIVETHFTFKELYFKMFDVGGQRSERKKWIHCFEG 217
Gi1α: DLDRIAQPNYIPTQQDVLRTRVKTTGIVETHFTFKDLHFKMFDVGGQRSERKKWIHCFEG 217
Goα : SLDRIGAADYQPTEQDILRTRVKTTGIVETHFTFKNLHFRLFDVGGQRSERKKWIHCFED 218
Gxα : DLERIAAPDYIPTVEDILRSRDMTTGIVENKFTFKELTFKMVDVGGQRSERKKWIHCFEG 218

Gsα : VTAIIFVVASSSYNMVIREDNQTNRLQEALNLFKSIWNNRWLRTISVILFLNKQDLLAEK 300
Gi2α: VTAIIFCVALSAYDLVLAEDEEMNRMHESMKLFDSICNNKWFTDTSIILFLNKKDLFEEK 278
Gi3α: VTAIIFCVALSDYDLVLAEDEEMNRMHESMKLFDSICNNKWFTDTSIILFLNKKDLFEEK 277
Gi1α: VTAIIFCVALSDYDLVLAEDEEMNRMHESMKLFDSICNNKWFTDTSIILFLNKKDLFEEK 277
Goα : VTAIIFCVALSGYDQVLHEDETTNRMHESLMLFDSICNNKFFIDTSIILFLNKKDLFGEK 278
Gxα : VTAIIFCVELSGYDLKLYEDNQTSRMAESLRLFDSICNNNWFINTSLILFLNKKDLLSEK 278

Gsα : VLAGKSKIEDYFPEFARYTTPEDATPEPGEDPRVTRAKYFIRDEFLRISTASGDGRHYCY 360
Gi2α: IT--QSPLTICFPEYTGANKYDE-------------AASYIQSKFEDLNKRKDTKEIY-- 321
Gi3α: IK--RSPLTICYPEYTGSNTYEE-------------AAAYIQCQFEDLNRRKDTKEVY-- 320
Gi1α: IK--KSPLTICYPEYAGSNTYEE-------------AAAYIQCQFEDLNKRKDTKEIY-- 320
Goα : IK--KSPLTICFPEYPGSNTYED-------------AAAYIQTQF-ESKNRSPNKEIY-- 320
Gxα : IR--RIPLSVCFPEYKGQNTYEE-------------AAVYIQRQFEDLNRNKETKEIY-- 321

Gsα : PHFTCAVDTENIRRVFNDCRDIIQRMHLRQYELL 394
Gi2α: THFTCATDTKNVQFVFDAVTDVIIKNNLKDCGLF 355
Gi3α: THFTCATDTKNVQFVFDAVTDVIIKNNLKECGLY 354
Gi1α: THFTCATDTKNVQFVFDAVTDVIIKNNLKDCGLF 354
Goα : CHMTCATDTNNIQVVFDAVTDIIIANNLRGCGLY 354
Gxα : SHFTCATDTSNIQFVFDAVTDVIIQNNLKYIGLC 355
```

Figure 4.1 Deduced amino acid sequences of α subunits of rat $G_s\alpha$, $G_i2\alpha$, $G_i3\alpha$, $G_i1\alpha$, $G_o\alpha$ and $G_x\alpha$ and bovine G_t1 and G_t2. The sequences of $G_s\alpha$ and $G_i2\alpha$ are from Itoh *et al.* [29]; $G_i3\alpha$ from Itoh *et al.* [5]; $G_i1\alpha$ from Jones and Reed [30]; $G_o\alpha$ from Itoh *et al.* [29] and Jones and Reed [30]; and $G_x\alpha$ from Matsuoka *et al.* [7]; $G_t\alpha1$ from Tanabe *et al.* [25], Medynski *et al.* [26] and Yatsunami and Khorana [27]; and $G_t\alpha2$ from Lochrie *et al.* [28]

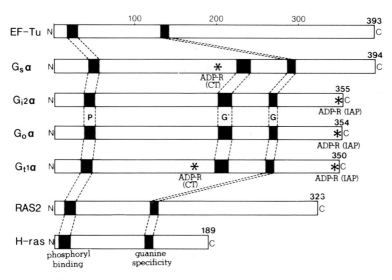

Figure 4.2 Schematic representation of structures of EF-Tu, $G_s\alpha$, $G_i2\alpha$, $G_o\alpha$, $G_t1\alpha$, RAS2, and H-*ras* p21

other GTP binding proteins except for ARF proteins [41]. Only the amino terminal sequence of the G' site, Asp–Xaa–Xaa–Gly (DXXG), is conserved in all GTP binding proteins. The DXXG sequence is found in p21 at residues 57 to 60, and in *E. coli* EF-Tu at residues 80 to 83. The deduced amino acid sequences of Gαs in the P, G',and G sites are shown in Figure 4.3(A), (B) and (C), respectively.

It must be noted that the predicted amino acid sequence of new $G_x\alpha$ was different from other Gα proteins at the P site. As shown in Figure 4.3(A), three amino acid residues in the consensus GTP hydrolysis site of $G_x\alpha$ (Thr–Ser–Asn, at positions 41–43) are different from the corresponding residues in other Gα proteins (Ala–Gly–Glu). It remains to be seen whether the kinetics of the $G_x\alpha$-mediated signal transduction may be different from other systems due to the replacement of Gly (which corresponds to Gly-12 of p21) by Ser.

ISOLATION OF HUMAN Gα GENES

We have screened human genomic libraries with above rat cDNA clones and isolated human genes coding for $G_s\alpha$, $G_i1\alpha$, $G_i2\alpha$, $G_i3\alpha$, and $G_o\alpha$. So far, we have determined the gene organization and nucleotide sequences of total exons of $G_s\alpha$, $G_i2\alpha$, and $G_i3\alpha$, and obtained the partial sequences (exons 1, 2 and 3) of $G_i1\alpha$ [4,5] (see Figure 4.4). The human $G_o\alpha$ gene is a huge gene spanning at least 90 kb, however, the organization of exons is completely identical to those of $G_i\alpha$ subfamily (Tsukamoto *et al.*, unpublished).

(A)

```
Gsα:     42-57:  R|L L L L G A G E S G K S T I V|
Giα:     35-50:  |K L L L L G A G E S G K S T I V|
Gi2α:    35-50:  |K L L L L G A G E S G K S T I V|
Gi3α:    35-50:  |K L L L L G A G E S G K S T I V|
Goα:     35-50:  |K L L L L G A G E S G K S T I V|
Gxα:     35-50:  |K L L L L G|T S N|S G K S T I V|
GP1α:    43-58:  |K L L L L G A G E S G K S T|V L
GP2α:  125-140:  |K|V|L L L G A G E S G K S T|V L
```

(B)

```
Gsα:    223-240:  |D V G G Q R|D|E R|R|K W I|Q|C F|N D
Giα:    200-217:  |D V G G Q R S E R K K W I H C F E G|
Gi2α:   201-218:  |D V G G Q R S E R K K W I H C F E G|
Gi3α:   200-217:  |D V G G Q R S E R K K W I H C F E G|
Goα:    201-218:  |D V G G Q R S E R K K W I H C F E|D|
Gxα:    201-218:  |D V G G Q R S E R K K W I H C F E G|
GP1α:   319-336:  |D|A|G G Q R S E R K K W I H C F E G|
GP2α:   296-313:  |D V G G Q R S E R K K W I H C F|D N
```

(C)

```
Gsα:    285-296:  I|S|V|I L F L N K|Q|D L
Giα:    262-273:  |T S I I L F L N K K D L
Gi2α:   263-274:  |T S I I L F L N K K D L
Gi3α:   262-273:  |T S I I L F L N K K D L
Goα:    263-274:  |T S I I L F L N K K D L
Gxα:    263-274:  |T S|L|I L F L N K K D L
GP1α:   381-392:  |T|P|F|I L F L N K|I|D L
GP2α:   358-369:  |T S|V V|L F L N K|I|D L
```

Figure 4.3 Conserved sequences of G-proteins. Sequences of P site (A), G' site (B) and G site (C) are shown.

Structure of the human $G_s\alpha$ gene and generation of 4 $G_s\alpha$ cDNAs by alternative splicing

The human $G_s\alpha$ gene isolated by Kozasa et al. [4] contained thirteen exons and twelve introns and spanned about 20 kb of genomic DNA (Figure 4.4A). It was known that there are two species of $G_s\alpha$ protein with different molecular masses (45 kDa and 52 kDa) [42]. Recently, Bray et al. [8] isolated four different $G_s\alpha$ cDNAs ($G_s\alpha1$ to $G_s\alpha4$) from human brain and characterized the partial structure. $G_s\alpha1$ and $G_s\alpha3$ are identical except that $G_s\alpha3$ lacks a single stretch of 45 nucleotides. $G_s\alpha2$ and $G_s\alpha4$ have three additional nucleotides (CAG) to $G_s\alpha1$ and $G_s\alpha3$ at the 5' end of exon 4. Robishaw et al. [19,43] isolated also two $G_s\alpha$ cDNAs from bovine adrenal that correspond to $G_s\alpha1$ and $G_s\alpha4$, and showed that these two cDNAs generated a 52-kDa and a 45-kDa protein when expressed in COS-m6 cells. Mattera et al. [14] also isolated two $G_s\alpha$ cDNAs from human liver that correspond to $G_s\alpha1$ and $G_s\alpha4$.

Comparison of the four types of human $G_s\alpha$ cDNAs [8] with the sequence of the

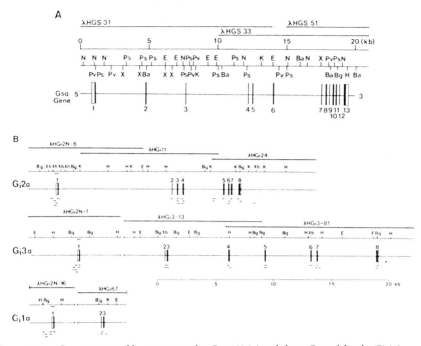

Figure 4.4 Organization of human genes for $G_s\alpha$ (A) [4] and three $G_i\alpha$ subfamily (B) [5]

human $G_s\alpha$ gene of Kozasa et al. [4] suggests that four types of $G_s\alpha$ mRNAs may be generated from a single $G_s\alpha$ gene by alternative splicing as shown in Figure 4.5. $G_s\alpha 1$ has a sequence identical to exons 2, 3 and 4, whereas $G_s\alpha 3$ lacks exon 3. $G_s\alpha 2$ and $G_s\alpha 4$ have three additional nucleotides (CAG) to $G_s\alpha 1$ and $G_s\alpha 3$, respectively, at the 5' end of exon 4. In the genomic sequence of the 3' splice site of intron 3, this CAG sequence is found. Although the 5' adjacent nucleotides to the CAG are TG and do not match with the 3' splice consensus sequence AG, this 3' splice site may be used for the production of $G_s\alpha 2$ and $G_s\alpha 4$. One additional serine residue in $G_s\alpha 2$ and $G_s\alpha 4$ may be the potential site for

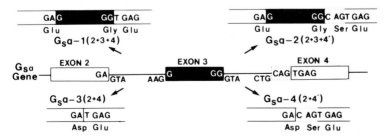

Figure 4.5 Generation of four different $G_s\alpha$ mRNAs by alternative splicing. The $G_s\alpha$ gene is shown in the centre. $G_s\alpha$ mRNAs are indicated by $G_s\alpha - 1$, -2, -3 and -4. (For details see [4])

phosphorylation by protein kinase C, and the alternative use of these splice sites may confer $G_s\alpha$ proteins with differential regulatory properties.

Human genes for $G_i\alpha$ subtypes

The coding region of the human $G_i2\alpha$ and $G_i3\alpha$ genes splits into eight exons and seven introns [5] (Figure 4.4B). There is an additional exon (exon 9) in the 3' non-coding region of $G_i2\alpha$ and $G_i3\alpha$, but this is not included in the figure. For human $G_i1\alpha$, we only have the sequences of exons 1 to 3 at present. Remarkably, the positions of the splice junctions on the sequence of cDNA for $G_i2\alpha$ and $G_i3\alpha$ were completely identical (Figure 4.6), although the lengths of introns are different [5]. The same splice sites are also conserved in the partial sequence (exons 1, 2 and 3) of the human $G_i1\alpha$ gene as well as in the human $G_o\alpha$ gene (T. Tsukamoto, unpublished). From the southern blot analysis, it appears that each of the three $G_i\alpha$ genes occurs as a single copy per haploid human genome.

Organization of human $G\alpha$ genes

The exon–intron organization of the $G_s\alpha$, $G_i2\alpha$, $G_i3\alpha$, and $G_o\alpha$ genes was compared with the predicted functional domain structure of proteins (Figure 4.6). The spatial orientation of each domain on the tertiary structure model will be discussed in a succeeding section (see Figures 4.8 and 4.9). The NH_2-terminal domain encoded by exon 1 is hydrophilic and contains the site for limited tryptic digestions. Although this region may be involved in interaction with $\beta\gamma$ subunits, its precise function has not yet been shown. Exon 2 encodes a short length region (24 and 14 amino acid residues, respectively, for $G_s\alpha$ and $G_i\alpha s$), which is the most conserved among all $G\alpha$ proteins and responsible for GTP hydrolysis. Exon 3 of $G_s\alpha$ is the one which is unique to $G_s\alpha$. This exon is lost by alternative splicing in some of the subtypes of $G_s\alpha$. The domain encoded by exons 4 to 6 of $G_s\alpha$ and 3 to 4 of

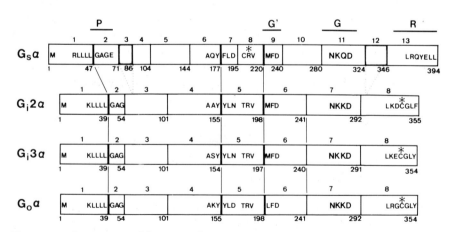

Figure 4.6 Organization of the exons of mammalian G-protein α subunits

$G_i\alpha$ is structurally divergent. Exon 8 of $G_s\alpha$ contains Arg-201, which is ADP-ribosylated in the presence of cholera toxin [44]. ADP-ribosylation of $G_s\alpha$ by cholera toxin causes a decrease of affinity for $\beta\gamma$ subunits [45]. Arg-179 in exon 5 of $G_i2\alpha$ corresponds to this arginine residue. The domain encoded by exons 9 to 11 of $G_s\alpha$, and 6 to 7 of $G_i\alpha$ is strongly conserved among all $G\alpha$ proteins. This domain is involved in formation of a core structure for GTP binding together with that coded by exon 2 [46]. The sequence, Asn–Lys–Xaa–Asp, consensus to all guanine nucleotide binding proteins, occurs in exon 11 of $G_s\alpha$ and exon 7 of $G_i\alpha$. The conserved Asp-223 in exon 9 of $G_s\alpha$ and Asp-201 in exon 6 of $G_i2\alpha$ may form a salt bridge to Mg^{2+}, which is linked to the β-phosphoryl group of GDP [38]. The exchange of GDP to GTP may result in displacement of the surrounding region residues 230–238 in exon 9 of $G_s\alpha$. A non-hydrolysable GTP analogue, but not GDP, prevents tryptic cleavage at Lys-210 in $G_o\alpha$ or Lys-205 in $G_t1\alpha$ [47].

Exon 12 of $G_s\alpha$ is unique to $G_s\alpha$, and exon 13 of $G_s\alpha$ and exon 8 of $G_i\alpha$ encode the COOH-terminal region. The domain may be involved in interaction with a receptor, since the Cys residue, which is ADP-ribosylated by pertussis toxin, is present in this region of $G_i\alpha$ and also the structure of this region is heterogeneous. In $G_x\alpha$, the Cys residue is replaced by Ile indicating that $G_x\alpha$ is probably refractory to modification by pertussis toxin. $G_s\alpha$ which is also resistant to pertussis toxin possessed Tyr instead of Cys in this position. It was shown that the replacement of Arg to Pro at the -6 position of $G_s\alpha$ gives rise to a mutant protein which is uncoupled with β-adrenergic receptor in S49 cells [32]. More recently, Masters et al. [48] have shown, using the chimaeric G_i/G_s construct, that the carboxy terminal domain of $G_s\alpha$ contains the structure specifying interactions with the effector enzyme, adenylate cyclase, as well as with the hormone receptor (see below).

Comparison of the exon organization of $G_i\alpha$ subfamily and $G_o\alpha$ with that of $G_s\alpha$ indicated that some of the exon junctions were conserved between $G_i\alpha$ subfamily and $G_s\alpha$. Thus, three out of twelve splice sites of the human $G_s\alpha$ gene are shared with the human $G_i\alpha$ genes, and exon 1 and exons 7 and 8 of $G_s\alpha$ correspond to exon 1 and exon 5 of $G_i\alpha$, respectively. More recent work by Raport et al. [49] revealed that the $G_t1\alpha$ and $G_i2\alpha$ genes possess the same organization as human $G_i\alpha/G_o\alpha$ genes.

Although not shown in Figure 4.6, human gene for $G_x\alpha$ consists of only two coding exons. Therefore, the gene organization of $G_x\alpha$ is quite different from other Gαs. The junction of exons 1 and 2 of $G_x\alpha$ is at the identical position as that of exons 6 and 7 of $G_i\alpha$ and $G_o\alpha$.

CONSERVATION OF PRIMARY STRUCTURE OF EACH Gα AMONG MAMMALIAN SPECIES

Table 4.3 shows that, in addition to the remarkable homologies of the overall structure, there is a strong conservation of the amino acid sequence in each subtype of G-protein α subunit. The amino acid sequence of $G_s\alpha$ is strongly conserved between human and rat;

Table 4.3 Conservation of G-protein α subunit sequences among different mammalian species

Species	Amino acid sequences	Nucleotide sequences
$rG_s\alpha$ vs $hG_s\alpha$	393/394 (99.7%)	1128/1182 (95.4%)
$bG_i1\alpha$ vs $hG_i1\alpha$	354/354 (100%)	998/1062 (94.0%)
$rG_i2\alpha$ vs $hG_i2\alpha$	350/355 (98.6%)	985/1065 (92.4%)
$rG_i3\alpha$ vs $hG_i3\alpha$	349/354 (98.6%)	981/1062 (92.4%)
$rG_o\alpha$ vs $bG_o\alpha$	348/354 (98.3%)	992/1062 (93.4%)
$rG_x\alpha$ vs $hG_x\alpha$	349/355 (98.3%)	977/1065 (91.7%)

h, human; r, rat; b, bovine

only one out of 394 amino acids being different. The sequence of $G_i1\alpha$ is completely identical between bovine and human. For $G_i2\alpha$, $G_i3\alpha$, $G_x\alpha$, and $G_o\alpha$, over 98% identity of amino acid sequences is maintained among different mammalian species.

The strong conservation of the amino acid sequence of each G-protein α subunit among distant mammalian species may reflect the presence of an evolutionary pressure to maintain the specific physiological function of each G-protein gene product.

An evolutionary tree of G-protein α subunits can be drawn, based on the homologies of the predicted amino acid sequences obtained from various mammalian sources (Figure 4.7). It is remarkable that the homologies among three $G_i\alpha$ species are higher than that between rod ($G_t1\alpha$) and cone ($G_t2\alpha$) transducin α subunits.

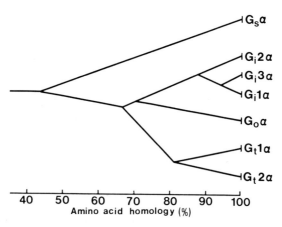

Figure 4.7 Relationships among the different mammalian Gα species. For details see [5]

STRUCTURAL MODEL AND MUTATIONAL ANALYSIS

Figure 4.8 shows a structural model of Gα [48] constructed by adopting the crystal structure of *E. coli* EF-Tu at the GDP binding domain [38,39]. As described in a previous section, the sequences conserved for all GTP binding proteins, i.e. P site, G' site and G site, are located in the close proximity of the bound guanine nucleotide.

The NH$_2$-terminal region encoded by exon 1 is followed by exon 2 containing the P site which is responsible for GTP hydrolysis. Gly-42 of G$_i$2α in this site (shown in Figure 4.8 as Gly-49) corresponds to Gly-12 of Ha-ras p21. However, the replacement of Gly-49 of G$_s\alpha$ by Val using the site-directed mutagenesis, did not give the expected phenotype of persistently activated adenylate cyclase in S49 cells [50]. Lys-46 of G$_i$2α (or Lys-53 of G$_s\alpha$) probably interacts electrostatically with the phosphoryl group of GDP.

The strongly conserved region encoded by exons 6 and 7 of G$_i\alpha$/G$_o\alpha$ of about 90 amino acids forms a guanine nucleotide binding pocket together with exon 2 containing P site. Asp-201 of G$_i$2α (shown as Asp-225· in Figure 4.8) interacts with phosphoryl group through Mg^{2+}. The consensus sequence Asp–Xaa–Xaa–Gly–Gln of this region corresponds to the transforming mutation site of p21, i.e. Ala-59 and Gln-61. As will be described below, the activating mutation of G$_s\alpha$ was obtained by replacement of Gln-227 by Leu, and H21a mutant of G$_s\alpha$ was found to be the Gly-226 → Ala replacement. It must be noted that the corresponding region in *E. coli* EF-Tu is Asp(80)–Cys–Pro–Gly–His(84),

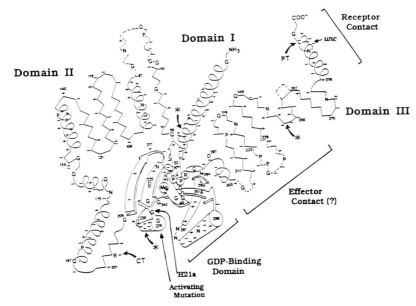

Figure 4.8 Structural model of Gα [48]. Numbers refer to positions of residues in αavg [46]. Asterisks with arrows are tryptic cleavage sites. Sites of point mutations and toxin modification sites are also indicated

where Cys-81 may be involved in interaction with aminoacyl-tRNA and ribosomes. A conformational change induced by the ligand change from GDP to GTP was detected at this region of EF-Tu [37]. Presumably, β strand containing Asp-201 may be displaced by the γ-phosphoryl group of GTP. Lys-210 of $G_o\alpha$, which is located in an α-helix downstream of the DXXGQ sequence, may also be the region where the conformational alteration is taking place.

The receptor interaction site of Gαs is located in the C-terminal region. As described above, modification of Cys at the -4 position by ADP-ribosylation with pertussis toxin abolished the receptor coupling of the toxin-sensitive Gαs and the *unc* mutation of $G_s\alpha$ had a replacement of Lys by Pro at the -6 position (for review see [50]). Furthermore, interaction of $G_t\alpha$ with rhodopsin was blocked by synthetic oligopeptides corresponding to the C-terminal sequences and also by monoclonal antibodies raised against this portion of $G_t\alpha$ [51,52].

The effector binding region was previously assumed to be located in Domain II of Figure 4.8, the region encoded by exons 3 and 4 of $G_i\alpha/G_o\alpha$. This was based on the molecular heterogeneity of various Gα species at this region and also on the location of the effector region of p21 around the NH$_2$-terminal region (residues 30–40 of p21, see Figure 4.9). However, the recent experiments in Bourne's lab [50] using the chimaeric constructs $G_i2\alpha/G_s\alpha$ or $G_t\alpha/G_s\alpha$ at the BamHI site (corresponding to amino acid residue 212 of $G_i2\alpha$) revealed that these chimaeras, when expressed in S49 cyc$^-$ cells, mediated stimulation of adenylate cyclase in response to β-adrenoceptor agonists. This implies that both receptor and effector binding sites of Gα may reside within the C-terminal 40% of the molecule (Domain III of Figure 4.8).

More recently, Holbrook and Kim [53] proposed the molecular model of Gα based on the crystal structure of Ha-ras p21 (Figure 4.9). As can be seen, the schematic diagrams of p21 and Gα are remarkably analogous. These authors defined three functional boxes, PO$_4$-Box, G-Box and S-Box, on the structure of p21 and Gα. PO$_4$-Box and G-Box correspond, respectively, to our P site and G site. On the other hand S-Box, which they consider as the potential conformational switch region, corresponds to our G' site. Although this region in *E. coli* EF-Tu is one of the candidates for the potential conformational switch site induced by GTP/GDP, the conformation of the effector region of p21 (L2 in Figure 4.2) may also be modulated.

Therefore, we propose that the conformational change may take place in several regions. The potential conformation switch sites may include a region of Gα at residues 201–205 (in terms of $G_i2\alpha$ sequence) which is homologous to residues 57–61 of p21, a region surrounding Lys-210 of which the sensitivity to trypsin was found to be modulated by the species of bound guanine nucleotide (L4 in Figure 4.9), a region around 167–177 where the sequence of Gα is homologous to the effector region (residues 31–41) of p21, and a certain sequence within a large region covering residues 212 to 315 in which the interaction with an effector was located by the use of two recombinant chimaeras. Further studies including site-directed mutagenesis and construction of chimaeric genes may throw more light on the structure–function relationship of Gα proteins.

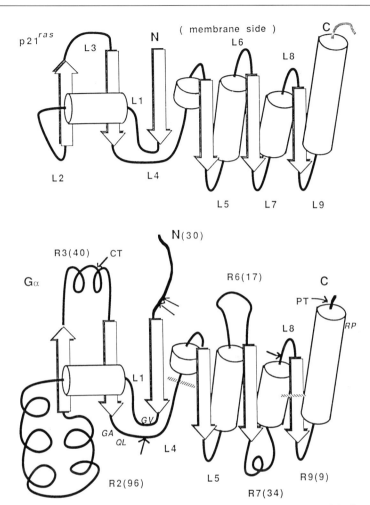

Figure 4.9 Schematic (topological) diagrams of the p21 (Upper) and the model of Gα (Lower) [53]. The number of amino acid residues in the loops is indicated in parentheses, and the sites of tryptic cleavage by arrows. The point mutations GV, GA, QL and AP indicate Gly→Val, Gly→Ala, Gln→Leu, and Ala→Pro substitutions, respectively. Hatched bars are junction sites for the chimaeric proteins. Reproduced by permission of Dr S.-H. Kim

BETA AND GAMMA SUBUNITS OF MAMMALIAN G-PROTEINS

Two subtypes of G-protein β subunit (Gβ1 and Gβ2) have been reported in mammalian cells [54,55]. Gβ1 and Gβ2 code, respectively, for 36 kDa and 35 kDa species of β subunit observed on NaDodSO$_4$ polyacrylamide gel electrophoresis [56]. Gβ1 is expressed in all kind of tissues including retina, whereas the 35 kDa species coded by Gβ2 was not seen in retina. Both Gβ1 and Gβ2 code for 340 amino acids, and they are about 90% identical.

The deduced amino acid sequence of Gβ1 and Gβ2 contains the repeated sequence of about 43 amino acids. The repetitive segmental structure similar to the β subunit was found also in yeast CDC4 [57], and STE4 ([58], see below).

The γ subunit of transducin was purified to a homogeneity and its amino acid sequence was determined. The cDNA coding for $G_t\gamma$ was isolated and its DNA sequence determined. The comparison of the deduced sequence of 74 amino acids of $G_t\gamma$ cDNA with that directly obtained from $G_t\gamma$ protein revealed that the NH_2-terminal methionine was removed. In addition, four C-terminal amino acids, Cys–Val–Ile–Ser, predicted from the DNA sequence were not found in $G_t\gamma$ protein. This might be due to the fatty acylation of the Cys residue, similar to p21 where the farnesylation of Cys at the -4 position is essential for its transforming activity.

The function of the $\beta\gamma$ subunits in the receptor-coupled G-protein activation cycle has been interpreted as shown in Figure 4.10. The Gα·GDP complex released from the amplifier (or effector) molecule is bound with $\beta\gamma$ to form a $\beta\gamma$·Gα·GDP complex which has an affinity towards receptor. In turn, receptor has a higher affinity to agonist when coupled with $\beta\gamma$·Gα·GDP. The interaction of agonist with the (R)($\beta\gamma$·Gα·GDP) complex induces the release of GDP so tightly bound to Gα that the exchange of GDP with external GTP is facilitated. The resulting (A)(R)($\beta\gamma$·Gα·GTP) complex dissociates to form Gα·GTP, G$\beta\gamma$, agonist and receptor. The Gα·GTP complex can now activate the amplifier (or effector) molecule. From this scheme, the major function of the $\beta\gamma$ subunit is to recognize the GDP bound form of Gα, to promote its binding to receptor and agonist, thus facilitating the recycling of Gα·GDP to Gα·GTP.

It is still controversial as to whether the $\beta\gamma$ complex has any specific role for activation

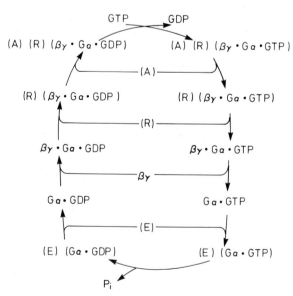

Figure 4.10 Function of G$\beta\gamma$ in recycling of Gα·GDP

of amplifier, in addition to the above described recycling process. Recently, it was reported that $\beta\gamma$ may be involved directly in activation of phospholipase A_2 [59] and K^+ channels [60] although the latter observation was questioned by others [61]. Katada et al. [62] also showed that $\beta\gamma$ interacts directly with calmodulin in the presence of Ca^{2+}, and suggested that $\beta\gamma$ may play some specific role in the calmodulin-modulated physiological processes.

G-PROTEINS FROM *SACCHAROMYCES CEREVISIAE*

A family of GTP binding protein, the *ras* family, is widely distributed among eukaryotes (see [63] for a review) including yeast *Saccharomyces cerevisiae* [64,65] and *Schizosaccharomyces pombe* [66]. It has been suggested that the *RAS2* gene in *S. cerevisiae* is involved in the activation of adenylate cyclase [67,68] and mimics the role of mammalian G_s.

However, in view of the strong conservation of the amino acid sequences of each G-protein species among different organisms (see Table 4.1), we speculated that G-protein may occur also in yeast. We have searched for G-protein homologous gene in yeast and isolated two genes *GPA1* [69] and *GPA2* [70] from *S. cerevisiae*, which are homologous with cDNAs for mammalian G-protein α subunits.

GPA1 and *GPA2* code for the sequences of 472 and 449 amino acid residues, respectively, with calculated M_rs of 54 075 and 50 516. When aligned with the α subunit of mammalian G-proteins to obtain maximal homology, GP1α (GPA1 encoded protein) and GP2α (GPA2 encoded protein) were found to contain the stretches of 110 and 83 additional amino acid residues, respectively, near the NH_2-terminus (Figure 4.11).

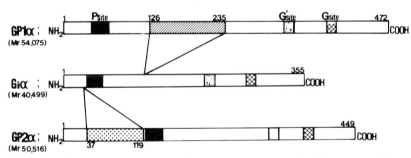

Figure 4.11 Schematic representation of the structure of yeast GP1α, yeast GP2α, and mammalian $G_i2\alpha$

COMPARISON OF THE AMINO ACID SEQUENCES OF YEAST GP1α AND GP2α WITH THOSE OF RAT BRAIN $G_i\alpha$ AND $G_o\alpha$

The deduced amino acid sequence of yeast GP1α and GP2α is highly homologous with those of rat brain $G_i\alpha$ and $G_o\alpha$. The homology is most remarkable in the region of GTP hydrolysis (P site, see Fig 4.3A) (amino acid residues 43–58 of GP1α, 125–140 of GP2α, and 35–50 of $G_i2\alpha$). As shown in Figure 4.3(C), the region responsible for GTP binding

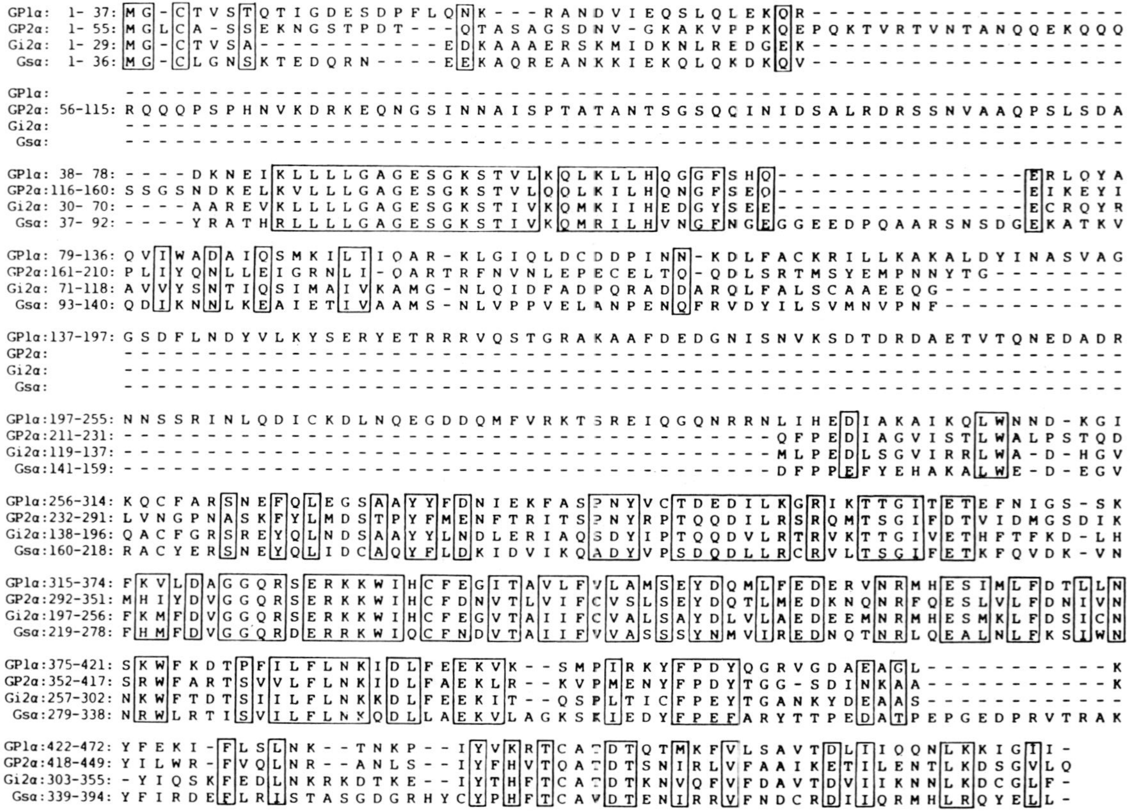

Figure 4.12 Alignment of the predicted amino acid sequence of yeast GP1α and GP2α with those of rat brain $G_i2α$ and $G_sα$. Identical or conservative amino acid residues are enclosed within solid lines [70]. NB. in GP2α, F at amino acid residue 435 should read P.

(G site) (amino acid residues 381–392 of GP1α, 358–369 of GP2α, and 263–274 of $G_i2α$) was also highly homologous. Another region of homology (G' site, see Figure 4.3B) was found in amino acid residues 319–336 of GP1α, 296–313 of GP2α, and 201–218 of Gi2α where a sequence of 14 contiguous amino acids was completely identical in yeast GP1α, GP2α and rat $G_i2α$.

The overall homology in nucleotide and amino acid sequences of yeast GP1α, yeast GP2α, rat $G_i2α$, and rat $G_sα$ is remarkable (Figure 4.12). Disregarding the unique sequences present in GP1α (residues 126–235) and GP2α (residues 37–119), the proteins are 60% homologous if conservative amino acid substitutions are considered to be homologous. The homology is smaller than that between rat $G_i2α$ and $G_oα$ (85%) but is comparable to that between rat $G_i2α$ and $G_sα$ (60%).

As is described elsewhere in detail [71,72], *GPA1* is a haploid specific gene and involved in the mating factor signal transduction. On the other hand, *GPA2* is expressed both in haploid and diploid cells, and may be involved in the regulation of cAMP levels in *S. cerevisiae* [70].

Recently, Whiteway *et al.* [58] made a remarkable observation that the *STE4* and *STE18* genes of *S. cerevisiae* encode β and γ subunits of GP1. Since both genes are required for the mating of yeast cells, and are expressed only in haploid cells, they are most likely the components of the mating-factor-coupled receptor signalling system. The nucleotide sequence analysis revealed that *STE4* codes for a protein of 423 amino acids highly homologous with mammalian Gβ1 and Gβ2. The deduced amino acid sequence of *STE18* exhibits also a homology to bovine $G_tγ$, except that *STE18* contains a poly Gln stretch near the NH_2-terminal region. From the genetic evidence, it appears that the products of *STE4* and *STE18* are involved, either directly or indirectly, in activation of the signals for the growth arrest at the late G1 phase.

The fact that the expression of *STE4* and *STE18* is limited in haploid cells immediately predicts that yeast cells may possess another set of genes coding for the potential β and γ subunits of G-protein GP2 which is present in both haploid and diploid cells. This also implies that *STE4* and *STE18* are probably involved specifically in the mating factor signalling system, since they are not replaced by the putative β and γ subunits expressed both in the diploid and haploid cells.

Finally, it is expected that demonstration of the occurrence of G-proteins in yeast may open the way for a detailed genetic analysis of the function of G-proteins in eukaryotic cells.

REFERENCES

1. Gilman, A. G. (1987) *Ann. Rev. Biochem.*, **56**, 615–649.
2. Stryer, L., and Bourne, H. R. (1986) *Ann. Rev. Cell Biol.*, **2**, 391–419.
3. Kaziro, Y., Itoh, H., Kozasa, T., Tsukamoto, T., Matsuoka, M., Nakafuku, M., Obara, T., Takagi, T., and Hernandez, R. (1988) *Cold Spring Harbor Symp. Quant. Biol.*, **53**, 209–220.
4. Kozasa, T., Itoh, H., Tsukamoto, T., and Kaziro, Y. (1988) *Proc. Natl. Acad. Sci. USA*, **85**, 2081–2085.

5. Itoh, H., Toyama, R., Kozasa, T., Tsukamoto, T., Matsuoka, M., and Kiziro, Y. (1988) *J. Biol. Chem.*, **263**, 6656–6664.
6. Weinstein, L. S., Spiegel, A. M., and Carter, A. D. (1988) *FEBS Lett.*, **232**, 333–340.
7. Matsuoka, M., Itoh, H., Kozasa, T., and Kaziro, Y. (1988) *Proc. Natl. Acad. Sci. USA*, **85**, 5384–5388.
8. Bray, P., Carter, A., Simons, C., Guo, V., Puckett, C., Kamholz, J., Spiegel, A., and Nirenberg, M. (1986) *Proc. Natl. Acad. Sci. USA*, **83**, 8893–8897.
9. Bray, P., Carter, A., Guo, V., Puckett, C., Kamholz, J., Spiegel, A., and Nirenberg, M. (1987) *Proc. Natl. Acad. Sci. USA*, **84**, 5115–5119.
10. Lavu, S., Clark, J., Swarup, R., Matsushima, K., Paturu, K., Moss, J., and Kung, H.-F. (1988) *Biochem. Biophys. Res. Commun.*, **150**, 811–815.
11. Beals, C. R., Wilson, C. B., and Perlmutter, R. M. (1987) *Proc. Natl. Acad. Sci. USA*, **84**, 7886–7890.
12. Didsbury, J. R., Ho, Y., and Snyderman, R. (1987) *FEBS Lett.*, **211**, 160–164.
13. Didsbury, J. R., and Snyderman, R. (1987) *FEBS Lett.*, **219**, 259–263.
14. Mattera, R., Codina, J., Crozat, A., Kidd, V., Woo, S. L. C., and Birnbaumer, L. (1986) *FEBS Lett.*, **206**, 36–41.
15. Suki, W. N., Abramowitz, J., Mattera, R., Codina, J., and Birnbaumer, L. (1987) *FEBS Lett.*, **220**, 187–192.
16. Codina, J., Olate, J., Abramowitz, J., Mattera, R., Cook, R. G., and Birnbaumer, L. (1988) *J. Biol. Chem.*, **263**, 6746–6750.
17. Fong, H. K. W., Yoshimoto, K. K., Eversole-Cire, P., and Simon, M. I. (1988) *Proc. Natl. Acad. Sci. USA*, **85**, 3066–3070.
18. Harris, B. A., Robishaw, J. D., Mumby, S. M., and Gilman, A. G. (1985) *Science*, **229**, 1274–1277.
19. Robishaw, J. D., Smigel, M. D., and Gilman, A. G. (1986) *J. Biol. Chem.*, **261**, 9587–9590.
20. Nukada, T., Tanabe, T., Takahashi, H., Noda, M., Haga, K., Haga, T., Ichiyama, A., Kangawa, K., Hiranaga, M., Matsuo, H., and Numa, S. (1986) *FEBS Lett.*, **197**, 305–310.
21. Nukada, T., Tanabe, T., Takahashi, H., Noda, M., Hirose, T., Inayama, S., and Numa, S. (1986) *FEBS Lett.*, **195**, 220–224.
22. Michel, T., Winslow, J. W., Smith, J. A., Seidman, J. G., and Neer, E. J. (1986) *Proc. Natl. Acad. Sci. USA*, **83**, 7663–7667.
23. Van Meurs, K. P., Angus, C. W., Lavu, S., Kung, H.-F., Czarnecki, S. K., Moss, J., and Vughan, M. (1987) *Proc. Natl. Acad. Sci. USA*, **84**, 3107–3111.
24. Ovchinnikov, Y. A., Slepak, V. Z., Pronin, A. N., Shlensky, A. B., Levina, N. B., Voeikov, V. L., and Lipkin, V. M. (1987) *FEBS Lett.*, **226**, 91–95.
25. Tanabe, T., Nukada, T., Nishikawa, Y., Sugimoto, K., Suzuki, H., Takahashi, H., Noda, M., Haga, T., Ichiyama, A., Kanagawa, K., Minamino, N., Matsuo, H., and Numa, S. (1985) *Nature (London)*, **315**, 242–245.
26. Medynsky, D. C., Sullivan, K., Smith, D., Van Dop, C., Chang, F.-H., Fung, B. K.-K., Seeburg, P. H., and Bourne, H. R. (1985) *Proc. Natl. Acad. Sci. USA*, **82**, 4311–4315.
27. Yatsunami, K., and Khorana, H. G. (1985) *Proc. Natl. Acad. Sci. USA*, **82**, 4316–4320.
28. Lochrie, M. A., Hurley, J. B., and Simon, M. I. (1985) *Science*, **228**, 96–99.
29. Itoh, H., Kozasa, T., Nagata, S., Nakamura, S., Katada, T., Ui, M., Iwai, S., Ohtsuka, E., Kawasaki, H., Suzuki, K., and Kaziro, Y. (1986) *Proc. Natl. Acad. Sci. USA*, **83**, 3776–3780.
30. Jones, D. T., and Reed, R. R. (1987) *J. Biol. Chem.*, **262**, 14241–14249.
31. Sullivan, K. A., Liao, Y.-C., Alborzi, A., Beiderman, B., Chang, F.-H., Masters, S. B., Levinson, A. D., and Bourne, H. R. (1986) *Proc. Natl. Acad. Sci. USA*, **83**, 6687–6691.
32. Sullivan, K. A., Miller, R. T., Masters, S. B., Beiderman, B., Heiderman, W., and Bourne, H. R. (1987) *Nature (London)*, **330**, 758–760.
33. Rall, T., and Harris, B. A. (1987) *FEBS Lett.*, **224**, 365–371.

34. Itoh, H., Katada, T., Ui, M., Kawasaki, H., Suzuki, K., and Kaziro, Y. (1988) *FEBS Lett.*, **230**, 85–89.
35. Lerea, C. L., Somers, D. E., Hurley, J. B., Klock, I. B., and Bunt-Milam, A. H. (1986) *Science*, **324**, 77–80.
36. Jones, D. T., and Reed, R. R. (1989) *Science*, **244**, 790–795.
37. Kaziro, Y. (1978) *Biochim. Biophys. Acta*, **505**, 95–127.
38. Jurnak, F. (1985) *Science*, **230**, 32–36.
39. LaCour, T. F. M., Nyborg, J., Thirup, S., and Clark, B. F. C. (1985) *EMBO J.*, **4**, 2385–2388.
40. DeVos, A. M., Tong, L., Milburn, M. V., Matias, P. M., Jankark, J., Noguchi, S., Nishimura, S., Miura, K., Ohtsuka, E., Kim, S.-H. (1988) *Science*, **239**, 888–893.
41. Sewell, J. L., and Kahn, R. A. (1988) *Proc. Natl. Acad. Sci. USA*, **85**, 4620–4624.
42. Northup, J. K., Sternweise, P. C., Smigel, M. D., Shleifer, L. S., Ross, E. M., and Gilman, A. G. (1980) *Proc. Natl. Acad. Sci. USA*, **77**, 6516–6520.
43. Robishaw, J. D., Russell, D. W., Harris, B. A., Smigel, M. D., and Gilman, A. G. (1986) *Proc. Natl. Acad. Sci. USA*, **83**, 1251–1255.
44. Van Dop, C., Tsubokawa, M., Bourne, H. R., and Ramachandran, J. (1984) *J. Biol. Chem.*, **259**, 696–698.
45. Kahn, R. A., and Gilman, A. G. (1984) *J. Biol. Chem.*, **259**, 6235–6240.
46. Masters, S. B., Stroud, R. M., and Bourne, H. R. (1986) *Protein Eng.*, **1**, 47–54.
47. Hurley, J. B., Simon, M. I., Teplow, D. B., Robishaw, J. D., and Gilman, A. G. (1984) *Science*, **226**, 860–862.
48. Masters, S. B., Sullivan, K. A., Miller, R. T., Beiderman, B., Lopez, N. G., Ramachandran, J., and Bourne, H. R. (1988) *Science*, **241**, 448–451.
49. Raport, C. J., Dere, B., and Hurley, J. (1989) *J. Biol. Chem.*, **264**, 7122–7128.
50. Bourne, H. R., Masters, S. B., Miller, R. T., Sullivan, K. A., and Heideman, W. (1988) *Cold Spring Harbor Symp. Quant. Biol.*, **53**, 221–228.
51. Deretic, D., and Hamm, H. E., (1987) *J. Biol. Chem.*, **262**, 10839–10847.
52. Hamm, H. E., Deretic, D., Arendt, A., Hargrave, P. A., Koenig, B., and Hofman, K. P. (1988) *Science*, **241**, 832–835.
53. Holbrook, S., and Kim, S.-H. (1989) *Proc. Natl. Acad. Sci. USA*, **86**, 1751–1755.
54. Sugimoto, K., Nukada, T., Tanabe, T., Takahashi, H., Noda, M., Minamino, N., Kangawa, K., Matsuo, H., Hirose, T., Inayama, S., and Numa, S. (1985) *FEBS Lett.*, **191**, 235–240.
55. Fong, H. K. W., Amatruda, T. T., Birren, B. W., and Simon, M. I. (1987) *Proc. Natl. Acad. Sci. USA*, **84**, 3792–3796.
56. Gao, B., Mumby, S., and Gilman, A. G. (1987) *J. Biol Chem.*, **262**, 17254–17257.
57. Fong, H. K. W., Hurley, J. B., Hopkins, R. S., Ryn, M.-L., Johnson, M. S., Doolittle, R. F., and Simon, M. I. (1986) *Proc. Natl. Acad. Sci. USA*, **83**, 2162–2166.
58. Whiteway, M., Hougan, L., Dignard, D., Thomas, D. Y., Bell, L., Saari, G. C., Grant, F. J., O'Hara, P., and MacKay, V. L. (1989) *Cell*, **56**, 467–477.
59. Jelsema, C. L., and Axelrod, J. (1987) *Proc. Natl. Acad. Sci. USA*, **84**, 3623–3627.
60. Logothetis, D. E., Kurachi, Y., Galper, J., Neer, E. J., and Clapham, D. E. (1987) *Nature (London)*, **325**, 321–326.
61. Birnbaumer, L. (1987) *Trends Pharmacol. Sci.*, **8**, 209–211.
62. Katada, T., Kusakabe, K., Oinuma, M., and Ui, M. (1987) *J. Biol. Chem.*, **262**, 11897–11900.
63. Barbacid, M. (1987) *Ann. Rev. Biochem.*, **56**, 779–827.
64. DeFeo-Jones, D., Scolnick, E. M., Koller, R., and Dhar, R. (1983) *Nature (London)*, **306**, 707–709.
65. Powers, S., Kataoka, T., Fasano, O., Goldfarb, M., Strathern, J., Broach, J., and Wigler, M. (1984) *Cell*, **36**, 607–612.
66. Fukui, Y., and Kaziro, Y. (1985) *EMBO J.*, **4**, 687–691.
67. Toda, T., Uno, I., Ishikawa, T., Powers, S., Katoaka, T., Broek, D., Cameron, S., Broach, J., Matsumoto K., and Wigler, M. (1985) *Cell*, **40**, 27–36.

68. Broek, D., Samiy, N., Fasano, O., Fujiyama, A., Tamanoi, F., Northup, J., and Wigler, M., (1985) *Cell*, **41**, 763–769.
69. Nakafuku, M., Itoh, H., Nakamura, S., and Kaziro, Y. (1987) *Proc. Natl. Acad. Sci. USA*, **84**, 2140–2144.
70. Nakafuku, M., Obara, T., Kaibuchi, K., Miyajima, I., Miyajima, A., Itoh, H., Nakamura, S., Arai, K., Matsumoto, K., and Kaziro, Y. (1988) *Proc. Natl. Acad. Sci. USA*, **85**, 1374–1378.
71. Miyajima, I., Nakafuku, M., Nakayama, N., Brenner, C., Miyajima, A., Kaibuchi, K., Arai, K., Kaziro, Y., and Matsumoto, K. (1987) *Cell*, **50**, 1011–1019.
72. Dietzel, D., and Kurjan, J. (1987) *Cell*, **50**, 1001–1010.

5
RECEPTOR-STIMULATED GTPASE ACTIVITY OF G-PROTEINS

Peter Gierschik and Karl H. Jakobs

Pharmakologisches Institut der Universität Heidelberg, Im Neuenheimer Feld 366, D-6900 Heidelberg, Federal Republic of Germany

INTRODUCTION

Guanine nucleotide binding regulatory proteins (G-proteins) play a pivotal role in signal transduction of a wide variety of plasma-membrane-located receptors for hormones, neurotransmitters, auto- and paracrine hormonal factors and even light. These signal transducing systems are composed out of at least three membrane-associated components, the hormone-specific receptor, the G-protein and an effector system, e.g. an intracellular signal forming enzyme or an ion channel. Based on the observations that poorly hydrolysable analogues of GTP such as guanylyl-5′-imidodiphosphate (Gpp(NH)p) and guanosine-5′-O-(3-thiotriphosphate) (GTP(S)) can cause, in contrast to GTP, persistent and, in many systems, receptor-independent alterations of the activity of the effector moieties, e.g. the adenylate cyclase or the retinal cyclic GMP hydrolysing phosphodiesterase, it was speculated that a GTPase activity, hydrolysing the active guanine nucleotide GTP to its inactive derivative GDP, is involved in control of G-protein-mediated signal transduction. One of the signal transducing G-proteins, i.e. the retinal G-protein transducin (G_t), was even termed 'GTPase' before its real function as transducing component between the 'light receptor', rhodopsin, and the effector moiety, the cGMP hydrolysing phosphodiesterase, has been recognized.

The basic demonstration of a receptor-controlled GTPase activity and its apparent function in termination of the hormonal response were first described by Cassel and Selinger [1] in the turkey erythrocyte membrane system. In these membranes, agonist activation of β-adrenoceptors simultaneously caused a GTP-dependent adenylate cyclase stimulation and an increase in high affinity GTPase activity. Although the ideas put forward by the authors were initially not corroborated in studies on purified G_s proteins (because of technical reasons), it is now fully accepted that any signal transducing G-protein, more exactly the G-protein α subunit, contains an inherent GTPase activity and that this GTP hydrolysing capacity can be controlled by receptor–G-protein interactions as well as by the bacterial toxins, cholera and pertussis toxin, ADP-ribosylating various G-protein α subunits. In this chapter, we will mainly consider the G-protein GTPase activity as a probe to analyse receptor–G-protein interactions and not the basic structural and functional properties of G-proteins, which have been dealt with in many recent reviews [2–7].

GTP HYDROLYSIS BY PURIFIED G-PROTEINS

Signal transducing G-proteins are composed of three subunits, termed α, β and γ, out of which the α subunits are the most distinct and best studied ones in the different G-proteins, while the functions of β and γ subunits, which occur under non-denaturing conditions as $\beta\gamma$ dimers and which are very similar or partially even identical in the different G-proteins, are far less clear. As shown with various purified G-proteins and as exemplified with the G-protein α subunit purified from bovine neutrophil membranes [8], the α subunits, which are the actual binding sites for guanine nucleotides, also contain the G-protein GTPase activity, hydrolysing bound GTP to GDP and inorganic phosphate (Figure 5.1). The steady state rate of GTP hydrolysis by G-protein α subunit GTPases is rather low, in the range of 0.3 min^{-1}. When G-protein α subunits are depleted of bound GDP before measurement of GTP hydrolysis or when G-proteins are reconstituted with agonist-liganded receptors, this rate of GTP hydrolysis is markedly increased and, most interestingly, to a similar extent by the two treatments [9,10]. These findings, together with the observation that agonist-activated receptors increase the dissociation rate of G-protein-bound GDP (2–7), strongly suggest that the rate of GTP hydrolysis by G-proteins is mainly determined by the dissociation of GDP from the G-protein α subunit. The actual rate of GTP hydrolysis was not found to be influenced by agonist-liganded receptors [9].

As analysed with the G-proteins G_i and G_o, the rate of GDP release from G-protein α subunits is under control by the $\beta\gamma$ subunits, apparently increasing the binding affinity for GDP [11]. An additional control of GTP hydrolysis is exerted by Mg^{2+} ions, which modulate and at high concentrations reverse the effect of $\beta\gamma$ subunits on GDP binding and subsequent GTPase activity of G-protein α subunits [11]. Thus, when high affinity GTPase activity and its stimulation by agonist-liganded receptors is determined with purified components or in membrane preparations, a rather complex reaction is examined. Although actually hydrolysis of GTP is measured, it is apparently not the rate of inactivation of G-proteins but the receptor-dependent binding of GTP to G-proteins which is studied. This binding of GTP to G-proteins is apparently primarily determined by the dissociation of bound GDP from the G-protein α subunits and, additionally, a receptor-catalysed isomerization of the GTP-bound G-protein from a low affinity to a high affinity binding state may be involved [12].

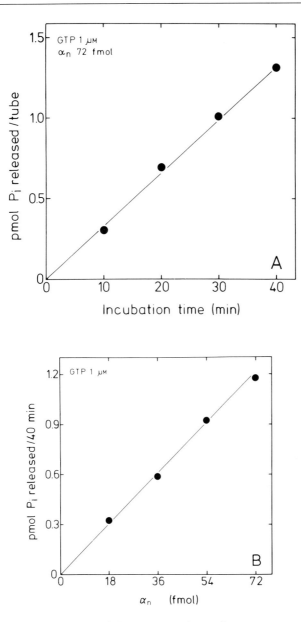

Figure 5.1 GTPase activity of a purified G-protein α subunit. The major pertussis-toxin-sensitive G-protein α subunit was purified from bovine neutrophil membranes [8]. The α subunit (α_n, 72 fmoles) was incubated for the indicated periods of time (A) or increasing amounts of the α subunit were incubated for 40 min (B) and the amount of the P_i released determined using 1 μM GTP as substrate

RECEPTOR-REGULATED GTP HYDROLYSIS IN MEMBRANES

Following the pioneering studies of Cassel and Selinger (1), receptor-regulated GTP hydrolysis has now been described in many different membrane systems and with a large variety of receptor agonists. Besides the retinal system, agonist stimulation of high affinity GTPase has been reported for receptors mediating stimulation and inhibition of adenylate cyclase via G_s and G_i proteins, respectively, stimulation of phospholipase C via pertussis-toxin-sensitive and -insensitive G-proteins and even for receptors for which the intracellular signal forming system has not been firmly established (for a review see [13]). On the other hand, for many receptors, agonist-stimulated GTP hydrolysis was not detected or at least not reported, although a coupling of these receptors to a G-protein has been definitely shown, either in agonist–receptor binding studies or in studying the agonist-dependent intracellular signal formation. This is particularly evident for receptors functioning via the cholera-toxin-sensitive G_s protein stimulation of adenylate cyclase. Only few examples of this type of receptor have been clearly shown to increase GTP hydrolysis in membrane preparations [13]. On the other hand, stimulation of GTP hydrolysis by receptors interacting with pertussis-toxin-sensitive G_i proteins, mediating inhibition of adenylate cyclase, stimulation of phospholipase C and/or alterations of ion channel activities, has been reported for many different membrane and receptor sytems. The reason for this discrepancy is not clear yet. It may be due to the generally smaller concentration of G_s proteins compared to G_i proteins in membranes (however, see [14]), thus precluding the determination of the possibly occurring small increase in GTP hydrolysis by the G_s protein within the relatively large 'noise' of GTP hydrolysis caused by G-protein-unrelated GTPase or NTPases. On the other hand, it has recently been reported that in membranes of S49 lymphoma cells, the agonist-liganded β-adrenoceptor can apparently activate about 10 pmol of G_s proteins per milligram of membrane protein [15]. Such a large activation, however, should also be measurable as β-adrenoceptor-stimulated GTPase activity. However, in contrast to the readily measurable GTPase stimulation by somatostatin (see also Figure 5.9), caused by G_i mediated inhibition of adenylate cyclase in these membranes, stimulation of GTP hydrolysis by the β-adrenoceptor agonist isoproterenol was not detected [16]. Thus, there may be additional, not yet recognized, possibly technical reasons, e.g. assay conditions for measuring GTPase activity, which preclude the detection of high affinity GTPase stimulation in membrane preparations by some receptor types in certain tissues.

REGULATION OF RECEPTOR-STIMULATED GTP HYDROLYSIS BY Mg^{2+} AND GUANINE NUCLEOTIDES

The apparent affinity of the receptor-regulated GTPase for its substrate GTP is in the range of 0.2 to 0.5 μM. No major differences in apparent substrate affinities between G_i and G_s protein-related GTPases were reported, although, as known from many different membrane systems, for agonist-induced inhibition of adenylate cyclase usually 10- to 20-fold higher concentrations of GTP are required than for agonist-induced stimulation of the enzyme [13,17]. Furthermore, the apparent affinity of the G-protein GTPase for its substrate GTP is not increased by agonist-activated receptors. Usually, only an increase in V_{max} is reported, while the K_m values for GTP observed in the presence of receptor agonists are either identical to the basal, unstimulated GTPase or even somewhat, but only slightly, higher in the presence than in the absence of receptor agonists [13]. These data suggest that the agonist-activated receptor increases the number of G-proteins, from which GDP is dissociated and to which, then, GTP can bind, followed by its hydrolysis, without inducing a major change in the affinity of the G-protein for GTP.

However, when other guanine nucleotides are added in addition to GTP to the GTPase assay, marked differences between basal and agonist-stimulated GTPase activity can be observed. As shown in Figure 5.2, when the hydrolysis-resistant GTP analogue GTPγ(S) was added to human platelet membranes and basal, epinephrine- and thrombin-stimulated GTPase activities were measured with 0.2 μM GTP as substrate, GTPγ(S) preferentially blocked the receptor-stimulated GTP hydrolysis. This differential sensitivity of basal and agonist-stimulated GTP hydrolysis to GTPγ(S) is even more evident when basal and agonist-stimulated GTP hydrolysis are plotted separately (Figure 5.2B). In addition, when comparing stimulation of GTP hydrolysis by epinephrine and thrombin and inhibition of these stimulated activities by GTPγ(S), it is evident that thrombin stimulation of GTPase, which is two- to three-fold larger than epinephrine-induced stimulation, is also about two- to three-fold more sensitive to inhibition by GTPγ(S) than epinephrine-stimulated GTP hydrolysis. Similar data were reported for inhibition of basal and opioid-stimulated GTP hydrolysis by GTPγ(S) in membranes of NG 108-15 cells [18]. One interpretation of these data could be that basal and agonist-stimulated GTP hydrolyses are due to distinct GTP hydrolysing enzymes, exhibiting different binding affinities for the GTP analogue GTPγ(S). However, treatment of differentiated HL-60 cells with pertussis toxin not only completely blocks agonist-dependent stimulation of GTP hydrolysis by chemotactic factors such as the formyl peptide, fMet–Leu–Phe (FMLP), and the leukotriene LTB$_4$, but this treatment also decreases basal, unstimulated GTPase activity by up to 80% [19,20]. These findings indicate, at a minimum, that the majority of GTP hydrolysis in these membranes is due to pertussis-toxin-sensitive G_i proteins, thus very similar or identical to the G-proteins activated by the agonist-liganded receptors. Nevertheless, in these membranes addition of GTPγ(S) to the GTPase assay preferentially blocks agonist-stimulated GTP hydrolysis, whereas inhibition of basal GTPase activity is only seen at concentrations of GTPγ(S) at which the agonist-stimulated GTP hydrolysis is almost completely blocked (Figure 5.3). These data, thus, suggest that the agonist-activated receptor, in addition to causing a release of G-protein-bound GDP, somehow decreases the binding affinity of the G-protein for GTP and/or increases the binding affinity for the GTP analogue GTPγ(S). The binding of this hydrolysis-resistant guanine nucleotide to the G-protein will prevent agonist-induced binding of GTP and, thus, inhibit preferentially agonist-stimulated GTP hydrolysis by the G-protein GTPase.

In complete contrast to the inhibition of GTP hydrolysis seen with the GTP analogue

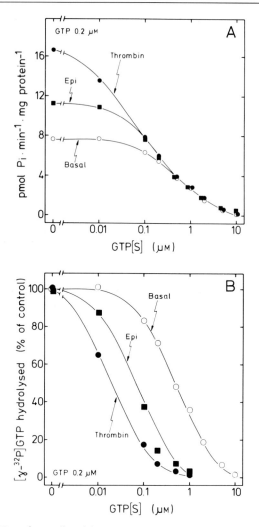

Figure 5.2 GTPγ(S) preferentially inhibits agonist-stimulated GTP hydrolysis in human platelet membranes. (A) High affinity GTPase activity was determined in human platelet membranes [30] in the absence (Basal, ○) and presence of 30 μM epinephrine (Epi, ■) or 0.1 U/ml thrombin (●) at the indicated concentrations of GTPγ(S). (B) Replot of basal, epinephrine- and thrombin-stimulated GTP hydrolysis at the indicated concentrations of GTPγ(S). Stimulated GTP hydrolysis was calculated as the difference between total stimulated minus basal activity

GTPγ(S) is the reduction of GTPase activity observed after addition of the nucleoside diphosphate GDP to the assay. This nucleotide, added in the absence of ATP and a nucleoside-triphosphate-regenerating system which are usually included in membrane GTPase assays [1,13], preferentially inhibits basal, unstimulated GTP hydrolysis, while the agonist-stimulated activity is far less sensitive. As shown in Figure 5.4, using 10 nM

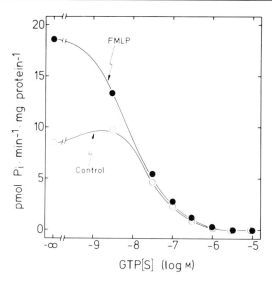

Figure 5.3 GTPγ(S) preferentially inhibits FMLP-stimulated GTPase activity in HL-60 membranes. High affinity GTPase activity was determined in membranes of HL-60 cells [19] without (Control, ○) and with (●) 10 μM FMLP at the indicated concentrations GTPγ(S)

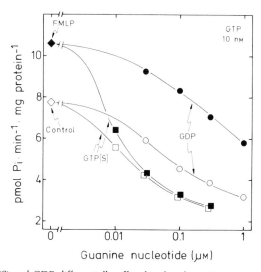

Figure 5.4 GTPγ(S) and GDP differentially affect basal and agonist-stimulated GTPase activity in HL-60 membranes. High affinity GTPase activity was determined in membranes of HL-60 cells with 10 nM GTP as substrate [19], but without ATP and a nucleoside triphosphate regenerating system, in the absence (□,○) and presence (■,●) of 10 μM FMLP at the indicated concentrations of GTPγ[S] (□,■) or GDP (○,●)

GTP as GTPase substrate, addition of 10 nM GTPγ(S) almost completely blocked agonist-stimulated GTPase activity in HL-60 membranes. In contrast GDP, which inhibited control, unstimulated activity with a five- to ten-fold lower potency than GTPγ(S), decreased GTPase activity measured in the presence of a maximally effective concentration of FMLP (10 μM) only at much higher concentrations. The agonist-stimulated portion of total GTP hydrolysis was not affected at all by up to 1 μM GDP, while at this same concentration GDP already reduced unstimulated activity by about 70%. When inhibition of basal and FMLP-stimulated GTP hydrolysis by GDP are plotted separately, as shown in Figure 5.2(B) for control and agonist-stimulated GTP hydrolysis in platelet membranes, the difference in sensitivity to GDP between basal and agonist-stimulated GTPase is even about 100-fold. Compared to GDP, inhibition of basal and FMLP-stimulated GTP hydrolysis in HL-60 membranes by the metabolically stable GDP analogue, guanosine-5'-O-(2-thiodiphosphate) (GDPβ(S)), exhibited, a much smaller difference in sensitivity. As illustrated in Figure 5.5, inhibition of total agonist-stimulated GTP hydrolysis by GDPβ(S) was slightly less sensitive than inhibition of basal GTP hydrolysis, while with GDP a large difference between unstimulated and stimulated activity is observed.

Similarly, as reported for purified G-proteins reconstituted with purified receptors [9], agonist-induced stimulation of GTP hydrolysis in membrane preparations requires only low, micromolar concentrations of Mg^{2+}, being half-maximally and maximally effective

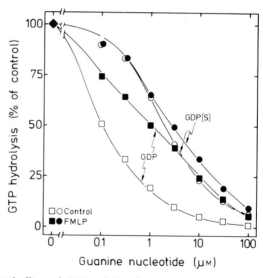

Figure 5.5 Differential effects of GDP and GDPβ(S) on agonist-stimulated GTP hydrolysis in HL-60 membranes. High affinity GTPase activity was determined in membranes of HL-60 cells as described in the legend to Figure 5.4 in the absence (□,○) and presence (■,●) of 10 μM FLMP at the indicated concentrations of GDP (□,■) or GDPβ(S) (○,●). Total GTP hydrolysis by high affinity GTPase measured in the absence or presence of FMLP is given in per cent of activity measured in the absence of GDP or GDPβ(S)

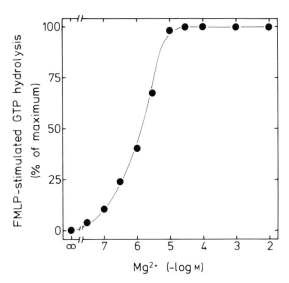

Figure 5.6 Mg^{2+} requirement of agonist-stimulated GTPase activity. High affinity GTPase activity was determined in membranes of HL-60 cells [19] in the absence and presence of 10 μM FMLP at the indicated concentrations of free Mg^{2+}. The Mg^{2+} dependence of agonist-stimulated GTP hydrolysis is given in per cent of maximal stimulated activity

at about 1 μM and 10 μM free Mg^{2+}, respectively [21,22] (Figure 5.6). However, in the presence of a 100-fold excess of GDP over GTP in the GTPase assay, a shift in the Mg^{2+} requirement for maximal agonist-stimulated GTP hydrolysis by a factor of about 100 is observed, with half-maximal and maximal activation now occurring at about 100 μM and 3 mM free Mg^{2+}, respectively (data not shown).

The large difference in GDP sensitivity between basal and agonist-stimulated GTP hydrolysis may be explained by the hypothesis that the activated receptor reduces the affinity of the G-protein for GDP and/or increases that for GTP (however, see above). Since with the GDP analogue GDPβ(S) only a rather small difference in sensitivity was observed, it may be speculated that additional processes are involved. The increase in Mg^{2+} requirement observed in the presence of GDP and the differential effects of GDP and GDPβ(S) suggest that the activated receptor may not only reduce the binding affinity of the G-protein for GDP but may, in addition, favour a transphosphorylation of GDP to GTP. This reaction may require millimolar rather than micromolar concentrations of Mg^{2+} [23] and is apparently only occurring with GDP but not with its metabolically stable analogue GDPβ(S).

INFLUENCE OF CHOLERA AND PERTUSSIS TOXIN ON RECEPTOR-STIMULATED GTP HYDROLYSIS

Cassel and Selinger [24] first described that cholera toxin treatment of turkey erythrocyte membranes inhibits receptor-stimulated G_s protein-related GTP hydrolysis. These data perfectly agreed with the simultaneously occurring GTP-dependent adenylate cyclase activation. Since that, it is generally assumed that the ADP-ribosylation of $G_s\alpha$ by cholera toxin specifically interferes with the GTP hydrolysis reaction of the G_s protein. Similar data were recently reported by us on the agonist-stimulated GTPase activity of G_i proteins ADP-ribosylated by cholera toxin in membranes of HL-60 cells [25]. However, if cholera-toxin-induced ADP-ribosylation of G-protein α subunits only interferes with the actual hydrolytic activity of the G-protein, then pretreatment of a G-protein with this toxin should have no effect on agonist-stimulated high affinity binding of GTP. However, as recently described in membranes of HL-60 cells, ADP-ribosylation of the G_i-proteins by cholera toxin largely suppressed agonist-stimulated binding of GTP(S) to membrane G-proteins [19]. Such an action is not compatible with just an inhibition of GTP hydrolysis. It rather suggests that the ADP-ribosylation induced by cholera toxin interferes with a receptor-regulated high affinity binding reaction of GTP to the G-protein and, therefore, the receptor-stimulated GTP hydrolysis is prevented. It is not known, although it has been mentioned that ADP-ribosylation of purified G_s protein by cholera toxin has no effect on its GTP hydrolysing activity [5], whether a similar action, i.e., a decrease in high-affinity binding of GTP(S), is also observed on the 'primary' target of cholera toxin, namely the G_s protein.

ADP-ribosylation of purified G_i proteins by pertussis toxin is reported to have no effects on the binding of GTPγ(S) to G-proteins and on the hydrolysis of GTP [26]. In contrast, the interaction of receptors with G_i proteins and, thus, the stimulation of GTP(S) binding to, and the GTP hydrolysis by, G-proteins induced by agonist-activated receptors is prevented by this treatment. However, as mentioned above, in some membrane preparations such as HL-60 and NG108-15 cells, pretreatment with pertussis toxin not only abolishes agonist-stimulated GTP hydrolysis but also largely reduces 'basal' GTPase activity [19,20,27]. Taking these data together with the findings obtained with purified G_i proteins, it has to be assumed either that, at least in some membrane preparations, ADP-ribosylation of G_i proteins by pertussis toxin also decreases receptor-independent G-protein GTPase activity or that unoccupied receptors, i.e. in the absence of agonists, are at least partially active and induce GTP binding to and subsequent GTP hydrolysis by G-proteins. ADP-ribosylation by pertussis toxin would prevent interaction of agonist-unoccupied receptors with G_i proteins and, thus, cause a decrease in 'basal' GTPase activity. This hypothesis that under some conditions unoccupied receptors can interact with G-proteins is supported by the observation that reconstitution of purified β-adrenoceptors with purified G_s proteins can cause an increased GTP hydrolysis by the G-protein even in the absence of a receptor agonist [28].

The two toxins, cholera and pertussis toxin, can also be used to analyse the G-protein(s) activated by specific receptors. An example of such a study is shown in Figure 5.7 [29,30]. Treatment of human platelet membranes with pertussis toxin prevents thrombin-induced and G_i protein-mediated adenylate cyclase inhibition, similarly to inhibition by epinephrine (adrenaline). However, stimulation of GTP hydrolysis induced by these two receptor agonists was differentially affected by the toxin treatment. While epinephrine-induced stimulation of GTPase activity was prevented by the toxin,

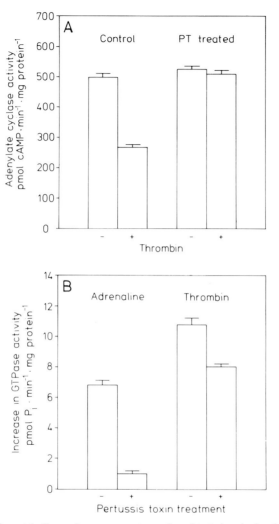

Figure 5.7 Differential effects of pertussis toxin on thrombin-induced adenylate cyclase inhibition and GTPase stimulation in human platelet membranes. (A) Thrombin-induced adenylate cyclase inhibition was determined in control and pertussis toxin (PT)-treated platelet membranes [29]. (B) Thrombin- and adrenaline-induced GTPase stimulation was determined in control and PT-treated platelet membranes [30]

thrombin-induced GTP hydrolysis was only partially decreased. These findings suggested that epinephrine-activated α_2-adrenoceptors interact with and activate essentially only G_i-like proteins in these membranes, while agonist-activated thrombin receptors activate both pertussis-toxin-sensitive and -insensitive G-proteins. However, since pertussis toxin can ADP-ribosylate many G-proteins, with the exceptions of G_s and G_z

(G_x), and since the target amino acid sequence for ADP-ribosylation by cholera toxin is present in all α subunits of G-proteins identified to date, the differential sensitivity to pertussis and cholera toxin is much less helpful than previously imagined to identify individual G-proteins interacting with and being activated by a specific receptor.

RELATIONSHIP BETWEEN AGONIST–RECEPTOR BINDING AND AGONIST-STIMULATED GTP HYDROLYSIS

Binding of agonists to receptors interacting with and activating G-proteins is regulated by several agents, leading to an increase or decrease of binding affinity. Guanine nucleotides such as GTP and GDP, by binding to G-protein α subunits, reduce agonist binding affinity of most of these receptors. On the other hand, divalent cations such as Mg^{2+} and Mn^{2+} at millimolar concentrations usually cause an increase in agonist affinity of G-protein-coupled receptors or are even required to observe the high affinity agonist binding state of the receptors. In contrast, monovalent cations, particularly sodium ions, have been shown to reduce agonist-binding affinity of many receptors interacting with G-proteins [2–7,31]. In order to see whether these changes observed in agonist–receptor binding affinity may also be reflected in agonist-induced GTPase stimulation, GTP hydrolysis was examined under almost identical conditions under which the changes in agonist–receptor affinity were observed.

First, formyl-peptide-stimulated GTP hydrolysis was studied in membranes of HL 60 cells at free Mg^{2+} concentrations leading either to the appearance of the high-affinity agonist-binding state of the receptor (4 mM free Mg^{2+}) or having no effect on agonist receptor binding (~ 0.45 μM free Mg^{2+}) [22]. As shown in Figure 5.8, at millimolar concentration of free Mg^{2+} not only high-affinity agonist receptor binding was induced (K_D for FMLP about 1 nM compared to about 30 nM in the absence or at the low, micromolar concentration of Mg^{2+}) [22], but the presence of this high, millimolar concentration of free Mg^{2+} not only high-affinity agonist receptor binding was induced (K_D for FMLP about 1 nM compared to about 30 nM in the absence or at the low, observed at high and low concentrations of Mg^{2+}, respectively, were in good agreement with the K_D values found under these conditions. Furthermore, at the high, millimolar concentration of Mg^{2+} most (60%–80%), but not all, of the receptors are in the high affinity agonist-binding state, while at the low, micromolar Mg^{2+} concentration almost all of the receptors are in a low affinity agonist-binding state [22]. Accordingly, at the high Mg^{2+} concentration, the Hill coefficient of FMLP-induced GTPase stimulation was only 0.58, suggesting that at this Mg^{2+} concentration two distinct receptor affinity states are involved in GTPase stimulation. On the other hand, at the low Mg^{2+} concentration, in addition to the reduction in agonist potency, the Hill coefficient was close to unity (0.83), indicating the involvement of essentially only one, namely a low affinity agonist-binding state of the receptor in GTPase stimulation.

In contrast to Mg^{2+}, sodium ions decrease agonist binding affinity of many G-protein-coupled receptors [31]. This decrease in agonist affinity is also reflected in

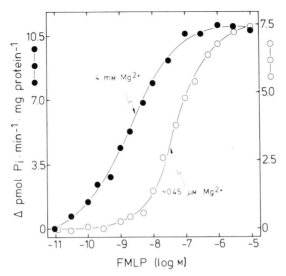

Figure 5.8 Mg^{2+} increases agonist potency in stimulating GTP hydrolysis in HL-60 membranes. High affinity GTPase activity was determined in HL-60 membranes at the indicated concentrations of FMLP with either 4 mM free Mg^{2+} (●, left ordinate) or about 0.45 μM free Mg^{2+} (○, right ordinate) [22]

agonist-induced GTPase stimulation. Similarly, as observed for bradykinin- and FMLP-induced GTPase stimulation in membranes of NG108-15 and HL-60 cells, respectively [20,32], the potency of somatostatin to stimulate GTP hydrolysis in membranes of S49 lymphoma cells [16] is significantly reduced by the presence of NaCl (100 mM) in the GTPase assay (Figure 5.9). Thus, divalent cations, at millimimolar concentrations increasing or inducing high affinity agonist receptor binding, simultaneously increase the agonist potency in stimulating GTP hydrolysis, whereas monovalent cations, particularly sodium ions, decrease the agonist receptor binding affinity and also decrease the potency of the agonist to stimulate GTP hydrolysis.

Finally, guanine nucleotides such as GTP and GDP added at increasing concentrations to the GTPase assay not only decrease agonist binding affinity of receptors or even reverse the high affinity agonist binding state induced by Mg^{2+} but also decrease the potency of the agonist to stimulate GTP hydrolysis. As exemplified in Figure 5.10, when GTPase stimulation was studied in membranes of HL-60 cells at increasing concentrations of FMLP using two distinct concentrations of GTP as GTPase substrate (30 nM and 1 μM), GTPase stimulation at the low GTP concentration (30 nM) exhibited a Hill coefficient of 0.42, suggesting an involvement of two distinct receptor affinity states in GTPase stimulation. At the high GTP concentration (1 μM), where most of the receptor sites are in a low affinity agonist binding state, the concentration response curve of FMLP was not only shifted to the right but also exhibited a Hill coefficient (0.80) much more close to unity, suggesting the involvement of essentially only one, low affinity binding

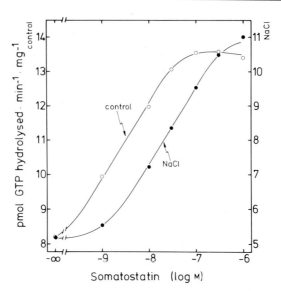

Figure 5.9 Na$^+$ decreases agonist potency in stimulating GTP hydrolysis in S49 lymphoma membranes. High affinity GTPase activity was determined in membranes of S49 lymphoma cells [16] in the absence of (Control, ○, left ordinate) or presence of (●, right ordinate) 100 mM NaCl at the indicated concentrations of somatostatin.

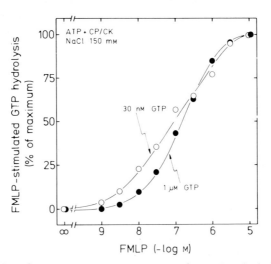

Figure 5.10 GTP decreases agonist potency in stimulating GTP hydrolysis in HL-60 membranes. GTPase activity was determined in HL-60 membranes with either 30 nM GTP (○) or 1 μM GTP (●) as substrate in the presence of ATP, the nucleoside triphosphate regenerating system, creatine phosphate plus creatine kinase (CP/CK), and NaCl [19] at the indicated concentrations of FMLP. Agonist-stimulated GTP hydrolysis is given in per cent of maximal stimulated activity

state of the receptor in agonist-induced GTPase stimulation. Thus, these data suggest, at a minimum, that the affinity state of the receptor for the agonist is also reflected in agonist-induced GTPase stimulation. Therefore, studies on stimulation of GTP hydrolysis in membrane preparations can be used not only to examine whether a given receptor interacts with and activates a G-protein, but apparently such studies are also useful to analyse the regulation of agonist binding affinity of the receptors.

CONCLUSIONS

GTP hydrolysis is apparently a function of any signal transducing G-protein α subunit. The rate of GTP hydrolysis, which is apparently required for termination of the active state of the G-protein in signal transduction, appears to be determined mainly by the release of previously bound nucleotide, usually GDP, from the G-protein and subsequent binding of GTP, while the actual rate of GTP hydrolysis itself may not be regulated. Since agonist-liganded receptors can induce the dissociation of G-protein-bound GDP and the subsequent binding of GTP, receptor agonists stimulate GTP hydrolysis by G-proteins. Therefore, studies on this receptor action are very helpful in analysing the initial steps in signal transduction both with regard to the underlying mechanisms of G-protein activation and the regulation of agonist–receptor binding. A further advantage of this method is that the natural nucleoside triphosphate GTP has to be used and not its hydrolysis-resistant analogues such as Gpp(NH)p and GTPγ(S), which can exhibit properties quite distinct from those of GTP (see, e.g. [25]). Finally, agonist-stimulated GTP hydrolysis can be used to analyse the type of G-protein involved in signal transduction by the receptor under study, although the discrimination by the two bacterial toxins, cholera and pertussis toxin, ADP-ribosylating various G-protein α-subunits, is apparently less powerful than previously imagined.

ACKNOWLEDGEMENTS

The author's studies reported herein were supported by the Deutsche Forschungsgemeinschaft and the Fritz Thyssen-Stiftung.

REFERENCES

1. Cassel, D., and Selinger, Z. (1986) *Biochim. Biophys. Acta*, **452**, 538–551.
2. Rodbell, M. (1980) *Nature (London)*, **284**, 17–22.
3. Schramm, M., and Selinger, Z. (1984) *Science*, **225**, 1350–1356.
4. Stryer, L. (1986) *Ann. Rev. Neurosci.*, **9**, 87–119.
5. Gilman, A. G. (1987) *Ann. Rev. Biochem.*, **56**, 615–649.
6. Birnbaumer, L., Codina, J., Mattera, R., Yatani, A., Scherer, N., Toro, M. J., and Brown, A. M. (1987) *Kidney International*, **32** (Suppl. 23), S14–S37.

7. Gierschik, P., Sidiropoulos, D., Dieterich, K., and Jakobs, K. H. (1989) In: *Growth Factors, Differentiation Factors and Cytokines* (Habenicht, A., ed.). Springer, Berlin–Heidelberg–New York (in press).
8. Gierschik, P., Sidiropoulos, D., Spiegel, A., and Jakobs, K. H. (1987) *Eur. J. Biochem.*, **165**, 185–194.
9. Brandt, D. R., Asano, T., Pedersen, S. E., and Ross, E. M. (1983) *Biochemistry*, **22**, 4357–4362.
10. Higashijima, T., Ferguson, K. M., Smigel, M. D., and Gilman, A. G. (1987) *J. Biol. Chem.*, **262**, 757–761.
11. Higashijima, T., Ferguson, K. M., Sternweis, P. C., Smigel, M. D., and Gilman, A. G. (1987) *J. Biol. Chem.*, **262**, 762–766.
12. Brandt, D. R., and Ross, E. M. (1986) *J. Biol. Chem.*, **261**, 1656–1664.
13. Aktories, K., Schultz, G., and Jakobs, K. H. (1984) In: *Neurotransmitter Receptors: Mechanisms of Action and Regulation* (Kito, S., Segawa, T., Kuriyama, K., Yamamura, H. I., and Olsen, R. W., eds). Plenum Press, New York–London, pp. 31–45.
14. Ransnäs, L. A., and Insel, P. A. (1988) *J. Biol. Chem.*, **263**, 9482–9485.
15. Ransnäs, L. A., and Insel, P. A. (1988) *J. Biol. Chem.*, **263**, 17239–17242.
16. Aktories, K., Schultz G., and Jakobs, K. H. (1983) *Mol. Pharmacol.*, **24**, 183–188.
17. Jakobs, K. H., Aktories, K., and Schultz, G. (1984) *Adv. Cyclic Nucleotide Protein Phosphoryl. Res.*, **17**, 135–143.
18. Vachon, L., Costa, T., and Herz, A. (1986) *J. Neurochemistry*, **47**, 1361–1369.
19. McLeish, K., Gierschik, P., Schepers, T., Sidiropoulos, D., and Jakobs, K. H. (1989) *Biochem. J.*, **260**, 427–434.
20. Gierschik, P., Sidiropoulos, D., Steisslinger, M., and Jakobs, K. H. (1989) *Eur. J. Pharmacol.* (Mol. Pharmacol. section; in press).
21. Hilf, G., and Jakobs, K. H. (1989) *Eur. J. Pharmacol.* (Mol. Pharmacol. section), **177**, 155–163.
22. Gierschik, P., Steisslinger, M., Sidiropoulos, D., Herrmann, E., and Jakobs, K. H. (1989) *Eur. J. Biochem.*, **183**, 97–105.
23. Parks, R. E., and Agrawal, R. P. (1973) In: *The Enzymes*, vol. VIII, Part A (Boyer, P. D., ed.). Academic Press, New York, pp. 307–333.
24. Cassel, D., and Selinger, Z. (1977) *Proc. Natl. Acad. Sci. USA*, **74**, 3307–3311.
25. Gierschik, P., and Jakobs, K. H. (1987) *FEBS Lett.*, **224**, 219–223.
26. Haga, K., Haga, T., Ichiyama, A., Katada, T., Kurose H., and Ui, M. (1985) *Nature (London)*, **316**, 731–733.
27. Klee, W. A., Koski, G., Tocque, B., and Simonds, W. F. (1984) *Adv. Cyclic Nucleotide Protein Phosphoryl. Res.*, **17**, 153–159.
28. Cerione, R. A., Codina, J., Benovic, J. L., Lefkowitz, R. J., Birnbaumer, L., and Caron, M. G. (1984) *Biochemistry*, **23**, 4519–4525.
29. Aktories, K., and Jakobs, K. H. (1984) *Eur. J. Biochem.*, **145**, 333–338.
30. Grandt, R., Aktories, K., and Jakobs, K. H. (1986) *Biochem. J.*, **237**, 669–674.
31. Gierschik, P., McLeish, K., and Jakobs, K. H. (1988) *J. Cardiovasc. Pharmacol.*, **12** (Suppl. 5), S20–S24.
32. Grandt, R., Greiner, C., Zubin, P., and Jakobs, K. H. (1986) *FEBS Lett.*, **196**, 279–283.

6
ON THE SPECIFICITY OF INTERACTIONS BETWEEN RECEPTORS, EFFECTORS AND GUANINE NUCLEOTIDE BINDING PROTEINS

Graeme Milligan

Molecular Pharmacology Group, Departments of Biochemistry and Pharmacology, University of Glasgow, Glasgow G12 8QQ, Scotland

INTRODUCTION

Suitable control of the integration and analysis of externally generated information by the signal transduction systems present in the plasma membrane of eukaryotic cells, would appear to require that strict fidelity of interaction be maintained between the appropriate components. As information amplification from the external to the internal environment of a cell requires a vectorial arrangement of receptor to G-protein to effector system, then the specificity (or otherwise) of contact between both receptor and G-protein and between G-protein and effector is likely to define the unique nature of the final response to the primary signal. A series of studies, largely using artificial liposome reconstitution systems, which have employed purified or partially purified receptors and guanine nucleotide binding proteins (G-proteins), have attempted to define the degree of selectivity of contacts between receptors and G-proteins. The results of many of these studies have indicated that a receptor, at least when placed in such an artificial environment, might be able to interact with a range of G-proteins [1–3]. Similarly, the recent observation that the stimulatory G-protein of the adenylate cyclase cascade (G_s) is also able to activate dihydropyridine-sensitive Ca^{2+} channels [4] has demonstrated that a receptor-activated G-protein, in the absence of spatial or other constraints, can in turn interact with more than a single effector system. The purpose of this chapter will thus be to assess critically recent experimental evidence pertaining to the degree of fidelity that is maintained in receptor–G-protein–effector interactions and to assess, in the light of these observations, the functional implications for hormonal modulation of cellular control.

HISTORICAL PERSPECTIVE

The original observations by Rodbell and co-workers [5] that hormonal stimulation of adenylate cyclase was critically dependent upon the presence of GTP led to the realization that the receptor and the catalytic moiety of adenylate cyclase were not the sole components of the signal transduction pathway. It took considerable time, however, before the coupling protein (G_s) was identified and obtained in a purified form [6]. This required the availability of the inappropriately named cyc$^-$ variant of the S49 lymphoma cell line (which is in fact $G_s\alpha^-$) [7], to act as an acceptor, and hence as an assay, for G_s in reconstitution studies on fractions derived from sodium cholate extracts of rabbit liver membranes (see Chapter 2 for details). Further, the recognition that the persistent activation of adenylate cyclase produced by a toxin elicited by *Vibrio cholerae* was due to the toxin catalysing an NAD-dependent ADP-ribosylation of the α subunit of G_s, provided another convenient means of detection of this polypeptide [8].

It had been realized from some of the earliest studies on the hormonal regulation of cAMP production that a number of agents were able to inhibit, rather than stimulate, adenylate cyclase. However, the fact that the magnitude of these effects was relatively small in comparison with those of the stimulatory agents, restricted the studies that were performed and thus hindered the development of the concept of a dual regulation of adenylate cyclase which was controlled by distinct G-proteins [9]. Thus, until the early part of this decade the notion of a separate G-protein, which would act to regulate hormonal control of the inhibition of adenylate cyclase, remained essentially a theoretical postulate. Again, it was the elucidation of the mechanism of action of a bacterial exotoxin which provided the necessary initial means of identification of the inhibitory G-protein. A series of elegant experiments by Ui and his collaborators [10–12], which had begun with the observation that a toxin produced by *Bordetella pertussis* could modulate α-adrenergic control of secretion from pancreatic islets [10] led to the recognition that this toxin (called either islet-activating protein or more usually pertussis toxin) attenuated hormone-mediated inhibition of adenylate cyclase. While pertussis toxin also catalyses an NAD-dependent ADP-ribosylation, the site of action (a cysteine residue compared to an arginine residue for cholera toxin) is different and the substrate, as originally defined, is the inhibitory G-protein of the adenylate cyclase cascade (G_i). Thus, the use of pertussis-toxin-catalysed ADP-ribosylation as an assay allowed the identification and purification of 'G_i' [13–15].

In contrast to the use of cyc$^-$ membranes as a assay for G_s, however, pertussis-toxin-catalysed ADP-ribosylation did not represent a functionally defined assay for G_i. While it originally appeared that there was but a single substrate for pertussis-toxin-catalysed ADP-ribosylation, then the assumption that this assay could thus be used to identify G_i could be justified. However, subsequent identification of other polypeptides with extremely similar mobilities in SDS–PAGE which were also substrates for this toxin [15–17] (see also Chapters 2 and 3) has required a reappraisal of this view. To date, the α subunits of some six G-proteins (TD1, TD2, G_i1, G_i2, G_i3 and G_o) are recognized as being potential substrates for ADP-ribosylation catalysed by pertussis toxin (Figure 6.1) by the presence of a cysteine residue located four amino acids from the C-terminus [18–19]. The extreme homology of this plethora of proteins, as well as of other identified G-proteins which are not substrates for pertussis toxin, as judged from their primary sequence, is the principal reason why it has been difficult to define specific functions for each of these proteins.

Depending upon the tissue studied, a considerable number of receptor-mediated signalling processes, including GTP-dependent activation of phospholipases of both the C [20] and A_2 [21] type, neurotransmitter inhibition of voltage-sensitive Ca^{2+} channels [22] and receptor regulated K^+ channels of pacemaking atrial cells and of certain secretory endocrine cells [23], as well as the receptor-mediated inhibition of adenylate cyclase have been shown to be disrupted by pretreatment with pertussis toxin. However, in the absence of further information, which can only be gained from carefully performed and well designed experimentation, it would be inappropriate to designate a specific function in any of these processes to a particular G-protein. Assessments of defined functionalities of particular pertussis-toxin-sensitive G-proteins is further complicated by the difficulties inherent in attempting to ascertain if a role of a G-protein is mediated by its α subunit or by the $\beta\gamma$ complex (see Chapter 1). As such, a major challenge in studies on G-protein function is to define the specificity or otherwise of the interactions of the pertussis-toxin-sensitive G-proteins with receptors and effectors.

TD1	K E N L K D C G L F
TD2	K E N L K D C G L F
G_i1	K N N L K D C G L F
G_i2	K N N L K D C G L F
G_i3	K N N L K E C G L Y
G_o	A N N L R G C G L Y
G_s	R M H L R Q Y E L L
G_z	Q N N L K Y I G L C

Figure 6.1 The C-terminal decapeptides of the currently identified signal transducing G-proteins. The cysteine residue located four amino acids from the C-terminus of the top six sequences is the site of pertussis-toxin-catalysed ADP-ribosylation. As such, each of these proteins could be a potential candidate to mediate pertussis-toxin-sensitive signalling events. The other two sequences lack this cysteine residue and hence are not substrates for this toxin

THE MULTIPLICITY OF PERTUSSIS-TOXIN-SENSITIVE G-PROTEINS

The concept that pertussis toxin was able to catalyse ADP-ribosylation of but a single substrate which represented the G-protein involved in receptor-mediated inhibition of adenylate cyclase was removed by the identification and partial purification of either two [15,16] or three [17] G-proteins, which were substrates for this toxin, from bovine brain. Production of complementary antisera able to discriminate between the two major forms of brain pertussis-toxin-sensitive G-proteins (G_i1 and G_o) [24,25] removed the possibility that the form with lower apparent molecular mass (G_o) was simply derived from G_i1 by proteolytic cleavage. Such observations, however, could not address questions relating to the function of each protein. The subsequent demonstration that antisera directed against either 'G_i' (now known to specifically identify G_i1) or G_o, did not cross-react with the major 40 kDa pertussis-toxin-sensitive G-protein present either in rat glioma cells [26] or in neutrophils [27] indicated that yet further levels of complexity in the expression of various forms of pertussis-toxin-sensitive G-proteins existed (see Chapter 3).

It has been the availability of information provided from the analysis of cDNA clones corresponding to the potential coding sequences of a range of G-proteins [18,19] which has allowed a beginning to be made on the problem of unravelling the roles of different pertussis-toxin-sensitive G-proteins. Production of anti-peptide antisera directed against specific sequences which can be predicted to be present in the individual G-proteins has allowed for the mapping of the tissue distribution of these G-proteins [28,29] (see Chapter 3). Such studies have confirmed that the inability of anti-G_i1 and G_o antisera to identify the major pertussis-toxin-sensitive G-protein of neutrophils and of glioma cells as noted above is because this polypeptide is the α subunit of G_i2 [29]. Further, the third pertussis-toxin-sensitive G-protein noted in one of the original purification of 'G_i-like' proteins from bovine brain [17], but not commented upon in the other two early purification schemes [15,16], has also been shown to have the immunological characterisitics of G_i2 [29]. While this information is not directly relevant to whether each of these individual gene products is present within a single cell, it does demonstrate that complex mixtures of pertussis-toxin-sensitive G-proteins can be expressed within a tissue. Given the physical and structural similarities of these proteins this fact raises particular problems for the purification of homogeneous populations of individual G-proteins both for reconstitution and for other functional studies. This will be discussed further at a later point.

RECONSTITUTION STUDIES AS AN ASSAY FOR G-PROTEIN FUNCTION

As noted above, it was the availability of the cyc$^-$ variant of the S49 lymphoma cell line, which contains neither relevant mRNA nor the polypeptide of $G_s\alpha$ [30], which provided a functional reconstitution system to detect exogenous G_s. This was based on the ability of the activated form of $G_s\alpha$ to stimulate the catalytic moiety of adenylate cyclase present in these membranes. Equivalent mutations of any of the pertussis-toxin-sensitive G-proteins are not available. However, as pertussis toxin treatment attenuates receptor-mediated inhibition of adenylate cyclase, then the restoration of hormonal-mediated reduction of cAMP generation in membrane systems derived from pertussis-toxin-pretreated cells theoretically offered an assay for G_i function. The relatively small effects of inhibitory hormones on adenylate cyclase function serve, however, to limit the sensitivity of such assays. Further, as noted above, the role of the individual α and $\beta\gamma$ subunits of any purified G-proteins which are employed in an assay of this form is difficult to assess. Both a mass-action-related effect to provide a pool of free $\beta\gamma$ subunit which would bind to and limit the subsequent activation of deactivated $G_s\alpha$ [31] and indications, at least in cyc¯ membranes, that the α subunit of G_i must play a direct role [32,33] in the inhibitory control of adenylate cyclase have received support.

Early experiments which analysed the reconstitution of 'G_i' from rabbit liver into membranes of pertussis-toxin-treated human platelets [31] appeared to provide a functional assay, while theoretically similar experiments which attempted to assess whether partially purified fractions enriched in either G_i1 or G_o from bovine brain were both able to reconstitute opioid-peptide-mediated inhibition of adenylate cyclase in membranes of pertussis-toxin-pretreated neuroblastoma × glioma hybrid, NG108-15, cells [15] were limited by the obvious lack of homogeneity of the G-proteins (a recurring problem in the field, as while technical refinements have steadily improved the resolution of various G-proteins they have also allowed the demonstration of the presence of hitherto undetected forms). It is thus likely to require the expression of defined G-proteins from either cDNAs [33–36] or genomic clones before such reconstitution experiments can be adequately performed.

While alterations in second messenger generation in response to exogenously provided G-protein have provided data pertaining to interactions between G-proteins and effector systems, different strategies have been used to assess interactions of hormone-activated receptors with G-proteins.

One obvious approach which has been applied to this question is to assess the affinity to interaction of agonist ligands with the receptor in the presence or absence of different G-proteins. This strategy is based on the observation that addition of analogues of GTP to ligand binding assays reduces the affinity of agonists, but not antagonists, at receptors which interact with G-proteins. As such, interaction of a G-protein with the receptor would be anticipated to increase the affinity of agonists at the receptor binding site. Using this approach, Florio and Sternweis [37] were able to demonstrate that partially purified muscarinic acetylcholine receptors were able to interact with both 'G_i' and with G_o when receptor and G-protein were co-introduced into artificial liposomes. These were both novel and intriguing results and while a succession of subsquent reports have essentially confirmed the basic premise of their report, both for muscarinic acetylcholine and other receptors [1–3], concerns as to the purity of the G-proteins used in these experiments as well as the heterogeneity of identified muscarinic receptor subtypes, in which it has recently been shown that both the M2 and M3 subtypes efficiently inhibit adenylate cyclase while the M1 and M4 subtypes strongly activate polyphosphatidylinositol hydrolysis [38], limit the conclusions which can usefully be drawn.

A second approach to the same question might be to ascertain if the addition of receptor agonists is able to promote the rate of binding and hence of the subsequent hydrolysis of GTP by various G-proteins. As all G-proteins undergo the cyclical binding and hydrolysis of GTP as part of their cycle of activation and deactivation, then this approach is theoretically amenable to studies of the interaction of receptors with any G-protein. Technical difficulties have been noted in many attempts to measure receptor-stimulated GTPase activities of certain G-proteins, such as G_s and the pertussis-toxin-insensitive form of the G-protein (G_p) which is involved in linking receptors to phospholipase C. This limitation is likely to reflect a combination of the expression of low levels of these proteins and their low turnover number [39].

Despite these caveats, addition of partially purified pertussis-toxin-sensitive G-proteins to membranes of pertussis-toxin-pretreated neuroblastoma × glioma hybrid cells allowed the reconstitution of an opioid-peptide-stimulated GTPase activity [40] (Figure 6.2), thus demonstrating functional interaction between the δ opiate receptor in

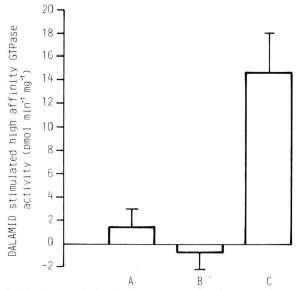

Figure 6.2 Reconstitution of δ opiate receptor stimulation of high affinity GTPase activity to membranes derived from pertussis-toxin-pretreated neuroblastoma × glioma, NG108–15, cells by pertussis-toxin-sensitive G-proteins from bovine brain.

Membranes were prepared from NG108–15 cells which had been treated with pertussis toxin (100 ng/ml) for 16 hours. These were incubated for 15 minutes at 30 °C with or without a mixture of pertussis-toxin-sensitive G-proteins isolated from bovine brain [15] and with DALAMID (10^{-5} M) in the presence of 0.04% sodium cholate. After this time high affinity GTPase activity was measured. Stimulation by DALAMID (A) of the G-protein preparation alone; (B) of membranes derived from pertussis-toxin-pretreated cells; and (C) following reconstitution of purified pertussis-toxin-sensitive G-proteins with membranes from pertussis-toxin-treated cells. Data modified from Milligan *et al.* [40]

these membranes and the exogenously provided G-proteins. As such, on the basis of a range of assays, it has been possible to demonstrate 'reconstitution' of receptor/pertussis-toxin-sensitive G-protein interactions.

THE SPECIFICITY OF RECEPTOR–G-PROTEIN–EFFECTOR INTERACTIONS

A number of recent experiments have attempted to define the specificity of either receptor–G-protein or G-protein–effector interactions. An elegant demonstration that G_k (the G-protein that mediates receptor control of K^+ channels) is equivalent to G_i3 has been provided by purification and sequencing of G_k activity and thus demonstrating that the obtained amino acid sequence is entirely compatible with the predicted sequence derived via the G_i3 cDNA [41]. It still remains for such definition to be provided for functions of other pertussis-toxin-sensitive G-proteins. In our own experiments on this topic we have used the neuroblastoma × glioma hybrid cell line, NG108-15, as a model system. This cell line expresses a considerable range of G-proteins including G_s, G_o, G_i2 and G_i3, but apparently no G_i1. As the cells represent a homogeneous population, then presumably all of these forms must be present in each cell and must at least be theoretically available to interact with receptors that generate pertussis-toxin-sensitive biochemical responses. Of the receptors on these cells, perhaps the most fully characterized is an opiate receptor of the δ subtype. This receptor mediates inhibition of adenylate cyclase, an event attenuated by pretreatment with pertussis toxin [42,43], hence indicating that the transduction of this signal must proceed via G_o, G_i2 or G_i3 or some combination thereof. As pertussis-toxin-catalysed ADP-ribosylation occurs on a cysteine residue, only four amino acids from the C-terminus of the α subunits of these proteins then we have generated a series of anti-peptide antisera able to identify the extreme C-terminus of each of these G-proteins with the hope that they might inhibit interactions between receptors and the relevant G-proteins [44]. Antiserum AS7 identifies both G_i1 and G_i2, as the C-termini of these two G-proteins are identical. However, as neither the polypeptide corresponding to G_i1 or relevant mRNA has been detected in these cells (unpublished), then in membranes of these cells this antiserum can be used as a useful probe for G_i2. Antiserum OC1 specifically identifies G_o [45]. Incubation of membranes of NG108-15 cells with antibodies derived from antiserum AS7 prevented the ability of a receptor-saturating dose of a synthetic enkephalin to stimulate high affinity GTPase activity [44], an assay which has been noted to provide strong correlation which opioid-mediated inhibition of adenylate cyclase in these cells [46]. In contrast, similar experiments using antiserum OC1 did not attenuate the opiate effect [47] (Figure 6.3). Further, a second receptor-mediated stimulation of high affinity GTPase in these cells, namely activation of a growth factor receptor by fetal calf serum was not abolished by antiserum AS7 [44]. These experiments provided strong support for the concept that the opioid-receptor-mediated event demonstrated specificity for the G-protein and that the G-protein in this system was G_i2.

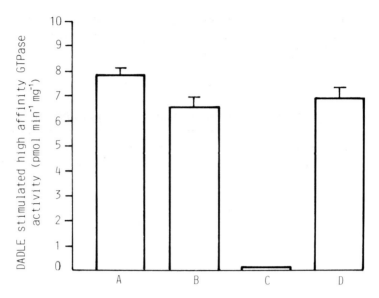

Figure 6.3 Antisera directed against the extreme C-termini of G_i and G_o. Modulation of opiate-stimulated GTPase activity in membranes of NG108–15 cells. Membranes of NG108–15 cells were incubated for 60 minutes at 37 °C with (A) water, (B) normal rabbit serum, (C) antibodies affinity-purified from antiserum AS7 (anti-G_i C-terminus) and diluted 1:100 into normal rabbit serum or (D) antiserum OC1 (anti-G_o C-terminus) diluted 1:100 into normal rabbit serum. The membranes were then assayed for opioid peptide (DADLE(10^{-6} M))-stimulated GTPase activity. The data are modified from McKenzie and Milligan [47] and updated from McKenzie et al. [44]

In this cell line, opioid peptides also function in a pertussis-toxin-sensitive manner to inhibit voltage-sensitive Ca^{2+} channel currents. Hescheler et al. [22] have demonstrated that addition of a purified preparation of brain G_o into pertussis-toxin-treated NG108-15 cells is more efficient in reconstituting opioid-peptide-mediated inhibition of Ca^{2+} currents than a preparation of G_i prepared from brain, suggesting that this function of the opiate receptor is transduced via G_o. However, it should be noted that brain G_i is predominantly G_i1, which is not expressed in detectable quantities in NG108-15 cells in which, as noted above, the predominant form of G_i is G_i2. While both sets of experiments were performed with NG108-15 cells, in the experiments reported by Hescheler et al. [22], the cells were pretreated with dibutyryl cAMP to 'differentiate' them, a procedure which is necessary before Ca^{2+} transients can be recorded. However, the experiments of McKenzie et al. [44,47] were performed on membranes of undifferentiated cells. Using pertussis-toxin-catalysed [^{32}P]ADP-ribosylation of membranes of both undifferentiated and dibutyryl cAMP 'differentiated' NG108-15 cells it is possible to make a rough assessment of the relative proportions of G_i and G_o in the membranes of these cells. In the undifferentiated cells greater than 90% of the radioactivity incorporated into the G-proteins in response to pertussis toxin is into G_i, suggesting that the relative ratio of G_i to

G_o in these membranes is high. By contrast the amounts of G_o are considerably higher in the 'differentiated' membranes, ranging in a series of experiments between 25% and 50% of the total pertussis-toxin-sensitive G-proteins (Figure 6.4). As such, the complete abolition of opiate-receptor-stimulated GTPase activity in membranes of undifferentiated NG108-15 cells, while defining an interaction of the opiate receptor with G_i2 in these membranes, cannot exclude an interaction with G_o, the contribution of which was unmeasurably small due to the low levels of this G-protein in the membrane.

As such, in experiments employing purified fractions of G-proteins for reconstitution studies it is vital that information is available on the nature of the endogenous G-proteins in the system before questions can be answered which address the specificity of fidelity of

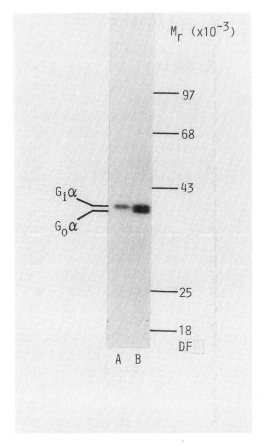

Figure 6.4 Pertussis-toxin-catalysed ADP-ribosylation of G-proteins in membranes of untreated and dibutyryl cAMP-'differentiated' NG108–15 cells. Membranes (25 µg) of untreated (A) of dibutyryl cAMP-'differentiated' (B) NG108–15 cells were ADP-ribosylated with [^{32}P]NAD and thiol-activated pertussis toxin, as described in Mullaney and Milligan [45]

interactions of G-proteins in the native system rather than an assessment as to whether particular receptor–G-protein interactions can occur under defined artificial conditions. A number of other reports have noted reconstitution of receptor-mediated alterations in Ca^{2+} currents by the addition of 'purified' brain G_o [48,49] but each report is limited by the same problems relating to the purity and the potential physiological relevance of the G-protein preparation employed.

Immunological techniques are also likely to provide information on the specificity of receptor–G-protein–ion channel effector system. Harris-Warwick et al. [50] have studied the involvement of a pertussis-toxin-sensitive G-protein in a dopamine-induced decrease in the voltage-dependent Ca^{2+} current in identified neurones of the snail *Helix aspersa*. Intracellular injection of $G_o\alpha$ to pertussis-toxin-treated snail neurones restored the effects of dopamine and intracellular injection of a polyclonal anti-$G_o\alpha$ antibody markedly reduced the effects of dopamine. Although the endogenous snail G-protein migrated differently from brain G_o it was identified by the anti-$G_o\alpha$ antibody, providing strong evidence that the snail homologue of G_o was indeed the coupling G-protein in this

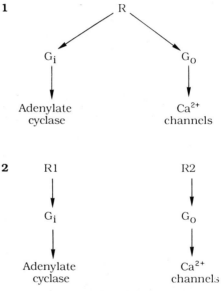

Figure 6.5 The nature of interactions of the δ opiate receptor with G-proteins and effector systems in neuroblastoma × glioma hybrid, NG108–15, cells. Pharmacological data indicate that the opiate receptor(s) expressed in NG108–15 cells are a single (δ) subtype. Opioid peptides mediate both inhibition of adenylate cyclase and inhibition of Ca^{2+} currents through voltage operated Ca^{2+} channels. (1) Evidence indicates that opioid effects on adenylate cyclase activity are transduced via G_i2 while effects on Ca^{2+} channels are mediated by G_o (see text). (2) As it remains to be ascertained whether the δ opioid receptor represents a single molecular gene product then the concept that the two effects of opioid peptides are produced by interaction with two distinct receptors, each of which is able to interact with but a single pertussis-toxin-sensitive G-protein, cannot currently be excluded.

system. We have used the anti-peptide antisera AS7 and OC1 described above to study opioid and α_2-adrenergic inhibition of Ca^{2+} currents in 'differentiated' NG108-15 cells and have demonstrated that the anti-G_o antiserum (OC1) was considerably more effective than the anti-G_i antiserum in attenuating the hormonal responses (unpublished). These data suggest that the opioid effect on Ca^{2+} currents in 'differentiated' NG108-15 cells is mediated via G_o, and that the opioid effect on GTPase and adenylate cyclase activity in the undifferentiated cells is transduced via G_i. Given the recent identification of a series of genes encoding information for muscarinic acetylcholine receptors which appear to have different functions [38], it must remain an open question as to whether a single molecular form of the δ opiate receptor is producing both the effects on adenylate cyclase and on Ca^{2+} channel current or whether greater diversity exists at the level of the receptor. Only when this has been defined will it be possible to assess if a single receptor can specifically activate multiple pertussis-toxin-sensitive G-proteins to produce differential effects (Figure 6.5).

ACKNOWLEDGEMENTS

I thank Fiona M. Mitchell for helpful comments and critical reading of this manuscript.

REFERENCES

1. Cerione, R. A., Staniszewski, C., Benovic, J. L., Lefkowitz, R. J., Caron, M. G., Gierschik, P., Somers, R., Spiegel, A. M., Codina, J. and Birnbaumer, L. (1985) *J. Biol. Chem.*, **260**, 1493–1500.
2. Ueda, H., Harada, H., Nozaki, M., Katada, T., Ui, M., Satoh, M., and Tagaki, H. (1988) *Proc. Natl. Acad. Sci. USA*, **85**, 7013–7017.
3. Kurose, H., Katada, T., Haga, T., Haga, K., Ichiyama, A., and Ui, M. (1986) *J. Biol. Chem.*, **261**, 6423–6428.
4. Yatani, A., Imoto, Y., Codina, J., Hamilton, S. L., Brown, A. M., and Birnbaumer, L. (1988) *J. Biol. Chem.*, **263**, 9887–9895.
5. Rodbell, M., Birnbaumer, L., Pohl, S., and Krans, H. M. J. (1971) *J. Biol. Chem.*, **246**, 1877–1882.
6. Sternweis, P. C., Northup, J. K., Smigel, M. D., and Gilman, A. G. (1981) *J. Biol. Chem.*, **256**, 11517–11526.
7. Johnson, G. L., Kaslow, H. R., and Bourne, H. R. (1978) *J. Biol. Chem.*, **253**, 7120–7123.
8. Gill, D. M., and Meren, R. (1978) *Proc. Natl. Acad. Sci. USA*, **75**, 3050–3054.
9. Rodbell, M. (1980) *Nature (London)*, **284**, 17–22.
10. Katada, T., and Ui, M. (1979) *J. Biol. Chem*, **254**, 469–479.
11. Katada, T., and Ui, M. (1982) *J. Biol. Chem.*, **257**, 7210–7216.
12. Katada, T., and Ui, M. (1982) *Proc. Natl. Acad. Sci. USA*, **79**, 3129–3133.
13. Bokoch, G. M., Katada, T., Northup, J. K., Ui, M., and Gilman, A. G. (1984) *J. Biol. Chem.*, **259**, 3560–3567.
14. Codina, J., Hildebrandt, J., Iyengar, R., Birnbaumer, L., Sekura, R. D., and Manclark, C. R. (1983) *Proc. Natl. Acad. Sci. USA*, **80**, 4276–4280.
15. Milligan, G., and Klee, W. A. (1985) *J. Biol. Chem.*, **260**, 2057–2063.
16. Sternweis, P. C., and Robishaw J. D. (1984) *J. Biol. Chem.*, **259**, 13806–13813.
17. Neer, E. J., Lok, J. M., and Wolf, L. G. (1984) *J. Biol. Chem.*, **259**, 14222–14229.
18. Jones, D. T., and Reed, R. R. (1987) *J. Biol. Chem.*, **262**, 14241–14249.

19. Suki, W. N., Abramowitz, J., Mattera, R., Codina, J., and Birnbaumer, L., (1987) *FEBS Lett.*, **220**, 187–192.
20. Ohta, H., Okajima, F., and Ui, M. (1985) *J. Biol. Chem.*, **260**, 15771–15780.
21. Jelsema, C. L., and Axelrod, J. (1987) *Proc. Natl. Acad. Sci. USA*, **84**, 3623–3627.
22. Hescheler, J., Rosenthal, W., Trautwein, W., and Schultz, G. (1987) *Nature (London)*, **325**, 445–447.
23. Codina, J., Grenet, D., Yatani, A., Birnbaumer, L., and Brown, A. M. (1987) *Science*, **236**, 442–445.
24. Pines, M., Gierschik, P., Milligan, G., Klee, W., and Spiegel, A. (1985) *Proc. Natl. Acad. Sci. USA*, **82**, 4095–4099.
25. Gierschik, P., Milligan, G., Pines, M., Goldsmith, P., Codina, J., Klee, W., and Spiegel, A. (1986) *Proc. Natl. Acad. Sci. USA*, **83**, 2258–2262.
26. Milligan, G., Gierschik, P., Spiegel, A. M., and Klee. W. A. (1986) *FEBS Lett.*, **195**, 225–230.
27. Gierschik, P., Falloon, J., Milligan, G., Pines, M., Gallin, J., and Spiegel, A. (1986) *J. Biol. Chem.*, **261**, 8058–8062.
28. Mumby, S. M., Kahn, R. A., Manning, D. R., and Gilman, A. G. (1986) *Proc. Natl. Acad. Sci. USA*, **83**, 265–269.
29. Goldsmith, P., Gierschik, P., Milligan, G., Unson, C., Vinitsky, R., Malech, H., and Spiegel, A. (1987) *J. Biol. Chem.*, **262**, 14683–14688.
30. Harris, B. A., Robishaw, J. D., Mumby, S. M., and Gilman, A. G. (1985) *Science*, **229**, 1274–1277.
31. Katada, T., Bokoch, G. M., Northup, J. K., Ui, M., and Gilman, A. G. (1984) *J. Biol. Chem.*, **259**, 3568–3577.
32. Katada, T., Bokoch, G. M., Smigel, M. D., Ui, M., and Gilman, A. G. (1984) *J. Biol. Chem.*, **259**, 3586–3595.
33. Jacobs, K. H., Aktories, K., and Schultz, G. (1983) *Nature (London)*, **303**, 177–178.
34. Graziano, M. P., Casey, P. J., and Gilman, A. G. (1987) *J. Biol. Chem.*, **262**, 11375–11381.
35. Olate, J., Mattera, R., Codina, J., and Birnbaumer, L. (1988) *J. Biol. Chem.*, **263**, 10394–10400.
36. Nukada, T., Mishina, M., and Numa, S. (1987) *FEBS Lett.*, **211**, 5–9.
37. Florio, V. A., and Sternweis, P. C. (1985) *J. Biol. Chem.*, **260**, 3477–3483.
38. Peralta, E. G., Ashkenazi, A., Winslow, J. W., Ramachandran, J., and Capon, D. J. (1988) *Nature (London)*, **334**, 434–437.
39. Milligan, G. (1988) *Biochem. J.*, **255**, 1–13.
40. Milligan, G., Simonds, W. F., Streaty, R. A., Tocque, B., and Klee, W. A. (1985) *Biochem. Soc. Trans.*, **13**, 1110–1113.
41. Codina, J., Olate, J., Abramowitz, J., Mattera, R., Cook., R. G., and Birnbaumer, L. (1988) *J. Biol. Chem.*, **263**, 6746–6750.
42. Burns, D. L., Hewlett, E. L., Moss, J., and Vaughan, M. (1983) *J. Biol. Chem.*, **258**, 1435–1438.
43. Klee, W. A., Milligan, G., Simonds, W. F., and Tocque, B. (1985) *Mol. Aspects Cell. Reg.*, **4**, 117–129.
44. McKenzie, F. R. Kelly, E. C. H., Unson, C. G., Spiegel, A. M., and Milligan, G. (1988) *Biochem. J.*, **249**, 653–659.
45. Mullaney, I., and Milligan, G. (1989) *FEBS Lett.*, **244**, 113–118.
46. Koski, G., and Klee, W. A. (1981) *Proc. Natl. Acad. Sci. USA*, **78**, 4185–4189.
47. McKenzie, F. R., and Milligan, G. (1989) In: *Receptors, Membrane Transport and Signal Transduction*, Vol. 29. Nato ASI series. Plenum Press, New York, pp. 65–74.
48. Ewald, D. A., Sternweis, P. C., and Miller, R. J. (1988) *Proc. Natl. Acad. Sci., USA*, **85**, 3633–3637.
49. Rosenthal, W., Hescheler, J., Trautwein, W., and Schultz, G. (1988) *FASEB J.*, **2**, 2784–2790.
50. Harris-Warwick, J., Hammond, C., Paupardin-Tritsch, D., Homburger, V., Rouot, B, Bockaert, J., and Gershenfeld, H. M. (1988) *Neuron*, **1**, 27–34.

7

TRANSDUCIN: THE SIGNAL TRANSDUCING G-PROTEIN OF PHOTORECEPTOR CELLS

Vijay N. Hingorani and Yee-Kin Ho

Department of Biological Chemistry, University of Illinois at Chicago, Health Sciences Center, Chicago, Illinois 60612, USA

TRANSDUCIN AND PHOTOTRANSDUCTION

Phototransduction in the retina serves an important role in converting light into neural signals. In vertebrate rod photoreceptor cells, this process occurs exclusively in the rod outer segment (ROS) and involves a light-activated cGMP enzyme cascade. Absorption of a photon by the receptor molecule rhodopsin on the ROS disk membrane, causes the photoisomerization of the 11-*cis*-retinal chromophore of rhodopsin to all-*trans*-retinal. Photolysed rhodopsin (R*) triggers a series of biochemical events that lead to the activation of a latent cGMP phosphodiesterase (PDE) which rapidly hydrolyses cytosolic cGMP. The coupling between R* and the latent PDE is mediated by a GTP binding protein called transducin (T) via a GTP binding and hydrolysis cycle. The activation signal is highly amplified; a single R* causes the hydrolysis of 10^5 cGMP molecules in a fraction of a second. The transient decrease of cGMP concentration causes the closure of the cGMP-sensitive cation channels on the plasma membrane of the cell. Reduction of the cations' permeability across the plasma membrane results in hyperpolarization of the cell [1,2].

The transducin-mediated signal coupling events of the cGMP cascade have been described in detail. In its latent form, transducin contains a bound GDP (T-GDP). R* binds T-GDP to form a complex with reduced affinity for guanine nucleotides. As a result, bound GDP is exchanged for GTP, converting transducin to an active form (T-GTP) that switches on the PDE [3]. The formation of T-GTP is accompanied by the release from R*, which can then interact with another T-GDP. This rapid recycling of R* generates hundreds of active T-GTP, and each is capable of activating PDE. Hence, signal amplification is achieved in two stages: a first stage gain of 10^2 due to the formation of hundreds of activated PDE; and a second stage gain of 10^3 due to the rapid hydrolysis of thousands of molecules of cGMP per second by an activated PDE. Deactivation of the T-GTP involves an intrinsic GTPase activity of transducin which hydrolyses the bound GTP and returns the transducin to its latent form of T-GDP.

G-Proteins as Mediators of Cellular Signalling Processes
Edited by Miles D. Houslay and Graeme Milligan. © 1990 John Wiley & Sons Ltd

THE COUPLING CYCLE OF TRANSDUCIN

All of the protein components of the retinal cGMP cascade have been purified. Details of the individual coupling events of transducin have been elucidated via reconstitution studies [4]. Both the transducin and the cGMP PDE are heterotrimers. Transducin is composed of three polypeptides: T_α (39 kDa), T_β (37 kDa) and T_γ (8.5 kDa). PDE is composed of two catalytic subunits, P_α (88 kDa) and P_β (84 kDa), and an inhibitor peptide, P_γ (14 kDa). T_α contains the GTP binding site and is the activator of the PDE. The GDP-bound form of T_α has a high affinity for $T_{\beta\gamma}$ and rhodopsin, and a low affinity for PDE. R* catalyses the exchange of GTP for the bound GDP on T_α. The T_α–GTP complex then dissociates from $T_{\beta\gamma}$/R* and acquires a high affinity for the effector enzyme, PDE. T_α–GTP interacts with P_γ removing the inhibition exerted by P_γ on the catalytic sites of PDE. T_β and T_γ form a tight complex. The $T_{\beta\gamma}$ complex apparently plays a role in presenting T_α to R*. Individually, T_α–GDP or $T_{\beta\gamma}$ does not bind well to R*. However, a complex of T_α–GDP–$T_{\beta\gamma}$ tightly associates with R*. Upon exchanging bound GDP for GTP, T_α–GTP dissociates from $T_{\beta\gamma}$ and R*. The dissociation of these three protein components ensures an irreversible step in the cascade. Hence, GTP cannot bind to the T_α site in the absence of R* and $T_{\beta\gamma}$ and the T_α-bound GTP will not release from the site easily which allows T_α to remain in its active GTP-bound state long enough to find its target, a latent PDE. T_α–GTP can only be converted back to its GDP-bound form by the hydrolysis of the bound GTP which terminates the interaction with PDE. Thus, T_α operates as a molecular switch which conveys the signal from a receptor to an effector enzyme which in turn elicits cellular responses. A schematic diagram depicting the coupling cycle of transducin is shown in Figure 7.1.

The basic mechanism of signal tranduction utilized by transducin is a subunit dissociation and association cycle regulated via guanine nucleotide binding and hydrolysis. By switching between GDP and GTP, T_α can exist in two comformations which differ in their interaction with the receptor and effector molecules. The GDP-bound form of T_α associates with $T_{\beta\gamma}$ which enchances its interaction with the receptor, R*. This interaction facilitates the exchange of GDP for GTP. The GTP-associated conformation of T_α favours the dissociation from $T_{\beta\gamma}$ and R*, making it accessible for the activation of the effector, PDE. The cycle is completed when the tightly bound GTP is hydrolysed to GDP. The hydrolysis of GTP is not an energy requirement for activation, but merely allows the active conformation of T_α to return back to its latent form. Non-hydrolysable analogues of GTP, such as Gpp(NH)p and GTPγS, have been found to be equally effective in activating PDE and in dissociating transducin subunits, indicating that the hydrolysis of GTP is not essential for the activation.

The purified T_α–GTPγS complex is capable of activating latent PDE presumably by removing the inhibitory P_γ peptide from the catalytic subunit of $P_{\alpha\beta}$. The P_γ–T_α–GTPγS complexes can be detected by ion-exchange chromatography as well as by immunoprecipitation [5,6] indicating that the active form of transducin is capable of interacting with P_γ. However, under physiological conditions the T_α–GTP may directly interact with the PDE complex removing the inhibitory action of P_γ without the formation of a soluble

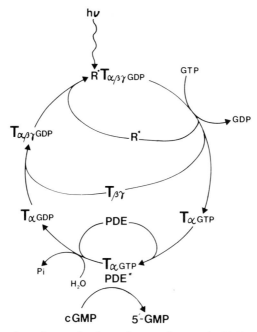

Figure 7.1 The signal coupling cycle of transducin in the retinal cGMP cascade. T represents transducin, R* is photolysed rhodopsin, and PDE is the cGMP phosphodiesterase

P_γ–T_α–GTP complex [7]. It has been demonstrated that binding of Cr(III)GTP to transducin activates PDE without dissociating the tranducin and PDE subunits from the ROS membrane [8]. Immunological studies using antibodies aganst PDE and T_α to follow the distribution of P_γ and T_α in the soluble and membrane-bound fractions during activation by GTPγS show that 90% of the T_α–GTPγS complex is released into solution, but all the P_γ remains bound to the membrane with less that 1% found in the soluble fraction [8]. These results imply that under physiological conditions, formation of a soluble P_γ–T_α–GTP complex is not an obligatory step for PDE activation and P_γ remains membrane bound during the activation cycle.

THE REACTION DYNAMICS OF THE TRANSDUCIN CYCLE

Purified transducin in solution does not exchange or hydrolyse GTP. The turnover of transducin requires the catalytic action of R*. The reaction cycle of tranducin can be separated into three reaction steps:

(1) Interaction with R* opens the GTP binding site of T_α for GTP exchange. This is followed by the dissociation of the R*–T_α–GTP–$T_{\beta\gamma}$ complex leading to the formation of T_α–GTP.

(2) Hydrolysis of the T_α-bound GTP at the GTP binding site of T_α generates an intermediate, T_α–GDP·P_i, containing tightly bound GDP and P_i.
(3) Turnover of the T_α–GDP·P_i occurs after the release of the tightly bound P_i. The regenerated T_α–GDP recombines with $T_{\beta\gamma}$ for another cycle of activation.

The relative rates of these reactions have been examined by pre-steady-state kinetic analyses [9]. The initial rates of P_i formation due to GTP hydrolysis have been studied using the rapid acid quenching method. It was found that the rate of P_i formation is biphasic with an initial rapid burst followed by a slow steady-state rate. This observation suggests that the hydrolysis of GTP at the site of T_α occurs rapidly and the rate-limiting step is on the release of P_i from the T_α–GDP·P_i intermediate which controls the steady-state rate of GTP hydrolysis of the transducin cycle. The initial burst of P_i formation, as well as the steady-state rate, increases proportionally with the amount of transducin sample. Moreover, substitution of H_2O with heavy water (D_2O) as solvent reduces the magnitude and rate of the initial burst, but has no effect on the steady-state rate of the GTPase activity. The lack of D_2O solvent isotope effect of the steady-state hydrolysis of GTP by transducin is not surprising since this rate is related to the release of P_i from T_α–GDP·P_i, a reaction that does not involve H_2O as a reactant.

Since the release of the tightly bound P_i from T_α–GDP·P_i is rate limiting. T_α–GDP·P_i can accumulate and become the most abundant species in the system at a steady state. The single turnover rate of P_i release from T_α–GDP·P_i has been measured by a filtration assay and shown to have a half-time of approximately 20–30 s [10]. The deactivation of PDE associated with the turnover of GTP has been estimated to have the same rate as the P_i release [8]. Moreover, the PDE deactivation rate also does not exhibit any D_2O solvent isotope effect. It may be concuded that the T_α–GDP·P_i is fully capable of activating PDE. After releasing the tightly bound P_i, T_α–GDP loses its ability to activate PDE as a stable complex. The tightly bound GDP stays at the site. The stable complex of T_α–GDP recombines with $T_{\beta\gamma}$, which interacts with R* again for another exhange cycle to occur. The reaction dynamics of the transducin cycle have been summarized in Figure 7.2.

ORGANIZATION OF THE TRANSDUCIN SUBUNITS

The subunit organization of transducin eluted from the ROS membranes has been investigated by chemical cross-linking [11]. Bifunctional reagents, *para*-phenyl dimaleimide and maleimidobenzoyl n-hydroxysuccinimide ester, were used to cross-link

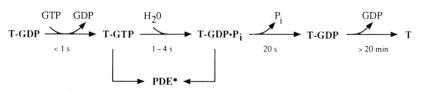

Figure 7.2 The dynamics of the transducin reaction cycle

transducin in an attempt to elucidate the topological arrangement of the transducin subunits and their proximal relationships. The composition of the cross-linked products was then the analysed according to their cross-linked molecular weights and identified with specific anti-T_α and anti-$T_{\beta\gamma}$ antisera by western blotting. The cross-linked products were digested with trypsin to define further which interacting regions of the T_α and $T_{\beta\gamma}$ subunit were bridged by the cross-linker. Based on these results, a topological model for the transducin subunits has been constructed as shown in Figure 7.3. An interesting aspect of this model is that the carboxyl terminal peptide of T_α may interact with the amino terminal region of T_β. Moreover, the close proximity of T_γ to T_α and T_β suggests that T_γ may actually play a role in conferring the specificity of the interaction between transducin and rhodopsin. The existence of oligomeric cross-linked products in the GDP-bound forms of transducin implies that transducin can form oligomeric aggregates in solution which may play a role in regulating the reaction dynamics of the transducin coupling cycle. The hydrodynamic properties of native transducin with GDP or Gpp(NH)p bound have been examined [12]. Sedimentation measurements of purified transducin–GDP reveal the existence of two protein components in solution. From the sedimentation coefficients, the molecular weights of 75 ± 7 kDa and 277 ± 16 kDa are deduced, corresponding to the expected monomeric and tetrameric forms of transducin. No dimers, trimers or oligomers larger than the tetramer were detected. Upon binding Gpp(NH)p, the two protein components observed in the sedimentation measurements are converted to a single protein component with an apparent molecular weight of

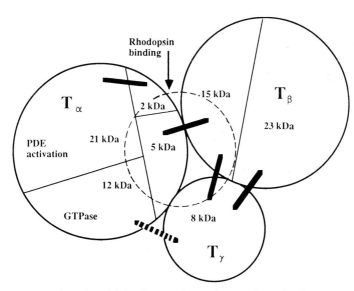

Figure 7.3 A topological model for the transducin subunits. The molecular sizes represent the tryptic fragments for each subunit. The heavy black lines indicate the observed cross-linking between various fragments of the subunits

37 ± 4 kDa, which is consistent with the dissociation of the putative tetrameric and monomeric forms of transducin into individual T_α and $T_{\beta\gamma}$ subunits.

The physiological role of the oligomeric transducin remains unclear. However, the observation is consistent with other studies. First, the kinetics of transducin activation determined by light scattering are best fitted with a scheme in which a single photolysed rhodopsin can interact with four transducin molecules simultaneously [13]. Second, an allosteric activation of transducin by R* has been postulated [14]. This effect can be accounted for if transducin exists as an oligomer. And third, freeze-etching electron microscopy has revealed a particle with the size of 80–120 Å for transducin bound to the surface of ROS membrane [15] which would be consistent with a tetrameric form of transducin.

THE PRIMARY SEQUENCE AND TRYPTIC PEPTIDE MAP OF TRANSDUCIN SUBUNITS

The amino acid sequence of each subunit of transducin has been deduced via molecular cloning [16–24]. Also, limited tryptic proteolysis has been used to dissect the native protein to generate a series of tryptic fragments. Comparison of the partial amino acid sequences of the peptide fragments with the deduced amino acid sequences of the subunits has allowed the development of a linear peptide map [23,25] as shown in Figure 7.4. T_α is first cleaved at Lys_{18} to produce a 38 kDa fragment which is then quickly cleaved at Arg_{310} to produce a transient 33 kDa fragment. The final cleavage at Arg_{204} results in a 21 kDa and a 12 kDa fragment which are stable even after prolonged treatment. This final cleavage is blocked if a non-hydrolysable analogue of GTP is incorporated into T_α. T_β has only one cleavage site at Arg_{129} and results in an amino terminal 15 kDa and a carboxyl terminal 23 kDa peptide. T_γ is not cleaved by trypsin. The linear tryptic map has been used extensively as a structural basis for analysing the functional domains of transducin.

Comparison of primary sequence homology reveals that T_α contains four regions that are similar to the GTP binding domains of EF-Tu, the ras p21 protein and other members of the G-protein family [26,27]. Since the tertiary structures of EF-Tu and ras p21 have been elucidated by X-ray crystallography, the role of these homologous regions in GTP binding can be predicted. The first homologous region with a consensus sequence of GXGXXGK, is located near the amino-terminus. In T_α, this loop (Gly_{36}–Ala–Gly–Glu–Ser–Gly–Lys_{42}) interacts with the γ-phosphate of GTP and may control the GTPase activity of the enzyme. The other three homologous regions cluster together between residues 190 and 286 as shown in Figure 7.4. These regions constitute the binding sites for the Mg^{2+} ion and the guanine ring structure of GTP. The Mg^{2+} ion is likely to associate with the loop containing residues Asp_{196}–Val–Gly–Gln_{200} with a salt bridge formed between the Mg^{2+} and the carboxyl group of Asp_{196}.

IDENTIFICATION OF THE FUNCTIONAL DOMAINS OF T_α

In order for T_α to undertake its varied activities it must contain interacting sites for R^*, $T_{\beta\gamma}$–GTP and PDE. Its coupling action is controlled by the coordinated interactions among these sites. Specific reconstitution assays have been developed to study the individual reaction steps of the retinal cGMP cascade shown in Figure 7.1. A [^{3}H]Gpp(NH)p binding assay is used to monitor the single turnover of the nucleotide exchange reaction. The binding of latent transducin–GDP to rhodospin membrane as well as the release of transducin subunits in the presence of Gpp(NH)p is monitored by a centrifugation method which separates the bound and free transducin [28]. For the multiple turnover of transducin catalysed by R^*, the continuous hydrolysis of [$\gamma-^{32}$P]GTP is followed. The PDE activity is measured with a pH electrode by monitoring the proton release due to the hydrolysis of cGMP [29]. Furthermore, PDE can be activated by reconstitution with the T_α–Gpp(NH)p complex or independent of transducin by brief trypsin treatment which cleaves the inhibitory P_γ peptide.

By combining the tryptic peptide map and the reconstitution assays, the structural and functional relationship of the transducin molecule has been characterized extensively. A variety of biochemical methods have been utilized to modify transducin. These include limited proteolysis, chemical modifications, immunochemical characterization, ADP-ribosylation and affinity labelling. In general, the purified T_α is first modified by a protease or affinity reagent, then the modified T_α is reconstituted with intact $T_{\beta\gamma}$, R^* and PDE for functional assays. Inhibition of individual reaction steps due to chemical modification can be easily identified and the site of the modification can also be localized on the tryptic peptide map. A large amount of information has been generated and is summarized in Table 7.1. Interestingly, all of the chemical modifications and limited proteolysis on T_α carried out so far inhibit only one of the many coupling activities of transducin, suggesting that T_α is composed of distinct domains with autonomic functions. For example, alteration of the amino- and carboxyl-terminal regions of T_α diminishes the interaction with rhodopsin and $T_{\beta\gamma}$, but not the binding of GTP and the activation of PDE. This suggests that these regions represent the receptor binding domain. Modifications of the mid-region of T_α, Leu_{19}–Arg_{204}, inhibit GTP hydrolysis and PDE activation but not the interaction with R^* and $T_{\beta\gamma}$. It is likely that this region contains the nucleotide and the effector binding sites. Based on the above results, a functional peptide map of T_α has been constructed as shown in Figure 7.4. In addition, two flexible hinge regions (Gly_{288}–Pro–Asn_{290} and Gly_{198}–Gly_{199}), which are located at the boundaries of the proposed functional domains, can be identified and may play an important role in allowing communication between the functional domains.

A THREE-DIMENSIONAL MODEL OF T_α

The large amount of biochemical data on transducin which has accumulated over the past few years has allowed the construction of a hypothetical three-dimensional model for the

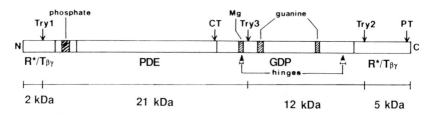

Figure 7.4 The tryptic peptide map of T_α. The tryptic cleavage sites are noted as Try-1, Try-2, and Try-3. The functional domains of T_α are incorporated. The four shaded regions represent the GTP binding site which is homologous to other GTP binding proteins such as elongation factor Tu. The CT and PT sites represent the cholera- and pertussis-toxin-catalysed ADP-ribosylation sites. The two black triangles represent the possible hinge regions that link the proposed funtional domains

T_α subunit [40] (Figure 7.5). The model serves to integrate the available biochemical data so as to provide new insights into the molecular mechanism of protein function. The following strategies have been employed to construct this model. First, the secondary structure and hydropathy of the 350 amino acid residues were predicted from the primary sequence of T_α with the use of well established algorithms. Then regions contained within the GTP binding domain were identified by comparison with similar regions in other G-proteins and were placed in the same spatial relationship as the folding topography of the GDP binding site seen in the crystal structure of EF-Tu. This constraint greatly defined and restricted the folding patterns of the putative PDE and rhodopsin binding sites which were predicted from the biochemical studies described previously. The PDE binding domain, encompassing residues Met_{49}–Val_{175}, was predicted to be made up of eight β strands in tandem, separated by turns and coils. This strongly suggests an anti-parallel β sheet topology. According to the results of the biochemical studies, the amino and carboxyl terminal regions were placed in close proximity to form the putative rhodopsin/$T_{\beta\gamma}$ interacting domain. A search was then made to identify flexible regions which might act as movable hinges to allow a conformational change in one domain of T_α to be conveyed to another. Two flexible regions located strategically between the interfaces of the proposed functional domains were identified. The flexibility of the Gly_{198}–Gly_{199} region provides a communicating link between the GTP and the PDE interacting domains and that of Gly_{288}–Pro_{289}–Asn_{290} links the rhodopsin/$T_{\beta\gamma}$ and the GTP binding domains.

MAPPING THE GTP BINDING SITE OF T_α

A hypothetical structure of the GTP binding site of transducin has been constructed by substituting the amino acid residues of the GDP binding site of EF-Tu obtained from X-ray crystallograhy [41] with the homologous residues of T_α (Figure 7.6). The GTP binding domain of G-proteins such as EF-Tu and the ras-p21 protein is composed of two separate peptide regions [26], a segment near the amino-terminus forms the site of

Table 7.1 Summary of results on the chemical modification of T_α

Chemical modification	Stoichiometry	Interacting or labelled site	Inhibitory effect
Fluorescein-5′-isothiocyanate labelling of lysine residues [30]	1 mole per T_α	21 kDa fragment (Major) (Leu_{19}–Arg_{204}) 12 kDa fragment (Minor) (Arg_{204}–Arg_{310})	GTP hydrolysis and PDE activation
N-Ethylmaleimide labelling of sulfhydryl group [28]	1 mole per T_α–Gpp(NH)p 3 moles per T_α–GDP	12 kDa fragment 2 moles on 12 kDa and 1 mole on 21 kDa fragment	$R^*/T_{\beta\gamma}$ binding
Ethyl dimethylamino-propyl carbodiimide labelling of carboxyl groups [31]	0.3 mole per T_α with [^{14}C]glycine ethylester trapping	Amino terminal 2 kDa fragment (Met_1–Lys_{18})	$R^*/T_{\beta\gamma}$ binding
Diethyl pyrocarbonate labelling of histidine [32]	1 mole per T_α–Gpp(NH)p	33 kDa fragment (Leu_{19}–Arg_{310})	PDE activation
8-azido-GTP [33]	1 mole per T_α	2 kDa fragment	GTP binding
Azido-anilido-GTP [33]	1 mole per T_α	21 kDa fragment	GTP binding
Pertussis-toxin-catalysed ADP-ribosylation [34,35]	1 site per T_α	5 kDa fragment (Cys_{347})	GTP exchange
Cholera-toxin-catalysed ADP-ribosylation [10,36]	1 site per T_α	21 kDa fragment (Arg_{174})	GTP hydrolysis
Monoclonal antibodies against amino-terminal peptides [37]	—	2 kDa fragment	$R^*/T_{\beta\gamma}$ binding
Antibodies against carboxyl terminal peptides [38]	—	5 kDa fragment	$R^*/T_{\beta\gamma}$ binding
Removal of amino terminal peptide with proteases [25,39]	—	2 kDa fragment	$R^*/T_{\beta\gamma}$ binding

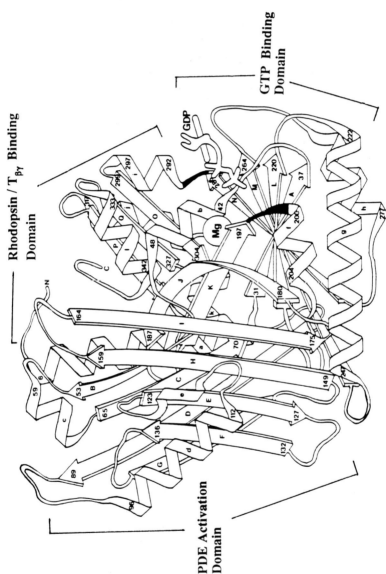

Figure 7.5 Proposed tertiary structure of T_α. The numbers represent the number of amino acid residues from the primary sequence. The arrows correspond to individual β strands. The guanine nucleotide binding domain (β strands N, M, L, A, K, J) and the PDE activation domains (β strands B, C, D, G, F, E, H, I) are deduced from the comparison of EF-Tu and the 'Greek key' β-barrel structure respectively. Two movable hinges (Gly_{198}–Gly_{199} and Gly_{288}–Pro_{289}–Asn_{290}) are located between the functional domains as shown in black

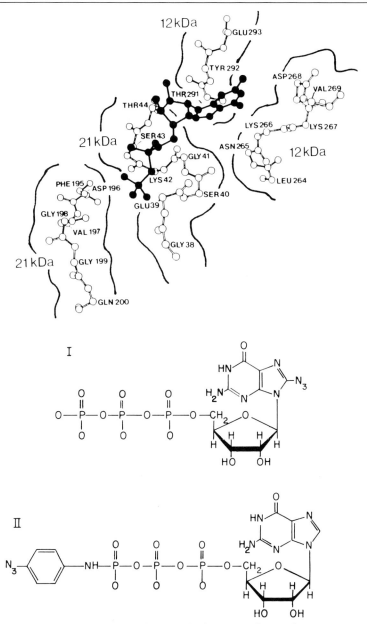

Figure 7.6 Proposed structure of the GTP-binding site of T_α. The tryptic peptides that contribute to the GTP binding site are indicated. A GDP molecule shown in black is incorporated in this model. Mg^{2+} is believed to associate with Asp_{196}, and a monodentate ligand is formed between the Mg^{2+} and the terminal phosphate of the bound guanine nucleotide. Photoaffinity analogues of GTP with the reactive azido group on the guanine ring (**I**) and phosphate moeity (**II**) are shown

interaction for the γ-P$_i$ of GTP and a larger segment closer to the carboxyl terminus forms the guanine ring binding site. These two regions fold together in close proximity to form the GTP binding site. These regions are located on different tryptic fragments. The γ-P$_i$ binding region is associated with the 21 kDa tryptic fragment of T$_\alpha$, whereas the guanine ring binding region is on the 12 kDa fragment (Figure 7.4). The specificity for the guanine ring may be regulated via interaction with Asp$_{265}$–Lys$_{266}$. The phosphate binding site is associated with residues Gly$_{36}$–Ala–Gly–Glu–Ser–Gly–Lys–Ser$_{43}$, which are located in the amino-terminal side of the 21 kDa tryptic fragment. The 2′-, 3′-hydroxy groups of the ribose are exposed to the solvent. Mg^{2+} ion binding occurs in a loop situated diametrically opposite the β-phosphate of GDP and a salt bridge may be formed between the Mg^{2+} and the carboxyl group of Asp$_{196}$. The Mg^{2+} ion is likely to form a monodentate ligand with the terminal phosphate of the guanine nucleotide.

The characteristics of the proposed structure have been experimentally mapped using a battery of affinity analogues of GTP with reactive groups attached to different parts of the GTP molecule [33]. Regions associated with the guanine ring have been probed with 8-azido-GTP (I) and the affinity labelling occurred on the 12 kDa tryptic fragment of T$_\alpha$. The sites which interact with the phosphate moiety of GTP have been probed with 4-(azidoanilido)-P^1-5′-GTP (II). The labelling site of this analogue was found to associate with the 21 kDa tryptic fragment. The ribose binding regions have been examined with 3′O-{3-[N-(4-azido-2-nitrophenyl)amino]propionyl}GTP and 2′,3′-dialdehyde derivative of GTP, but none covalently attached to T$_\alpha$. The overall result of mapping the nucleotide binding site of transducin with affinity labelling is in complete agreement with the proposed model of T$_\alpha$.

PROPOSED COUPLING MECHANISM OF T$_\alpha$

The structural information obtained from the proposed model along with the available biochemical data lead to a hypothetical mechanism for the coupling action of T$_\alpha$. Interactions with R* and T$_{\beta\gamma}$ occur at the two terminal regions of the T$_\alpha$ molecule. Binding of R* opens up the guanine nucleotide binding site of T$_\alpha$ for rapid GTP/GDP exchange. This can be accomplished through the movement of the flexible hinge region (Gly$_{288}$–Pro$_{289}$–Asn$_{290}$). Upon binding of GTP, two major conformational changes occur in the T$_\alpha$–GTP complex. First, it dissociates from photolysed R* and T$_{\beta\gamma}$. Second, the PDE activation site of the T$_\alpha$-GTP complex is exposed for interaction with PDE. In order to accommodate the γ-phosphate of GTP, additional space is needed in the GTP binding pocket of T$_\alpha$. Peptides located at the interface of the three functional domains are pushed away to open up the GTP binding site. As a result of this spatial rearrangement, the helix containing Arg$_{204}$, which is directly attached to the flexible hinge region (Gly$_{198}$–Gly$_{199}$), moves towards the PDE binding domain. Hence, the conformational changes originating from GTP binding are transmitted to the PDE binding site. The Mg^{2+} binding site bridges the GTP binding domain with the R*/T$_{\beta\gamma}$ interacting domain. Interaction between the bound Mg^{2+} and GTP may play an important role in regulating the dissociation of T_α from R* and $T_{\beta\gamma}$. The amino-terminal region of T_α has been suggested to form an amphipathic helix containing a hydrophobic groove with charge pairs surrounding it. This characteristic may play a role in regulating the binding and dissociation of transducin from the ROS membrane.

Several features of the model of T_α as well as the suggested coupling mechanism have been supported by experimental results. The movement of the flexible hinge (Gly_{198}–Gly_{199}) in conveying the signal from the GTP binding site to the PDE activating domain is supported by results obtained by proteolysis. Upon binding of Gpp(NH)p, the tryptic cleavage site at Arg_{204} located at the middle of the interfacial helix is no longer available for digestion, indicating that movement of this helix occurs during the activation of T_α [25]. Interaction of the amino- and carboxyl-terminal peptides of T_α forming the rhodopsin/$T_{\beta\gamma}$ binding site has been confirmed by fluorescence spectroscopy [42] and chemical cross-linking studies [11]. The involvement of interaction between the bound Mg^{2+} and GTP in dissociating T_α from R* and $T_{\beta\gamma}$ was tested by substituting GTP with Cr(III) β,γ-bidentate GTP analogues. Cr(III)GTP complex is a stable exchange-inert complex that eliminates the interaction between the β,γ phosphates of GTP with the bound Mg^{2+}. Indeed, Cr(III)GTP is capable of activating PDE but fails to dissociate the transducin subunits from the ROS membrane [8].

COMPARISON OF T_α OF ROD AND CONE PHOTORECEPTOR CELLS

It is known that a similar but not identical transduction system exists in the cone photoreceptor cells for colour perception [43]. A distinct PDE which interacts with the cone T_α has been suggested. The primary sequence homology between the rod and cone T_α molecules is more than 85% [20] as shown in Figure 7.7. Also, the secondary structures are predicted to be the same, indicating that they share a similar folding pattern. The guanine nucleotide binding domain is essentially identical except for a variation of the suggested hinge region for the rod T_α (Gly_{288}–Pro–Asn_{290}). The change in cone T_α to Gly–Asn–Asn may have functional significance such as differences in the affinity and exchange rates for guanine nucleotides. There are several variations in the proposed PDE activation domain which may represent part of the PDE interacting site. Indeed, these variable locations are on the surface of the molecule and could easily interact with the PDE complex. The receptor binding domain shows interesting variations. There are four additional residues (Glu–Leu–Ala–Lys) on the amino-terminal helix of cone T_α. However, the hydrophobic groove as well as the change pair characteristics surrounding the groove remain unchanged. This observation implies that in spite of the additional residues the nature of the interaction of cone transducin with colour opsin remains the same. It has been suggested that both T_α molecules use the same T_β subunit as their modulator indicating that the T_β interacting site of cone and rod T_α may be similar. The carboxyl terminal ends of the two T_α molecules are identical. These comparisons provide additional support for the model.

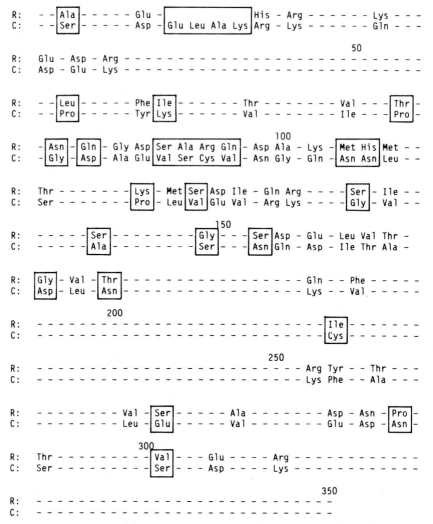

Figure 7.7 Comparison of the primary sequence of rod and cone T_α. R and C represent the sequences of rod and cone T_α, respectively. Only the non-identical residues are indicated. The boxed residues are those that are not conserved

FUNCTIONAL ROLE OF $T_{\beta\gamma}$

Information concerning the structural and functional aspects of the $T_{\beta\gamma}$ subunit of transducin is still limited. Most of the biochemical methods used in characterizing T_α have also been applied to $T_{\beta\gamma}$ in an attempt to elucidate its functional domains and essential residues. Surprisingly, all of the chemical modifications, including N-ethylmaleimide labelling of the sulfhydryl groups [28] and proteolysis with trypsin [25] have no effect on $T_{\beta\gamma}$ with respect to coupling of T_α to R*. It is clear that $T_{\beta\gamma}$ is required for the binding of T_α to R* in the initial activation of the cGMP cascade. However, whether the dissociated $T_{\beta\gamma}$ carries out additional functions in the rod outer segment is still unclear. The $\beta\gamma$ subunits of other G-proteins have been suggested to have an inhibitory role through their interaction with the α subunit thereby keeping it associated with the receptor and avoiding the interaction with the effector enzyme [44]. This does not seem to be the regulatory function of $T_{\beta\gamma}$, since excess $T_{\beta\gamma}$ does not affect the activation of the T_α or the PDE in a reconstitution system. Recently a retinal 33 kDa phosphoprotein has been isolated and shown to form a specific complex with $T_{\beta\gamma}$ [45]. It is likely that the interaction of $T_{\beta\gamma}$ and the 33 kDa protein will decrease the availability of $T_{\beta\gamma}$ needed for the binding of T_α to R*, which could indirectly inhibit the activation of the cGMP cascade. Further experiments are needed to elucidate the regulatory role of $T_{\beta\gamma}$ in rod outer segments.

PERSPECTIVES

This chapter summarizes the large amount of biochemical data on the structure and function of transducin which has accumulated over the past few years. Most of the experiments carried out thus far have been solution studies; these form a basis for the application of other techniques such as site directed mutagenesis and X-ray crystallography. Some questions which remain to be answered include the physiologic role played by the oligomeric structure of the protein, regulation of the cascade cycle by rod outer segment components which have not been identified, and precise structural determinations of the molecules and their mode of interaction.

Transducin belongs to a family of GTP binding proteins (the G-proteins) which are now known to play a major role in regulating a diversity of cellular signalling events. All members of the G-protein family share a common structural motif as a heterotrimer and appear to exhibit a similar mode of action. The parallel between the light activation of cGMP cascade in ROS and the hormonal regulation of adenylyl cyclase is overwhelming. It is generally believed that the biochemical pathway elucidated in the visual system will help to provide a better understanding of the role of G-proteins in other less well defined signal transduction processes.

The basic mechanism of signal transduction utilized by transducin is a subunit dissociation and association cycle regulated via guanine nucleotide binding and hydrolysis. Several well characterized enzymes also use this general nucleotide-mediated switching mechanism for mechanical coupling and informational transfer in biopolymer synthesis and degradation. It has been implicated in the elongation factor Tu (EF-Tu) in protein synthesis [46], the actin-myosin ATPase [47], the rec A protein in DNA recombination [48] and the ATP-dependent La protease [49]. A comparison of the regulatory mechanism of transducin with other biological coupling enzymes could elucidate the common motif of bound nucleotides as molecular switches in cellular regulation [50].

REFERENCES

1. Liebman, P. A., Park, K. R., and Dratz, E. A. (1987) *Ann. Rev. Physiol.*, **49**, 765–791.
2. Applebury, M. L., and Hargrave, P. A. (1986) *Vision Research*, **26**, 1881–1895.
3. Fung, B. K.-K., Hurley, J. B., and Stryer, L. (1981) *Proc. Natl. Acad. Sci. USA*, **78**, 152–156.
4. Fung, B. K.-K. (1983) *J. Biol. Chem.*, **258**, 10495–10502.
5. Deterre, P., Bigay, H. J., Robert, M., Pfister, C., Kuhn, H., and Chabre, M. (1986) *Protein*, **1**, 188–193.
6. Fung, B. K.-K., and Griswold-Prenner, I. (1989) *Biochemistry*, **28**, 3133–3137.
7. Sitaramayya, A., Harkness, J., Parkes, J. H., Gonzalez-Oliva C., and Liebman, P. A. (1986) *Biochemistry* **25**, 651–656.
8. Frey, S. E., Hingorani, V. N., Su-Tsai, S.-M., and Ho, Y.-K. (1988) *Biochemistry*, **27**, 8209–8218.
9. Ting, T. D., Ho, Y.-K., (1989) *Investigative Opthalmology and Visual Science*, **30** (3), Supplement Abs. p. 172.
10. Navon, S. E., and Fung, B. K.-K. (1984) *J. Biol. Chem.*, **259**, 6686–6693.
11. Hingorani, V. N., Tobias, D. T., Henderson, J. T., and Ho, Y.-K. (1988) *J. Biol. Chem.*, **263**, 6916–6926.
12. Goldin, S., Mazar, A., and Ho, Y.-K. (1989) Unpublished results.
13. Bennett, N., and Dupont, Y. (1985) *J. Biol. Chem.*, **260**, 4156–4168.
14. Wessling-Resnick, M., and Johnson, G. L. (1987) *J. Biol. Chem.*, **262**, 3696–3706.
15. Roof, D. J., Korenbrot, J. I., and Heuser, J. E. (1982) *J. Cell. Biol.*, **95**, 501–509.
16. Medynski, D. C., Sullivan, K., Smith, D., Van Dop, C., Chang, F. H., Fung, B. K.-K., Seeburg, P. H., and Bourne, H. R. (1985) *Proc. Natl. Acad. Sci. USA*, **82**, 4311–4315.
17. Yatsunami, K., and Khorana, H. G. (1985) *Proc. Natl. Acad. Sci. USA*, **82**, 4316–4320.
18. Tanabe, T., Nukada, T., Nishikawa, Y., Sugimoto, K., Suzuki, H., Takahashi, H., Noda, M., Haga, T., Ichiyama, A., Kangawa, K., Minamino, N., Mutsuo, H., and Numa, S. (1985) *Nature (London)*, **315**, 242–245.
19. Ovchinnikov, Y. A., Lipkin, V. M., Shuvaeva, T. M., Bogachuk, A. P., and Shemyakin, V. V. (1985) *FEBS Lett.*, **179**, 107–110.
20. Lochrie, M. A., Hurley, J. B., and Simon, M. I. (1985) *Science*, **228**, 96–99.
21. Fong, H. K. W., Hurley, J. B., Hopkins, R. S., Miake-Lye, R., Johnson, M. S., Doolittle, R. F., and Simon, M. I. (1986) *Proc. Natl. Acad. Sci. USA*, **83**, 2162–2166.
22. Sugimoto, K., Nukada, T., Tanabe, T., Takahashi, H., Noda, M., Minamino, N., Kangawa, K., Matsuo, H., Hirose, T., Inayama, S., and Numa, S. (1985) *FEBS Lett.*, **191**, 235–240.
23. Hurley, J. B., Simon, M. I., Teplow, D. B., Robishaw, J. D., and Gilman, G. (1984) *Science*, **226**, 860–862.
24. Yatsunami, K., and Khorana, H. G. (1985) *Proc. Natl. Acad. Sci. USA*, **82**, 4316–4320.
25. Fung, B. K.-K., and Nash, C. R. (1983) *J. Biol. Chem.*, **258**, 10503–10510.
26. Halliday, K. R. (1984) *J. Cyclic Nucleotide Res.*, **9**, 435–448.
27. Leberman, R., and Egner, U. (1984) *EMBO J.*, **3**, 339–341.
28. Ho, Y.-K., and Fung, B. K.-K. (1984) *J. Biol. Chem.*, **259**, 6694–6699.
29. Yee, R., and Liebman, P. A. (1978) *J. Biol. Chem.*, **253**, 8902–8909.
30. Hingorani, V. N., and Ho, Y.-K. (1987) *Biochemistry*, **26**, 1633–1639.
31. Hingorani, V. N. (1988) Ph.D. Thesis, University of Illinois at Chicago.
32. Tobias, D. T., Chang, L. F., Ho, Y.-K. (1989) Unpublished results.
33. Hingorani, V. N., Ho-Chan, L.-F., and Ho, Y.-K. (1989) *Biochemistry*, **28**, 7424–7432.
34. Watkins, P. A., Moss, J., Burns, D. L., Hewlett, E. L., and Vaughan, M. (1984) *J. Biol. Chem.*, **259**, 1378–1381.
35. Van Dop, C., Tsubokawa, M., Bourne, H. R., and Ramachandran J. (1984) *J. Biol. Chem.*, **259**, 696–698.
36. Abood, M. E., Hurley, J. B., Pappone, M.-C., Bourne, H. R., and Stryer, L. (1982) *J. Biol. Chem.*, **257**, 10540–10543.

37. Navon, S. E., and Fung, B. K.-K. (1987) *J. Biol. Chem.*, **262**, 15746–15751.
38. Cerione, R. A., Kroll, S., Rajaram, R., Unson, C., Goldsmith, P., and Spiegel, A. M., (1988) *J. Biol. Chem.*, **263**, 9345–9352.
39. Navon, S. E., and Fung, B. K.-K. (1988) *J. Biol. Chem.*, **263**, 489–496.
40. Hingorani, V. N., and Ho, Y.-K. (1987) *FEBS Lett.*, **220**(1), 15–22.
41. laCour, T. F. M., Nyborg, J., Thirup, S., Clark, B. F. C. (1985) *EMBO J.*, **4**, 2385–2388.
42. Hingorani, V. N., and Ho, Y.-K (1988) *J. Biol. Chem.*, **263**, 19804–19808.
43. Lerea, C. L., Somers, D. E., Hurley, J. B., Klock, I. B., and Bunt-Milam, A. H. (1986) *Science*, **234**, 77–80.
44. Northup, J. K., Sternweis, P. C., Gilamn, A. G. (1983) *J. Biol. Chem.*, **258**, 11361–11368.
45. Lee, R. H., Lieberman, B. S., and Lolley, R. N. (1987) *Biochemistry*, **26**, 3983–3990.
46. Kaziro, Y. (1978) *Biochim. Biophys. Acta.*, **505**, 95–127.
47. Hibberd, M. G., and Trentham, D. R., (1986) *Ann. Rev. Biophys. Chem.*, **15**, 119–161.
48. Kowalczykowski, T. (1987) *Trends Biochem. Sci.*, **12**, 141–145.
49. Menon, A. S., Waxman, L., and Goldberg, A. L. (1987) *J. Biol. Chem.*, **262**, 722–726.
50. Ho, Y.-K., Hingorani, V. N., Navon, S. E., and Fung, B. K.-K. (1989) In *Current Topics in Cellular Regulation*, **30**, 171–201, (Horecker, B. L., and Stadtman, E. R., eds). Academic Press, New York.

8

G-PROTEIN-MEDIATED SIGNALLING IN OLFACTION

Richard C. Bruch

Department of Neurobiology and Physiology, Northwestern University, 2153 Sheridan Road, Evanston, IL 60208, USA*

and

Geoffrey H. Gold

Monell Chemical Senses Center, Philadelphia, PA, USA

INTRODUCTION

In vertebrates, peripheral chemosensory neurones in the olfactory epithelium detect olfactory stimuli, transduce stimulus-encoded information into ion channel activity and transmit these ionic signals directly to the central nervous system. Peripheral olfactory neurones are bipolar cells with single, unbranched and non-myelinated axons collectively forming the olfactory nerve (cranial nerve I) which synapses with second-order neurones in the olfactory bulb. The single, unbranched dendrites of the receptor cells project towards the external environment and terminate at the apical end in cilia or microvilli [1–3]. Stimulus interaction with the apical dendritic cilia and microvilli of the primary chemosensory neurones initiates the subsequent transmembrane signal transduction events that lead to membrane depolarization, action potential generation and synaptic transmission [1]. Although these general features of olfactory stimulus–response coupling are well established from electrophysiological studies, consensus conclusions have not been reached regarding the molecular basis of stimulus detection and the sequence of transductional events that link stimulus recognition to the ion channels underlying membrane depolarization. Although stimulus detection and recognition are generally regarded as receptor-mediated events, alternative mechanisms that do not require stimulus–receptor interaction have also been proposed to mediate stimulus activation of olfactory neurones [3–6]. The receptor hypothesis and the extent to which receptor-independent processes may contribute to the overall mechanism of olfactory transduction, remain unresolved problems since the receptors have not been identified at the molecular level.

* Address to which correspondence should be addressed.

Several members of the heterotrimeric family of signal transducing G-proteins [7] have been identified in the olfactory epithelium [4]. A role for these proteins in mediating the initial events of olfactory reception and signal transduction has also been indicated by recent biochemical and neurophysiological evidence [3,4,8–10]. This chapter summarizes the major findings regarding the role of G-proteins in mediating transmembrane signalling events in peripheral olfactory neurones. Many aspects of G-protein involvement in olfaction have been recently reviewed (3,4,8–12). However, in this chapter we have also considered the additional studies available to us during 1988, regarding the identification, subcellular localization and functional roles of G-proteins in olfactory neurones. The evidence implicating G-protein involvement in coupling of stimulus to second messenger pathways is described, and compared to additional studies suggesting a role for alternative signalling mechanisms, not linked to G-proteins, in olfactory stimulus–response coupling.

G-PROTEIN IDENTIFICATION IN OLFACTORY NEURONES

Five members of the heterotrimeric family of signal transducing G-proteins have been identified in the olfactory epithelium [4]. Both G_s and G_i, the G-proteins generally associated with dual regulation of adenylate cyclase [7], and G_o, were initially identified in isolated olfactory cilia from frog by bacterial toxin-catalysed ADP-ribosylation [13]. These observations were subsequently confirmed and further supported by immunoblotting analysis with subunit-specific antisera [14]. Similar cholera and pertussis toxin substrates were also identified in isolated cilia from toad, rat and cow [15,16]. In catfish, although a cholera toxin substrate corresponding to G_s was identified in isolated cilia preparations, immunoreactive G_o was not detected in these preparations, or in membranes derived from intact olfactory epithelium [17]. In addition, a 40 kDa pertussis toxin substrate that cross-reacted with antisera to a common amino acid sequence of G-protein α subunits, was also detected in isolated cilia from catfish [17]. The electrophoretic mobility of the 40 kDa pertussis toxin substrate compared to the purified α subunits of G_i (G_{i1}) [7] and G_o, and its immunoreactivity, suggested homology with an alternative form of G_i, G_{i2} [7].

Five cDNA clones encoding the α-subunits of G_s, G_o and three forms of G_i were also isolated from rat olfactory epithelium [18]. The novel G_{i3}, initially identified in olfactory epithelium, was found to be distributed in a variety of additional rat tissues by northern analysis [18]. G_{i3} was also identified in HL-60 cells by in vitro translation experiments [19], and genes encoding homologues of rat G_{i3} were identified by cDNA cloning in human genomic libraries [20]. In addition, cDNA cloning experiments indicated that G_{i3} corresponded to G_k, a G-protein coupled to potassium channels [21]. Thus, a G-protein unique to the olfactory epithelium, analogous to transducin in visual cells, was not detected by biochemical or molecular cloning techniques. In this regard, it should also be noted that immunoreactive transducin was not detected in isolated olfactory cilia preparations by immunoblotting analysis [4]. The functional roles of G_o and the three forms of G_i identified by molecular cloning in olfactory signal transduction, and their cellular localization in the olfactory epithelium, have not yet been investigated.

Immunohistochemical localization of G-proteins at the cellular level in mature animals, and initial investigations of the developmental aspects of G_s expression in fetal olfactory neurones, were also reported. Subunit-specific antisera were used to evaluate the cellular localization and distribution of G-proteins in frog olfactory epithelium [12,14].

Immunohistochemical studies using antiserum to the β-subunit indicated that these subunits were prominantly distributed along the dendritic ciliary surface of the tissue and in the axons of olfactory neurones. Antiserum to a common amino acid sequence of G-protein α-subunits showed preferential localization of these subunits in both receptor cells and glandular cells. Although G_o was detected in isolated olfactory cilia by western blotting, immunohistochemical localization of G_o in the olfactory epithelium was not feasible. Immunohistochemical methods, at the light and electron microscopic levels, were also used to study the ontogeny of G_s expression in olfactory neurones [22,23]. In fetal rats, G_s and G-protein β subunits were first detectable in olfactory cilia at E16, coinciding with the developmental expression of cilia and stimulus-evoked action potentials.

A putative 56 kDa glycosylated GTP binding protein was also isolated from skate olfactory epithelium [24]. A potential role for this polypeptide in olfactory signal transduction was suggested based on its co-purification as a complex with putative olfactory receptors and the dissociation of the complex by GTPγS. In addition, stimulus enhancement of GTPase activity associated with the putative receptor-GTP binding protein complex, but not with the dissociated GTP binding protein alone, was also reported. However, the identity of this polypeptide, its cellular localization and its functional role in olfaction remain unclear. Direct demonstration of guanine nucleotide binding to this polypeptide and its functional coupling to G-protein-linked effectors (page 117) have also not been reported.

G-PROTEIN-MEDIATED SIGNALLING EVENTS

Stimulus detection and recognition

The molecular nature of stimulus detection and recognition and the transductional events associated with stimulus-evoked olfactory responses are generally assumed to be receptor-mediated processes involving stimulus binding to receptor proteins and subsequent activation of appropriate receptor-linked effectors [1–4,8–10]. However, an alternative, but not mutually exclusive, hypothesis regarding the molecular nature of stimulus activation of olfactory neurones has also been proposed that does not require stimulus interaction with macromolecular receptors [3–5,10]. Recent studies have also suggested that both receptor-mediated and receptor-independent mechanisms may operate simultaneously during stimulus detection [6,25].

The molecular identity of the anticipated olfactory receptor proteins remains perhaps the most fundamental unresolved problem in olfaction. Radioligand binding studies in aquatic species have shown that stimulus amino acids interact with binding sites in isolated olfactory cilia in a manner consistent with generally accepted criteria for ligand–receptor interaction [3,4,10,11,26]. However, the low affinity of this interaction (micromolar K_D values) has generally been assumed to prohibit identification and isolation of the putative receptors by traditional ligand binding affinity techniques [9]. An additional complication of receptor identification in olfaction arises from the limited availability of comparable ligand binding data in air-breathing vertebrates [27], thereby limiting confident prediction of receptor specificity across species. The wide diversity and selectivity of stimulus recognition in olfaction has also encouraged speculation that olfactory receptors may be similar to other multi-substrate proteins, such as the immunoglobulins [9] or the cytochrome P-450 oxygenases [28]. Ligand binding studies in aquatic species support the hypothesis that stimulus detection is mediated by a limited number of, probably glycoprotein [29], receptor subtypes [30]. Stimulus binding in isolated cilia preparations paralleled the specificity, potency and stereoselectivity characteristics expected for the receptors from electrophysiological studies [31,32]. Coupling of stimulus amino acid receptors to G-proteins was also demonstrated in isolated olfactory cilia by radioligand binding [17]. The affinity of these receptors for their ligands was significantly reduced when stimulus binding was measured in the presence of GTP or a hydrolysis-resistant analogue, reflecting an establishd pharmacological criterion associated with G-protein-linked receptors [7].

Although ligand binding studies suggest that olfactory stimulus detection and recognition are mediated by G-protein-linked receptors, this conclusion requires rigorous confirmation by isolation of the putative receptors. It should be noted, however, that several candidate receptor proteins have been isolated by affinity-based chromatography on immobilized stimulus columns [3,4,10,33–35]. In addition, gp 95, a transmembrane glycoprotein localized exclusively in olfactory cilia, was also suggested as a candidate olfactory receptor [9]. Confirmatory evidence has not yet been reported to establish that these candidate receptors exhibit stimulus binding activity or the ability to interact with other components of the signal transduction pathway(s). A soluble odorant-binding protein, associated with secretory glands, was also isolated from bovine olfactory epithelium [12,36]. The same or similar polypeptide was cloned from frog olfactory epithelium and shown to exhibit some sequence homology to serum transport proteins [37]. Since the soluble nature and cellular localization of this polypeptide are unexpected characteristics of an olfactory receptor, a role for the odorant-binding protein in stimulus solubilization and transport has been suggested [12,36].

An additional model, which does not require stimulus–receptor interaction or G-protein involvement [3–5,10], has also been proposed for detection and discrimination of olfactory stimuli. Since several stimuli depolarized non-chemosensory neuroblastoma cells, at concentrations that also altered membrane fluidity, it was suggested that specific receptor proteins were not required for olfactory reception [10,38]. Several stimuli were also shown to depolarize azolectin liposomes and alter membrane fluidity at concentrations that correlated with frog and porcine olfactory thresholds [39]. Since these effects of stimuli on membrane properties were also modulated by manipulation of the lipid composition of the phospholipid vesicles, it was suggested that stimulus adsorption on hydrophobic membrane domains was sufficient to describe stimulus reception. The specificity and sensitivity of olfactory reception were therefore postulated to arise from cell-to-cell differences in membrane lipid composition among individual receptor cells [39,40]. It is not clear, however, to what extent the phospholipid composition of azolectin liposomes reflects that of the cellular membranes of olfactory neurones since lipid composition data have not been reported. The influence of membrane proteins at appropriate protein:lipid ratios on these stimulus effects on membrane properties has also not been investigated.

Taken together, the combined evidence regarding the molecular nature of olfactory reception suggests that it is likely that both receptor-mediated and receptor-independent processes contribute to the overall mechanism of receptor cell activation [3,4]. A similar conclusion was also suggested recently, since olfactory stimuli activated melanosome dispersion and elevated intracellular cAMP levels in non-olfactory melanophores [6]. These stimulus-induced effects were demonstrable with the same types and concentrations of stimuli previously shown to activate adenylate cyclase in isolated olfactory cilia (page 117). Two mechanisms were suggested to be involved in stimulus detection, involving specific receptors and receptor-independent events, with both pathways acting to elevate cAMP [6]. Partial inhibition of electrophysiological responses by monoclonal antibodies to polypeptides isolated by chromatography on immobilized stimulus columns, was also interpreted to indicate that multiple mechanisms were involved in olfactory reception [25].

Cyclic nucleotide second messengers

A second messenger role for both cAMP and cGMP in olfactory signal transduction has been suggested by biochemical and electrophysiological evidence [3,4,8–12]. About half of the total adenylate cyclase activity in the olfactory epithelium was retained in isolated cilia preparations. In these preparations, stimuli elevated cAMP levels, but only in the presence of guanine nucleotides, suggesting that the olfactory adenylate cyclase was G-protein-linked [13]. Adenylate cyclase was also solubilized from isolated olfactory cilia and reconstituted in proteoliposomes [41]. In the reconstituted membranes, forskolin and GTPγS stimulated the enzyme, suggesting that a G-protein-linked adenylate cyclase complex had been reconstituted. Since the olfactory adenylate cyclase also exhibited additional characteristics associated with the classical hormone-regulated enzyme [7], it was proposed that an analogous G_s-linked cyclic nucleotide cascade mediated olfactory signal transduction [8,9,13]. Snyder and co-workers subsequently confirmed these initial observations, and also tested a large number of stimuli for the ability to enhance guanine-nucleotide-dependent adenylate cyclase activity in isolated cilia [12,42]. Differential stimulation of adenylate cyclase by distinct classes of stimuli, corresponding to floral, fruity, herbaceous and minty odorants, but not putrid and solvent-like stimuli, was observed. Since not all stimuli elicited enhanced cAMP formation, it was suggested that at least one additional transduction mechanism must be available to account for responses to these stimuli. By contrast, in isolated cilia from catfish, representative stimuli from each amino acid receptor subtype similarly enhanced Gpp(NH)p-dependent adenylate cyclase activity [43]. In this regard, it should also be noted that a positive correlation was obtained between electrophysiological response magnitudes and the ability of a large number of stimuli to enhance cAMP formation in isolated cilia [44]. This correlation suggested that many, if not all, stimuli activated adenylate cyclase. It was therefore proposed that the inability to detect stimulus-dependent elevation of cAMP levels in vitro [42] may reflect a smaller number of receptor cells responsive to some stimuli [44].

This combined evidence suggested that the olfactory adenylate cyclase was likely to be similar to the G-protein-linked enzyme associated with hormone and neurotransmitter receptors [7]. Thus, it is generally assumed that olfactory receptors are coupled to the G-proteins, G_s and G_i, generally associated with dual regulation of adenylate cyclase (Figure 8.1). In this regard, it should be noted that, although stimulatory and apparently non-stimulatory odorants have been reported, no stimuli associated with G_i-linked inhibition of adenylate cyclase have been identified. G_s may mediate at least some olfactory responses in humans, since type Ia pseudohypoparathyroid patients, genetically deficient in G_s activity, also exhibited impaired ability to identify several common stimuli [45].

That stimulus-elicited elevations of cAMP in olfactory neurones, particularly in olfactory cilia, are linked to regulation of the ion channels underlying membrane depolarization is now supported by biochemical and neurophysiological evidence [3,4,8–11,46]. Cyclic-AMP may regulate ion channel activity in olfactory neurones by at least two mechanisms (Figure 8.1). Cyclic-AMP-dependent protein kinase activity and endogenous protein substrates for the enzyme were identified in isolated olfactory cilia [16,47]. Activation of the protein kinase by cAMP may lead to direct phosphorylation of ion channels or phosphorylation of an intermediary protein that subsequently modulates ion channel activity. Although evidence has not been reported to distinguish between these possible regulatory mechanisms, it should be noted that ATP-dependent decreases in mean channel open time were observed in rat olfactory epithelium homogenates incorporated into phospholipid bilayers; the modulation of channel activity in the presence of ATP was also antagonized by protein kinase inhibitor [48]. The ATP-dependent, and presumably phosphorylation-dependent, modulation of channel activity was associated with channels similar to those underlying the cyclicnucleotide-gated conductance in olfactory cilia described previously [49]. The cyclicnucleotide-gated conductance was initially observed in excised membrane patches from individual cilia on dissociated olfactory neurones from toad [49]. Unlike the GMP-gated conductance mediating phototransduction [50], both cAMP and cGMP reversibly modulated membrane conductance in voltage-clamped membranes with similar potency. Since exogenous nucleotide triphosphates were also not required to observe the cyclicnucleotide-gated conductance, it was concluded that the underlying cation channels were directly gated by cyclic nucleotides. The same or similar cyclicnucleotide-gated conductance has also been observed in olfactory membranes from catfish [43] and rat [48], suggesting that the cyclic-

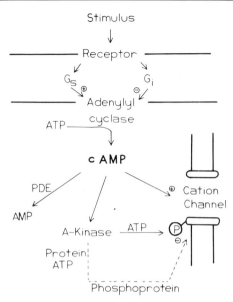

Figure 8.1 Second messenger role of cAMP in olfaction. Stimulus-activated elevation of cAMP in olfactory cilia leads to increases in membrane conductance by direct gating of cation channels. These channels may also be regulated by cAMP-dependent protein kinase (A-kinase), either by phosphorylation of channel proteins or by phosphorylation of an intermediary regulatory protein (dashed line). Dephosphorylation by, as yet unidentified, protein phosphatases, and degradation of cAMP by cyclic nucleotide phosphadiesterase, would lead to signal termination. (Reproduced by permission of Academic Press from R. C. Bruch (1989) *G-Proteins*, pp. 411–427)

nucleotide-gated channels are common transduction components in vertebrate olfaction. In addition, cAMP, but not cGMP, enhanced stimulus-evoked transepithelial current transients recorded from olfactory epithelium explants [46]. Small potentiation of stimulus-evoked current transients by cAMP were also recorded from oocytes injected with total mRNA from frog olfactory epithelium [51].

The combined biochemical and electrophysiological evidence is thus consistent with Sutherland's original criteria for a second messenger role for cyclic nucleotides, particularly cAMP, in olfactory signal transduction [9]. However, in general, high stimulus concentration and prolonged exposure to stimulus were required to observe significant enhancement of cAMP formation in isolated cell-free preparations from the olfactory epithelium [13,42,43,52]. In isolated olfactory cilia from catfish, the dose–response curve for stimulus activation of adenylate cyclase was signficantly shifted to the right of the corresponding receptor occupation curve (Figure 8.2). Good correlation was not obtained between stimulus-dependent cAMP formation and electrophysiological potency. In contrast to other G_s-linked receptors, stimuli also did not accelerate the rate of cAMP formation over that observed with Gpp(NH)p alone [43]. The rate of cAMP formation elicited by stimulus was also not related to activation of cyclic nucleotide phosphodiesterase. The slow rate of cAMP formation and the high levels of receptor

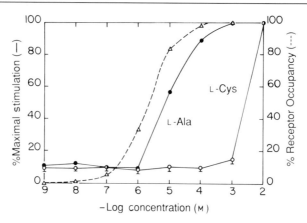

Figure 8.2 Amino acid stimulation of adenylate cyclase in isolated cilia. Cyclic AMP formation was measured in the presence of 10 μM Gpp(NH)p and the indicated concentrations of L-alanine and L-cysteine (solid lines). Per cent receptor occupancy for L-[3H]alanine binding was calculated from competitive binding curves measured under the conditions of the adenylate cyclase assay

occupancy required to observe stimulus effects on adenylate cyclase in vitro suggested that cAMP may mediate tonic or adaptive olfactory responses [43]. Since decreased adenylate cyclase activity was observed following prior exposure to stimulus, Pace et al. [53] also concluded that cAMP may be involved in adaptive olfactory responses. Additional experiments with intact olfactory neurones are required to evaluate the potential role of cAMP in receptor desensitization, signal termination and mediation of adaptive olfactory responses.

A potential second messenger role for cGMP in olfaction has also been investigated [10,11]. Initial electrophysiological recordings indicated that membrane-permeable cAMP analogues reversibly attenuated stimulus-evoked responses, while the corresponding cGMP analogues were ineffective [54]. Although less potent than cAMP, cGMP also activated cyclic-nucleotide-dependent protein kinase activity in isolated olfactory cilia [47]. Endogenous protein substrates for cGMP-dependent protein kinase have not yet been identified in olfactory cilia or in other domains of the receptor cells. In contrast, both cAMP and cGMP reversibly modulated membrane conductance with similar potency in excised voltage-clamped olfactory cilia membranes [49] and in isolated cilia incorporated into artificial phospholipid bilayers [43]. However, stimulus-induced transepithelial current transients recorded from the olfactory epithelium were enhanced by cAMP, but were inhibited by cGMP [46]. Guanylate cyclase activity was identified in both soluble and particulate fractions from the olfactory epithelium, although only about 2% of the total activity was retained in isolated cilia preparations [43]. In isolated cilia, exogenous calcium stimulated Mn^{2+}-dependent guanylate cyclase activity, an effect that was markedly potentiated by the calcium ionophore A23187. Stimuli did not affect Mn^{2+}-dependent cGMP formation, irrespective of the presence of stimulatory levels of calcium, in these cell-free preparations. Although direct coupling of stimulus to cGMP production has not been established, it is likely that cGMP levels in olfactory neurones may be regulated by intracellular calcium. Elevation of intracellular calcium, perhaps within localized subcellular domains [55], as a result of calcium release from intracellular stores and/or influx of extracellular calcium, could act to stimulate guanylate cyclase [56,57] and simultaneously inhibit adenylate cyclase [42,52]. Further studies in intact olfactory neurones are required to determine the relevance of cGMP-mediated signalling mechanisms in olfaction, particularly for those stimuli that may not activate adenylate cyclase [42,44].

Phosphoinositide-derived second messengers

In many cells, G-proteins have been implicated in coupling cell-surface receptors to phospholipases [58]. Receptor-mediated activation of phospholipase C has also been investigated in isolated olfactory cilia [3,4,10,11]. The majority of the enzyme activity was recovered in soluble fractions from the olfactory epithelium of catfish, rat and mouse. However, a small, but reproducible, fraction of the activity was retained in isolated cilia [59,60]. In isolated cilia from catfish, stimulus amino acids, at concentrations an order of magnitude lower than those required to stimulate adenylate cyclase, elicited increases in inositol phosphate production. Stimulus-dependent inositol phosphate formation was also observed only in the presence of guanine nucleotides, suggesting that a G-protein mediated stimulus activation of phospholipase C (Figure 8.3). The identity of the G-

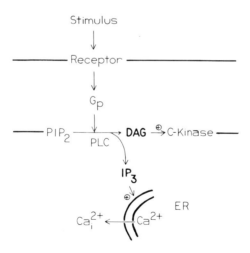

Figure 8.3 Phosphoinositide-derived second messengers in olfaction. Stimulus activation of phospholipase C (PLC) is mediated by a G-protein (G_p) and catalyses the hydrolysis of phosphatidylinositol-4, 5-bisphosphate (PIP_2). IP_3 subsequently releases calcium from the endoplasmic reticulum (ER), thereby elevating intracellular calcium levels. Calcium, together with diacylglycerol (DAG), may stimulate protein kinase C (C-kinase). (Reproduced by permission of Academic Press from R. C. Bruch (1989) *G-Proteins*, pp. 411–427)

protein involved in coupling olfactory receptors to phosphoinositide hydrolysis has not yet been determined. In contrast to stimulus-dependent cAMP formation, stimuli elicited rapid increases in inositol phosphate production within several seconds of exposure to stimulus [61,62]. Anion-exchange chromatography of the inositol phosphate products formed under both basal and stimulated conditions showed that inositol trisphosphate (IP_3) accounted for at least 90% of total inositol phosphate products formed.

The second messenger role of IP_3 in mediating calcium release from intracellular stores [63] was subsequently evaluated in subcellular fractions from the olfactory epithelium and in dissociated olfactory neurones [4,64]. Microsomes isolated from the olfactory epithelium sequestered calcium in a Mg^{2+}/ATP-dependent manner. In the presence of oligomycin, to block mitochondrial calcium uptake, IP_3 rapidly elicited the transient release of about 30% of the sequestered calcium that could be released by the calcium ionophore A23187. In dissociated olfactory neurones, preloaded with the fluorescent calcium indicator Fura-2, stimulus amino acids evoked significant elevations of intracellular calcium levels [64]. It is likely therefore that IP_3-mediated calcium release from intracellular stores contributes to stimulus-evoked elevations of intracellular calcium in these cells. As noted above (page 121), elevations of intracellular calcium may be linked to cyclic nucleotide metabolism in olfactory cilia. Although the further role of intracellular calcium in olfactory signal transduction has not been extensively investigated, it may also function, together with diacylglycerol, to stimulate protein kinase C (Figure 8.3). Protein kinase C has been identified in isolated olfactory cilia by immunoblotting analysis and phorbol ester binding [14]. However, tissue-specific protein substrates for the kinase and the role of these substrates in olfactory stimulus–response coupling remain to be identified.

CONCLUSIONS

A role for members of the heterotrimeric family of signal transducing G-proteins in mediating transmembrane signalling in olfactory neurones is clearly indicated by the evidence summarized in this chapter. Two stimulus-activated G-protein-linked second messenger systems have been identified in isolated olfactory cilia. The potential involvement of cyclic nucleotides and phosphoinositide-derived second messengers in regulation of ion channels and intracellular ion fluxes associated with neuronal excitability has also been demonstrated in olfaction. However, it is also likely that additional mechanisms participate in olfactory stimulus–response coupling that may not be directly regulated by G-proteins. Stimulus interaction with hydrophobic membrane domains [30,40] and direct stimulus gating of ion channels [65] have been proposed as alternative olfactory signalling mechanisms. Additional studies are now needed to determine the molecular events underlying stimulus regulation of these multiple signalling mechansims in olfactory neurones.

ACKNOWLEDGEMENTS

Work in the authors' laboratories was supported by the National Science Foundation and the National Institutes of Health.

REFERENCES

1. Getchell, T. V. (1986) *Physiol. Rev.*, **66**, 772–818.
2. Getchell, T. V., Margolis, F. L., and Getchell, M. L. (1985) *Prog. Neurobiol.*, **23**, 317–345.
3. Bruch, R. C., Kalinoski, D. L., and Kare, M. R. (1988) *Ann. Rev. Nutr.*, **8**, 21–42.
4. Bruch, R. C. (1989) *G-proteins and Calcium Signalling* (Naccache, P. H., ed.). CRC Press, Boca Raton, in press.
5. Kurihara, K., Yoshii, K., and Kashiwayanagi, M. (1986) *Comp. Biochem. Physiol.*, **85A**, 1–22.
6. Lerner, M. R., Reagan, J., Gyorgyi, T., and Roby, A. (1988) *Proc. Natl. Acad. Sci. USA*, **85**, 261–264.
7. Gilman, A. G. (1987) *Ann. Rev. Biochem.*, **56**, 615–649.
8. Lancet, D. (1986) *Ann. Rev. Neurosci.*, **9**, 329–355.
9. Lancet, D. (1988) *Molecular Neurobiology of the Olfactory System* (Margolis, F. L., and Getchell, T. V., eds). Plenum Press, New York, pp. 25–50.
10. Bruch, R. C. (1989) *G-proteins* (Birnbaumer, L., and Iyengar, R., eds), pp. 411–427. Academic Press, New York.
11. Bruch, R. C., and Teeter, J. H. (1989) *Chemical Senses: Receptor Events and Transduction in Taste and Olfaction* (Brand, J. G., Teeter, J. H., Kare, M. R., and Cagan, R. H., eds), pp. 283–298. Marcel Dekker, New York.
12. Snyder, S. H., Sklar, P. B., and Pevsner, J. (1988) *Molecular Neurobiology of the Olfactory System* (Margolis, F. L., and Getchell, T. V., eds). Plenum Press, New York. pp. 3–24.
13. Pace, U., Hanski, E., Salomon, Y., and Lancet, D. (1985) *Nature (London)*, **316**, 255–258.
14. Anholt, R. R. H., Mumby, S. M., Stoffers, D. A., Girard, P. R., Kuo, J. F., and Snyder, S. H. (1987) *Biochemistry*, **26**, 788–795.
15. Pace, U., and Lancet, D. (1986) *Proc. Natl. Acad. Sci. USA*, **83**, 4947–4951.
16. Kropf, R., Lancet, D., and Lazard, D. (1987) *Soc. Neurosci. Abstr.*, **13**, Part 2, 1410.
17. Bruch, R. C., and Kalinoski, D. L. (1987) *J. Biol. Chem.*, **262**, 2401–2404.
18. Jones, D. T., and Reed, R. R. (1987) *J. Biol. Chem.*, **262**, 14241–14249.
19. Goldsmith, P., Rossiter, K., Carter, A., Simonds, W., Unson, C. G., Vinitsky, R., and Spiegel, A. M. (1988) *J. Biol. Chem.*, **263**, 6476–6479.
20. Itoh, H., Toyama, R., Kozasa, T., Tsukamoto, T., Matsuoka, M., and Kaziro, Y. (1988) *J. Biol. Chem.*, **263**, 6656–6664.
21. Codina, J., Olate, J., Abramowitz, J., Mattera, R., Cook, R. G., and Birnbaumer, L. (1988) *J. Biol. Chem.*, **263**, 6746–6750.
22. Mania-Farnell, B., and Farbman, A. I. (1987) *Chem. Senses*, **12**, 678.
23. Farbman, A. I., and Mania-Farnell, B. (1989) *Chem. Senses*, **14**, 185.
24. Novoselov, V. I., Krapivinskaya, L. D., Krapivinsky, G. B., and Fesenko, E. E. (1988) *FEBS Lett.*, **234**, 471–474.
25. Price, S., and Willey, A. (1988) *Biochim. Biophys. Acta*, **965**, 127–129.
26. Rhein, L. D., and Cagan, R. H. (1981) *Biochemistry of Taste and Olfaction* (Cagan, R. H., and Kare, M. R., eds). Academic Press, New York, pp. 47–68.
27. Price, S. (1981) *Biochemistry of Taste and Olfaction* (Cagan, R. H., and Kare, M. R., eds). Academic Press, New York, pp. 69–84.

28. Margolis, F. L. (1987) *Sensory Transduction: Discussions in Neuroscience IV* (Hudspeth, A. J., Macleish, P. R., Margolis, F. L., and Wiesel, T. N., eds), pp. 47–52 FESN, Geneva.
29. Kalinoski, D. L., Bruch, R. C., and Brand, J. G. (1987) *Brain Res.*, **418**, 34–40.
30. Bruch, R. C., and Rulli, R. D. (1988) *Comp. Biochem. Physiol.*, **91B**, 535–540.
31. Caprio, J. (1978) *J. Comp. Physiol.*, **123**, 357–371.
32. Caprio, J., and Byrd, R. P., Jr (1984) *J. Gen. Physiol.*, **84**, 403–422.
33. Fesenko, E. E., Novoselov, V. I., and Bystrova, M. F. (1987) *FEBS Lett.*, **219**, 224–226.
34. Fesenko, E. E., Novoselov, V. I., and Bystrova, M. F. (1988) *Biochim. Biophys. Acta*, **937**, 369–378.
35. Novoselov, V. I., Krapavinskaya, L. D., and Fesenko, E. E. (1988) *Chem. Senses*, **13**, 267–278.
36. Bignetti, E., Cattaneo, P., Cavaggioni, A., Damiani, G., and Tirindelli, R. (1988) *Comp. Biochem. Physiol.*, **90B**, 1–5.
37. Lee, K. H., Wells, R. G., and Reed, R. R. (1987) *Science*, **235**, 1053–1056.
38. Kashiwayanagi, M., and Kurihara, K. (1985) *Brain Res.*, **359**, 97–103.
39. Nomura, T., and Kurihara, K. (1987) *Biochemistry*, **26**, 6135–6140.
40. Nomura, T., and Kurihara, K. (1987) *Biochemistry*, **26**, 6141–6145.
41. Anholt, R. R. H. (1988) *Biochemistry*, **27**, 6464–6468.
42. Sklar, P. B., Anholt, R. R. H., and Snyder, S. H. (1986) *J. Biol. Chem.*, **261**, 15538–15543.
43. Bruch, R. C., and Teeter, J. H. (1989) (submitted for publication).
44. Lowe, G., Nakamura, T., and Gold, G. H. (1989) *Proc. Natl. Acad. Sci. USA*, **86**, 5641–5645.
45. Weinstock, R. S., Wright, H. N., Spiegel, A. M., Levine, M. A., and Moses, A. M. (1986) *Nature*, **322**, 635–636.
46. DeSimone, J. A., Persaud, K., and Heck, G. L. (1988) *Molecular Neurobiology of the Olfactory System* (Margolis, F. L., and Getchell, T. V., eds). Plenum Press, New York, pp. 159–181.
47. Heldman, J., and Lancet, D. (1986) *J. Neurochem.*, **47**, 1527–1533.
48. Vodyanoy, V., and Vodyanoy, I. (1987) *Soc. Neurosci. Abstr.*, **13**, Part 2, 1410.
49. Nakamura, T., and Gold, G. H. (1987) *Nature (London)*, **325**, 442–444.
50. Stryer, L. (1986) *Ann. Rev. Neurosci.*, **9**, 87–119.
51. Getchell, T. V. (1988) *Neurosci. Lett.*, **91**, 217–221.
52. Shirley, S. G., Robinson, C. J., Dickinson, K., Aujla, R., and Dodd, G. H. (1986) *Biochem. J.*, **240**, 605–607.
53. Pace, U., Heldman, J., Shafir, I., Rimon, G., and Lancet, D. (1987) *Soc. Neurosci. Abstr.*, **13**, Part 1, 362.
54. Menevse, A., Dodd, G., and Poynder, T. M. (1977) *Biochem. Biophys. Res. Commun.*, **77**, 671–677.
55. Alkon, D. L., and Rasmussen, H. (1988) *Science*, **239**, 998–1005.
56. Rasmussen, H., Apfeldorf, W., Barrett, P., Takuwa, N., Zawalich, W., Kreutter, D., Park, S., and Takuwa, Y. (1986) *Phosphoinositides and Receptor Mechanisms* (Putney, J. W., Jr, ed.), Alan R. Liss, New York, pp. 109–147.
57. Ui, M. (1986) *Phosphoinositides and Receptor Mechanisms* (Putney, J. W., Jr, ed.), Alan R. Liss, New York, pp. 163–195.
58. Fain, J. N., Wallace, M. A., and Wojcikiewicz, R. J. H. (1988) *FASEB. J.*, **2**, 2569–2574.
59. Huque, T., and Bruch, R. C. (1986) *Biochem. Biophys. Res. Commun.*, **137**, 36–42.
60. Boyle, A. G., Park, Y. S., Huque, T., and Bruch, R. C. (1987) *Comp. Biochem. Physiol.*, **88B**, 767–775.
61. Bruch, R. C., Kalinoski, D. L., and Huque, T. (1987) *Chem. Senses*, **12**, 173.
62. Bruch, R. C., Rulli, R. D., and Boyle, A. G. (1987) *Chem. Senses*, **12**, 642–643.
63. Putney, J. W., Jr (1987) *Amer. J. Physiol.*, **252**, G149–G157.
64. Restrepo, D., and Bruch, R. C. (1989) *Chem. Senses*, **14**, 740.
65. Labarca, P., Simon, S. A., and Anholt, R. R. H. (1988) *Proc. Natl. Acad. Sci. USA*, **85**, 944–947.

9
G-PROTEINS AND THE REGULATION OF ION CHANNELS

Annette C. Dolphin

Department of Pharmacology, St George's Hospital Medical School, London SW17 ORE, UK

INTRODUCTION

Ion channels of many classes are present in every cell plasma membrane, although they are particularly prevalent in concentration and variety in the membranes of excitable tissues. Ion channels may be divided into voltage-operated, second-messenger-operated and receptor-operated subtypes, although increasingly any classification has exceptions and overlaps.

Ion channels are involved in homeostasis of the intracellular environment, in transducing signals from the external world to the inside of the cell, and in mediating and limiting cell excitability. They are involved as an essential intermediary in many specialized functions including secretion, synaptic transmission and transduction of sensory signals.

Ion channel activity may be modulated in several ways at the single channel level. Changes may occur in the following:

(1) The number of functional channels, and at the extreme a new type of channel may be expressed.
(2) The probability of opening.
(3) The voltage sensitivity of channel gating.
(4) The mode of opening.

Examples of all these are known.

When considering the interaction between G-proteins and ion channels, it is not an easy task to provide incontrovertible evidence that a direct interaction is occurring rather than a response mediated via a second messenger. However, in a cell-free patch of membrane attached to the recording pipette, soluble second messengers can be ruled out, and it may be possible to show that the effect of an activated G-protein on an ion channel does not involve a phosphorylation process. Nevertheless, it is possible that cytoskeletal elements and membrane-associated enzymes remain attached to the patch, or that processes such as dephosphorylation are occurring. Thus to further rule out enzyme-mediated events, it should ideally be demonstrated that the effect of an agonist on the ion channel is reversible in the patch. The extent to which the responses of various ion channels to receptor activation have been shown clearly to be either a direct effect of an activated G-protein or to involve a second messenger is discussed in the following sections.

MODULATION OF ION CHANNEL FUNCTION BY CYCLIC AMP

The first examples of ion channel modulation whose mechanism was described in detail involve cAMP-dependent phosphorylation of Ca^{2+} channels in cardiac myocytes and K^+ channels in *Aplysia* sensory neurones. In these cases the GTP binding proteins G_s and G_i clearly modulate the channel at the level of second messenger production. Modulation may also occur of the dephosphorylation process, for example by calcium-dependent phosphatases.

Calcium channel modulation by cyclic-AMP-dependent phosphorylation

Beta adrenergic receptor activation has effects on several currents observed in heart cells, but one of the most prominent is its enhancement of the slow inward calcium current in ventricular myocytes. This response is mimicked by membrane-permeable analogues of cAMP [1] and by the catalytic subunit of cAMP-dependent protein kinase, providing evidence that protein phosphorylation is involved [2]. Many studies of single channel kinetics have examined the molecular basis for this enhancement. 8-Bromo cyclic AMP [1] was observed to increase the opening probability of the Ca^{2+} channels underlying this current (see page 136). The channels open in bursts, and the increased probability of opening is due to a reduction in the interburst interval. However, Lipscombe et al. [3] have also reported an increased open time, and Bean et al. [4] have shown that the number of functional channels is also increased. The conclusion of this work was that a channel had to be phosphorylated to be functional within the normal range of activation potentials, and phosphorylation shifted the voltage for activation of the channels to more hyperpolarized potentials. Biochemical studies have provided evidence that it is the channel itself which is phosphorylated. The channel has been isolated by virtue of its specific binding of dihydropyridine Ca^{2+} channel ligands, and has been shown to consist of at least five subunits, α_1, α_2, β, γ and δ; of which the α_1 and β subunits are both substrates for cAMP-dependent protein kinase [5]. The α_1 subunits will reconstitute Ca^{2+} channel activity in lipid bilayers, and phosphorylation increases Ca^{2+} flux through these channels [5,6].

All the effects of activation of muscarinic receptors on Ca^{2+} currents in heart are due to a reduction in cAMP, both by inhibition of adenylate cyclase and by activation of a phosphodiesterase [7]. By these means, a reduction in β-adrenergic-induced phosphorylation is produced [8]. No effect of muscarinic activation is observed unless the current has been enhanced by prior phosphorylation. Similarly, inhibitors of proten kinase, while completely abolishing the effect of β-adrenergic stimulation, do not markedly inhibit basal calcium currents. These findings suggest that even in the dephosphorylated state the channels are available to open [9].

The nature, specificity and regulation of the phosphatase involved in dephosphorylating the channel remain to be determined. In addition, there may also be other intracellular factors or metabolic events which are required for these channels to remain functional. Although single channel events may be recorded with stability in cell-attached patches in which the recording pipette is placed on the intact cell, they are rapidly silenced if the pipette and membrane patch are subsequently detached from the cell to form cell-free patches [10].

The involvement of G-proteins in the mediation of the responses in cardiac cells to β-adrenergic and muscarinic activation has been examined by Breitwieser and Szabo [11]. The non-hydrolysable GTP analogue guanylylimidodiphosphate (GMP-PNP), which permanently activates G-proteins, causes the effects of both isoprenaline and acetylcholine to become irreversible by activating G_s and G_i respectively. It is interesting that although isoprenaline normally reverses the effect of acetylcholine, in the presence of GMP-PNP, when both G-proteins are activated, it is unable to do so.

Potassium channel modulation by cyclic-AMP-dependent phosphorylation

S CHANNELS

A class of serotonin-sensitive potassium channels has been characterized in *Aplysia* sensory neurones. It is open at the resting membrane potential and is relatively voltage- and Ca^{2+}-insensitive. Serotonin causes channel closure and subsequent neuronal depolarization. This process has been implicated as one of the mechanisms of synaptic facilitation [12] involved in a short-term learning process in *Aplysia*. It results from cAMP-dependent phosphorylation reducing the number of functional channels available to open [13]. An investigation of the proteins phosphorylated by serotonin is likely to provide a route for identification of the K^+ channel [14].

CALCIUM-DEPENDENT POTASSIUM CHANNELS

There are at least two classes of Ca^{2+}-activated K^+ currents, one of which, I_c, is ubiquitously found. This current is blocked by charybdotoxin and carried by large conductance (BK) channels. It has recently been shown that the open-state probability of BK channels in tracheal myocytes is increased by cyclic AMP-dependent phosphorylation

[14a]. The other Ca^{2+}-activated K^+ current has been described in hippocampal pyramidal cells [15] and other neurones [16]. It has slower decay, and has been termed I_{AHP}, since it is responsible for the slow after hyperpolarization which determines action potential discharge rate [17]. Isoprenaline, by activating cAMP-dependent protein kinase, blocks I_{AHP} and reduces the ability of hippocampal pyramidal cells to limit action potential firing [16]. The channels underlying the slow AHP have yet to be identified, and thus it remains unclear whether phosphorylation affects the channel itself.

Chloride channel modulation by cyclic-AMP-dependent phosphorylation

Chloride channels are of particular importance in epithelia and glandular tissue because of their role in secretion. A chloride channel has been identified in airways epithelium that requires phosphorylation by cAMP-dependent protein kinase to open [18]. This channel is relatively voltage-insensitive and is opened physiologically by β-adrenergic agonists [19]. It appears to be a fault in this channel which is the underyling cause of the genetic defect in cystic fibrosis. In this condition, the channel does not respond to activation by β-adrenergic agonists or cAMP-dependent protein kinase [18]. It remains to be determined whether the channel itself is no longer a target for phosphorylation, or whether phosphorylation no longer results in channel opening.

Cyclic-AMP-dependent modulation of other channels

Several other voltage-sensitive ion channels are influenced by cAMP. These include the hyperpolarization-activated pacemaker current in the sino-atrial node, which is inhibited by muscarinic receptor activation, an effect that can be blocked by cAMP analogues [20]. Agonist-operated channels may also be phosphorylated by cAMP-dependent (as well as other) protein kinases. For example, the nicotinic receptor/channel complex is a substrate for this kinase. Channel phosphorylation does not affect the initial rate of ion flux, but it increases the rate of channel desensitization [21].

MODULATION OF ION CHANNELS BY PROTEIN-KINASE-C-DEPENDENT PHOSPHORYLATION

Protein kinase C (PKC) is a soluble protein kinase which depends on low levels (10^{-7} M) of Ca^{2+} for its activation, and can be activated physiologically by diacylglycerol produced in the membrane by phospholipase C (PLC) activity (for review see [22]). It can also be activated by arachidonic acid, which may be produced either by phospholipase A_2 (PLA_2) or by further metabolism of arachidonate-containing diacyglycerols by DAG lipase [23]. For this reason, PKC activation may be indirectly under the control of more than one G-protein: G_p, the putative G-protein that couples receptors to PLC [24] and a G-protein whose identity is unclear, which appears to couple receptors to activation of PLA_2 [25]. The former G-protein, G_p, does not appear to be sensitive to pertussis toxin in most cell types, except in haemopoietic cells [26]. In contrast, pertussis toxin sensitivity has been reported for the latter G-protein associated with PLA_2 [25]. Several membrane-permeable agents may be used to activate PKC. These include synthetic diacylglycerols including oleoylacetylglycerol (OAG) or dioctanoylglycerol (DOG), which may be metabolized intracellularly, and the persistent activators of the phorbol ester family, including tetradecanoylphorbol acetate (TPA), and phorbol dibutyrate (PDBu). PKC activation may involve translocation to the inner surface of the plasma membrane, where membrane proteins will be phosphorylated [22], although the substrates of PKC, particularly when persistently activated by phorbol esters, are many and ubiquitous.

Several experimental approaches have indicated that PKC activation leads to modulation of the activity of many classes of ion channel, and these will be described in the following sections. However, in many cases, the evidence that this is a pathway utilized by neurotransmitters to modulate ion channel activity is incomplete. This point is an important one with regard to protein kinase C activators, since they may have other effects in the membrane [27]. In addition, PKC may phosphorylate other components of the signal transduction system, including G-proteins [28] and receptors [22], leading to down-regulation of the pathway.

Calcium channel modulation by PKC

ACTIVATION OF CALCIUM CHANNELS IN APLYSIA NEURONES BY PKC

Aplysia neurosecretory bag cells, either treated with phorbol esters or injected with protein kinase C, show an increased Ca^{2+} current [29]. Subsequently, using the cell-attached patch technique, it has been shown that this is due to the appearance of a novel Ca^{2+} channel with a greater single channel conductance, there being no change in the number or characteristics of the Ca^{2+} channels of smaller conductance which are present under control conditions [30]. It has been observed more recently that this effect is extremely long lasting, and that PKC activation is likely to be responsible for some physiological events associated with hormone secretion from these cells [31]. A similar effect on Ca^{2+} currents has also been observed in sensory neurones of *Aplysia* [32].

ACTIVATION OF OTHER CALCIUM CHANNELS BY PKC

Angiotensin II enhances Ca^{2+} currents in ventricular myocytes and this response is mimicked by phorbol esters [33]. In agreement with this, the PKC activators TPA and DOG increase the probability of L-type Ca^{2+} channel opening in cell-attached patches on these cells [34]. It is interesting that parallel effects were not observed on whole cell Ca^{2+} currents in this study, suggesting the loss of cytoplasmic factors responsible for the enhancement. Prolonged application of TPA but not DOG resulted in a decrease in the probability of opening of Ca^{2+} channels, possibly due to down-regulation of PKC itself, which occurs by enhanced proteolysis of the activated PKC [35]. In agreement with the results on ventricular myocytes, PDBu rapidly enhances single Ca^{2+} channel activity in cell-attached patches of frog sympathetic neurones [3]. TPA also enhances Ca^{2+} currents in oocytes injected with brain mRNA [36,37], although in this case corresponding single channel activity has not been recorded.

INHIBITION OF CALCIUM CHANNELS BY PKC

There are as yet no examples of rapid inhibition of single Ca^{2+} channels by PKC activation. However, there are many studies which show inhibition of components of whole cell Ca^{2+} current by PKC. These will be discussed in detail on page 139, where the evidence in favour of a direct coupling of G-proteins to Ca^{2+} channels to mediate inhibition by various neuromodulators will be compared with the evidence in favour of second messenger mediation of the transduction process.

INHIBITION OF M CURRENTS BY PKC

The M current is a K^+ current which is activated at membrane potentials more depolarized than about -40 mV and deactivates slowly upon hyperpolarization. This current is inhibited by acetylcholine, acting at muscarinic receptors, and by LHRH and bradykinin, among other hormones and neurotransmitters (for review see Brown [35]). The effect of muscarine can be mimicked by GTP analogues, but the G-protein is not sensitive to pertussis toxin [39,40]. Muscarinic receptors have been cloned by Fukuda et al. [41], and expressed in NG108-15 cells, where it appears that the same muscarinic receptor subtype (M_1) stimulates phospholipase C, activates the Ca^{2+}-dependent Cl^- and K^+ conductances and inhibits M currents. In contrast, the M_2 receptor is associated, presumably via a different G-protein, with adenylate cyclase inhibition, and activation of Na^+ and K^+ currents [41]. The PKC activator dioctanoylglycerol partially inhibits M currents in bullfrog smooth muscle cells [42], and phorbol esters partially inhibit M currents in rat sympathetic neurones [40]. However, Pfaffinger et al. [43] and Bosma and Hille [43a] have reported that although phorbol esters partially inhibit M current in frog sympathetic ganglion, and prevent subsequent inhibition by substance P, the effect of LHRH is not occluded. PKC inhibitors prevent the effect of phorbol esters but do not affect the response to agonists. Because of this, they suggest that inhibition of the substance P response by phorbol esters is due to interference in the signal transduction process, for example by receptor phosphorylation. This result illustrates the general principle that to implicate a second messenger in a response it is not sufficient to show that the effect of an agonist is mimicked by a second messenger analogue, but it must also be shown that the effect of the agonist is prevented by inhibition, either of the second messenger producing enzyme or of the second messenger activated enzyme.

In hippocampal pyramidal cells, PKC activators have no effect on M currents, whereas IP_3 injection into the neurones causes their attenuation, possibly by a Ca^{2+}-independent mechanism [44].

INHIBITION OF CALCIUM-DEPENDENT POTASSIUM CURRENTS BY PKC

Muscarinic inhibition of the slow AHP in hippocampal pyramidal neurones appears to involve PKC activation [45]. Thus, activation of cAMP-dependent protein kinase and PKC has similar effects on this current, as it does on cardiac calcium channels (compare pages 126 and 129). It is not clear whether the same phosphorylation site is involved in the action of both kinases, although PKC can phosphorylate some of the same substrates as cAMP-dependent protein kinase.

INHIBITION OF OTHER CHANNELS BY PKC

Xenopus oocytes injected with mRNA from chick forebrain express Na^+ and GABA-activated channels, both of which are inhibited by phorbol esters [37], although the GABA receptor–ion channel complex has been reported to require phosphorylation to be functional [46].

OTHER SECOND MESSENGER EFFECTS ON ION CHANNELS

Phospholipase A_2

Arachidonic acid production by PLA_2 has been shown in several systems to be receptor stimulated, and under the control of a pertussis-toxin-sensitive G-protein [47,47a]. The activation of PLA_2 by activated transducin in rod outer segments is unusual in that it occurs via the $\beta\gamma$ subunits [47a]. In *Aplysia* sensory neurones the peptide FMRF-amide increases the opening of the same K^+ channels (S channels) that are closed by serotonin (see Belardetti and Siegelbaum [48] for review). The effect can be mimicked by GTP analogues and arachidonic acid or its lipoxygenase metabolites particularly 12-HPETE, and is prevented by lipoxygenase inhibitors [49]. FMRF-amide overrides the effect of serotonin on the S channel, causing reopening of channels that have been closed by phosphorylation. The mechanism of channel modulation does not appear to involve arachidonic acid activation of PKC, and a recent study has shown that one point of interaction may be at the level of the phosphorylation process, since FMRF-amide reduced basal and serotonin-stimulated phosphorylation in *Aplysia* neurones [14].

The controversy concerning whether K^+ channels in atrial myocytes are opened by α_i or by $\beta\gamma$ (see page 133) has recently been at least partially resolved: $\beta\gamma$ subunits are known to stimulate PLA_2 [47a], and it appears that this is the route by which they activate cardiac K^+ channels [49a,49b]. Leukotrienes A_4 and C_4, and also some other arachidonic acid metabolites are active. Their effect may not be on the channel itself but rather at the level of G-protein activation, because observation of the response in inside-out patches requires GTP [49a]. This may represent a positive feedback loop, since the effect of acetylcholine on the potassium channel occurs primarily by direct interaction with activated α_i subunits (see page 133) and is not prevented by lipoxygenase inhibitors or antibodies to PLA_2 [49b].

Modulation of calcium-activated currents

Indirect modulation of Ca^{2+}-dependent ionic currents resulting from G-protein activation can occur by several means. Firstly, G-proteins can directly, or indirectly via second messengers, influence Ca^{2+} influx through voltage-gated channels. Secondly, they will stimulate PLC and increase IP_3 production, causing release of internal Ca^{2+}. Thirdly, there are several lines of evidence implicating a guanine-nucleotide- dependent process in the release of internal Ca^{2+} from internal stores including endoplasmic reticulum.

There are three major classes of ion channel that have an obligatory role for Ca^{2+} for their opening: Ca^{2+}-activated K^+, Cl^- and non-specific cation channels. Several studies have shown Ca^{2+}-activated K^+ and Cl^- channels to be regulated by release of internal Ca^{2+} [50]. It has also been observed that their run-down is reduced, or their amplitude enhanced by GTP analogues [50–52]. This may indicate a role for GTP analogues in maintaining Ca^{2+} release from internal stores. However, the GTP-mediated release of Ca^{2+} from endoplasmic reticulum has been shown to require GTP hydrolysis and to be inhibited by non-hydrolysable analogues of GTP [53,54]. Thus the existence of a G-protein on endoplasmic reticulum associated with the Ca^{2+} release channel remains unclear. Other work has shown that GTPγS induces the appearance of Ca^{2+}-activated Cl^- currents in oocytes injected with rat brain mRNA, when de novo expression of rat brain G_i and G_o. This results in G-protein-mediated PLC stimulation and a consequent enhancement of IP_3 production. In these experiments a similar response is also mediated by 5-HT, and this effect is sensitive to pertussis toxin [55].

Parallel findings have been obtained by Bolton and Lim [56] examining the effect of GTP analogues on spontaneous transient outward currents in smooth muscle cells. These are thought to be due to periodic release of stored Ca^{2+}, activating Ca^{2+}-activated K^+ channels. They are rapidly inhibited by GTPγS, which apparently depletes internal Ca^{2+}, since no effect of caffeine, which normally releases Ca^{2+} from internal stores, can then be induced. Similar results indicating that GTPγS can release internal Ca^{2+} in smooth muscle have been obtained by Kobayashi et al. [57].

DIRECT INTERACTION OF ION CHANNELS WITH GTP BINDING PROTEINS

The first evidence that G-proteins might directly modulate ion channels in the membrane without an intervening second messenger came from a comparison of the behaviour of Ca^{2+} and K^+ channels to modulation by muscarinic agonists in cardiac myocytes. These findings initiated the idea that an ion channel could be considered as an effector protein, just like the enzymes adenylate cyclase or phospholipase C whose activity could be altered by interaction with an activated G-protein.

Direct interaction of potassium channels with G-proteins

AGONIST-ACTIVATED POTASSIUM CHANNELS IN CARDIAC MYOCYTES

Muscarinic activation of the inwardly rectifying K^+ current in atrial myocytes is rendered irreversible by GTP analogues, but the activation is not prevented by isoprenaline or cAMP analogues, indicating that inhibition of adenylate cyclase is not involved [11,58]. However, the evidence that no other second messenger is involved required demonstration of the effect in cell-free patches [59], where responses to second messengers could, in principle, be eliminated. Soluble second messengers would rapidly diffuse away from the patch (Figure 9.1), and second messenger responses requiring protein phosphorylation can be eliminated by the absence of ATP in the medium. Under these conditions, opening of the K^+ channel can be achieved by a non-hydrolysable GTP analogue applied to the cytoplasmic side of the patch. Alternatively, GTP/Mg^{2+} may be applied to the cytoplasmic side, together with agonist (either acetylcholine or adenosine) in the patch pipette exposed to the external side of the membrane [60]. In this case, K^+ channel activation is reversible upon removal and replacement of Mg^{2+}, providing evidence that the initial response is not due to activation of membrane-associated phosphatases. The response could be blocked by pertussis-toxin-A subunits [61], and mimicked by addition of a G_i-type protein purified from human erythrocytes and termed G_K [59]. This G-protein, preactivated with $GTP\gamma S$, added to the inner face of the patch is thus able to associate with the K^+ channel and apparently exert a direct effect on channel opening.

Similar effects have been observed in the pituitary cell line GH_3, where the ability of somatostatin and acetylcholine to activate K^+ channels is mimicked by activated G_K and by the activated α subunit of G_K [62,63]. These results indicate that the α subunit, and not a contaminating species, is the mediator of the response. In addition, a monoclonal antibody generated against this α subunit blocks muscarinic activation of atrial K^+ channels when carbachol is present in the patch pipette [64].

The finding by Logothetis et al. [65] that the $\beta\gamma$ subunits of G-proteins will activate chick embryo atrial K^+ channels has been disputed by Kirsch et al. [63]. This issue may now be resolved by the finding that whereas α_i can act directly on the channel, $\beta\gamma$ acts indirectly on PLA_2 [49a,49b] (see page 131). This has previously been described in retina, when transducin is activated and its α subunit activates phosphodiesterase [47a]. Thus, in general, $\beta\gamma$ subunits generated by receptor activation of a G-protein may perform two physiological functions. As well as turning off other activated α subunits, which may be a major mechanism in adenylate cyclase inhibition [66], they may also themselves activate enzymes [47a].

The kinetics of muscarinic-induced activation of G_K to activate the K^+ channel has been examined by Breitwieser and Szabo [67]. The catalysis which results from receptor activation by agonist has been estimated to involve an increase in GDP off-rate from the endogenous G-protein from 0.44 $\min^{-1}$ to 135 $\min^{-1}$. Further mechanistic considerations have been examined by Kurachi et al. [68], concerning short-term desensitization. Agonist activation of K^+ currents results in a peak of current which decays to a plateau phase. During this plateau, further activation will not occur to the same or another

Second messenger mediated process

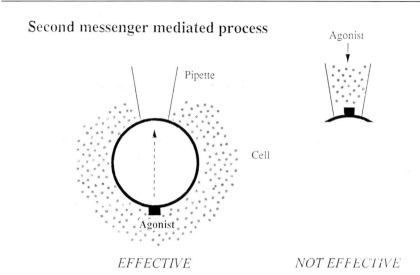

Direct or G-protein mediated process

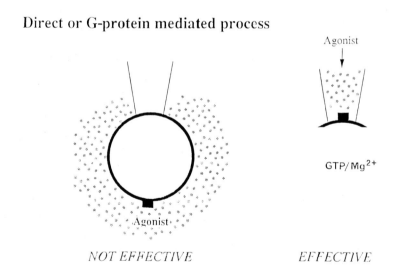

Figure 9.1 Modulation of ion channels by neurotransmitters and hormones: a means of distinguishing agonist effects on ion channels mediated by a second messenger from those mediated directly via a G-protein. If a response is observed in the cell-attached patch mode when agonist is applied outside the pipette, it is likely to be mediated by a second messenger; if not, it is likely to be a direct effect and should still be observed in an inside-out patch. If a G-protein is involved, GTP–Mg^{2+} will be essential on the cytoplasmic surface of the isolated patch

agonist. Similar desensitization to agonist occurs when the K^+ channel is activated by a GTP analogue, suggesting that the desensitization process involves partial inactivation of the channel, when it is associated with activated G-protein.

Agonist-activated potassium channels in neuronal tissue

In central and peripheral neuronal tissue, many neurotransmitters and neuromodulators have been observed to activate a K^+ conductance, resulting in hyperpolarization. Such neurotransmitters include GABA acting at $GABA_B$ receptors, serotonin acting at $5\text{-}HT_{1a}$ receptors, noradrenaline acting at α_2 receptors, somatostatin, dopamine acting at D_2 receptors, adenosine acting at A_1 receptors and opiates acting at μ and δ receptors (for review see Nicoll [16]). Several of these studies in slices and isolated cells have shown the involvement of a G-protein [69–73]. By injection of pertussis toxin intracerebroventrically or into discrete brain areas, before preparation of slices, the G-protein has been shown to be sensitive to this toxin [70,71,74,75].

Sasaki and Sato [76] showed that dopamine, histamine and acetycholine produced similar maximal effects on a K^+ current in single isolated *Aplysia* neurones. Several other studies have shown that in tissues where more than one agonist activates the K^+ current, if the maximal effect is produced by one agonist, no further activation can be produced by the second agonist [77]. This indicates that the same K^+ channels can be coupled to more than one receptor, either by the same G-protein, specific for each channel or by different subsets of G-proteins, specific for each receptor, all of which would have access to every channel. It is likely that the G-proteins are present in excess over both receptors and channels.

With respect to synaptic responses, Thalmann [71] has observed that pertussis toxin blocks the late slow i.p.s.p. recorded in hippocampal CA_3 neurones following synaptic activation of mossy fibres, a conductance mediated by activation of $GABA_B$ receptors. Similarly, pertussis toxin blocks the α_2-adrenergic-mediated K^+-dependent i.p.s.p. recorded in submucous plexus neurones following sympathetic activation [78].

Van Dongen *et al.* [79] have examined the activation of K^+ channels by G_o in inside-out patches of cultured hippocampal neurones. Under the conditions used, in the absence of Ca^{2+} and with an antagonist to block ATP-sensitive K^+ channels, no K^+ channels were observed in control patches, but were induced in about half of the patches by either purified activated G_o or activated recombinant α_o. Four types of K^+ channel were induced in this way, they were distinct in terms of single channel conductance and by the presence or absence of inward rectification. In this study, G_o was not observed to modulate other K^+ channels, including Ca^{2+} activated K^+ channels, ATP-sensitive K^+ channels and transient A channels.

AGONIST INHIBITION OF POTASSIUM CURRENTS

In several tissues the inhibition of a K^+ current has been reported to occur via a G-protein that is insensitive to pertussis toxin. This current is distinct from the M current in that it is active at the resting membrane potential. For example, substance P suppresses an inwardly rectifying K^+ current in nucleus basalis and locus coeruleus neurones [80]. Whether this is a direct effect remains unclear, although a role for adenylate cyclase has been ruled out. The same group have recently shown that muscarinic activation inhibits a resting K^+ current, producing an inward current in cultured hippocampal neurones, and again a pertussis-toxin-insensitive G-protein is involved [81]. In contrast, noradrenaline, acting at an α_1 receptor decreases a background K^+ conductance in Purkinje myocytes, but in this case the G-protein involved is sensitive to pertussis toxin [82].

Modulation of calcium channels by G-protein activation

Many investigations have studied the interaction between Ca^{2+} currents and GTP binding proteins. In most cases, the unequivocal demonstration of a direct interaction has not been achieved but in this section the possibility of direct coupling of Ca^{2+} channels to the different classes of G-protein will be assessed.

Voltage-activated Ca^{2+} channels have been classified in several ways, firstly according to the voltage threshold of their activation, into low and high threshold channels [83,84]. Low threshold channels or 'T' channels are of small conductance (about 8 pS in Ba^{2+}) and inactivate rapidly with sustained depolarization above about -50 mV. They give rise to transient whole cell currents. High threshold channels recorded in cardiac myocytes are of a single very slowly inactivating class with a single channel conductance of about 24 pS; these have been termed 'L' channels. In neurones an additional class of high threshold channel has been described, of intermediate conductance about 13 pS, which is inactivating and has been termed 'N' [85,86].

T currents are not present in all cell types, but when present they can be clearly delineated because of their low threshold for activation and inactivation [87], and because of their sensitivity to Ni^{2+}, and other blockers including octanol and amiloride (for review see Tsien et al. [86]). However, the contribution of N and L channels to the whole cell currents in neurones is far less clear-cut [10,88]. The problem arises partly because of their similar threshold for activation. A major drawback remains the lack of antagonist which is specific for N channels. In addition, although L currents are sensitive to inhibition by dihydropyridine Ca^{2+} channel antagonists, these may also increase the rate of current inactivation [89]. In DRGs, N channels inactivate quite rapidly, but in sympathetic neurones they are less easily distinguished from L channels by their rate of inactivation [86].

ACTIVATION OF CALCIUM CHANNELS

In cell-attached patches from ventricular myocytes Ca^{2+} channel activity is increased by β-adrenergic-receptor-mediated cAMP-dependent phosphorylation, as discussed on page 126. In cell-free patches the channels are normally rapidly lost, but it has recently been found that activation of G_s increases the probability of opening of L-type channels in these patches [90]. It was initially observed that the survival of L channels in inside-out patches could be enhanced by GTP analogues. This only occurred following previous exposure of the cell to a β-adrenergic agonist; there was a requirement for channel phosphorylation prior to patch excision. Subsequently it was observed that the effect of β agonists could be mimicked by the dihydropyridine agonist Bay K 8644. This result suggests that when this agonist is bound it can also maintain the channel in a state that allows the activated G-protein to interact with it. The response of the Ca^{2+} channels in the patch to GTP analogues is not mimicked by agents inducing cAMP-dependent phosphorylation, but can be reproduced by preactivated G_s or α_s. In contrast, no effect of activated G_k was observed in this system.

Similar effects have been observed using skeletal muscle t-tubule Ca^{2+} channels purified and incorporated into lipid bilayers. Unlike cardiac Ca^{2+} channels, these channels show stable basal activity, and stimulation of Ca^{2+} channel activity by activated G_s can occur in the absence of Bay K 8644 or prior phosphorylation. Under these conditions the involvement of second messengers is completely ruled out [91]. The functional role of this direct action of G_s has yet to be determined since all the effects of isoprenaline on Ca^{2+} currents in intact cells can be prevented by inhibitors of cAMP-dependent phosphorylation [9].

There are several other reports of enhancement of high threshold Ca^{2+} currents by G-protein activation. Angiotensin enhances Ca^{2+} currents in heart cells [33] and Y1 cells (an adrenocortical cell line) and by LHRH in GH_3 cells [92]. The former response is thought to occur by PKC activation, but both of the latter two responses are pertussis-toxin sensitive, whereas the ability of these hormones to activate polyphosphoinositide turnover is not prevented by this toxin. Nevertheless, it is possible that DAG production from non-inositol-containing lipids occurs by a pertussis-toxin-sensitive route [93].

INHIBITION OF HIGH THRESHOLD CALCIUM CHANNELS

The initial observations on which these studies are based showed that in many cells that exhibit Ca^{2+} action potentials, or in which there is a substantial calcium-dependent plateau phase of the action potential, neurotransmitters and neuromodulators decrease this component [94]. Subsequently it has been observed that calcium currents in many cell types including neurones and gland cells are inhibited by a variety of hormones and neurotransmitters. In cardiac myocytes all the inhibitory effects on Ca^{2+} currents occurring via a pertussis-toxin-sensitive G_i appear to result from a reduction in cAMP levels. However, in many other tissues there is little evidence that Ca^{2+} current inhibition by neurotransmitters and hormones involves inhibition of adenylate cyclase.

The initial evidence that a GTP binding protein is involved was suggested by the finding that the inhibition of Ca^{2+} currents by GABA, the $GABA_B$ analogue baclofen and noradrenaline in chick and rat dorsal root ganglion neurones could be enhanced and mimicked by GTP analogues [95]. It could be inhibited by GDP analogues [96] and pertussis toxin. Similarly, inhibition of Ca^{2+} currents in the pituitary cell line AtT-20 is inhibited by somatostatin and by GTP analogues [97]. The ability of Ca^{2+} currents to be modulated by neurotransmitters varies widely, but in some cell types this can reach complete inhibition [98]. However, in many systems a differential inhibition of the transient component of the whole cell current occurs, which has been suggested by some workers to represent inhibition of N current (for review see Tsien et al. [86]). In some cases this interpretation must be regarded with caution, particularly where Ca^{2+} is the charge carrier, since a reduction of the initial current may reduce Ca^{2+}-dependent inactivation. This would give an apparently greater reduction in the transient component of the current. However, in experiments using Ba^{2+} as the charge carrier, GTP analogues clearly inhibit the transient component of the whole cell current in rat DRGs [95] and slow current activation [98, 99], although this has also been observed in AtT-20 cells that do not possess N channels [97]. From these studies it is clear that L currents can also be inhibited by neurotransmitters and hormones and by GTP analogues [99]. Neuropeptide Y (NPY) has been shown by Walker et al. [100] to inhibit both the transient and the sustained portion of the calcium current in rat sensory neurones. In addition, NPY inhibits Ca^{2+} flux into these cells and the release of substance P, which is also inhibited by dihydropyridines. In this case it is clear that release is dependent on Ca^{2+} influx through L channels, although in many systems transmitter release has not been found to be reduced by dihydropyridine antagonists.

Many of the neurotransmitters which have been observed to inhibit Ca^{2+} currents belong to a similar set to those which activate K^+ currents by the direct mediation of a pertussis-toxin-sensitive G-protein. However, it has rarely been demonstrated that both inhibition of Ca^{2+} currents and activation of K^+ currents will occur simultaneously following activation of the same neurotransmitter receptor, and it is unknown whether the same subtype of G-protein is involved in the two events.

Definitive evidence for the involvement of a G-protein in coupling receptors to Ca^{2+} current inhibition comes from experiments in which G-proteins are included in the patch pipette when recording Ca^{2+} currents in pertussis-toxin-treated cells. It has been found that these exogenously added G-proteins can restore the ability of neurotransmitters to inhibit Ca^{2+} currents. These experiments have generally suggested that G_o or its α subunit is more effective than G_i in restoring coupling [101–102]. An interesting finding of Ewald et al. [103] is that, whereas responses to NPY require G_o for full restoration of coupling, the response to bradykinin could be restored partially by three pertussis-toxin-sensitive G-proteins including G_o. The authors suggest that it is the receptor that dictates specificity of coupling, and that bradykinin receptors have less stringent recognition requirements for coupling. In several systems, the use of anti-G-protein antibodies which inhibit function has confirmed that a G_o protein is able to mediate inhibition of calcium currents by neurotransmitters [102–104].

There is no unequivocal evidence that the Ca^{2+} channels modulated by neurotransmitter are directly influenced by interaction with G-proteins. However, most evidence suggests that neither activation nor inhibition of adenylate cyclase markedly affects the whole cell Ca^{2+} current in neurones [105–110]. This may indicate that channels are fully phosphorylated under the whole cell clamp conditions used, or that the activity of neuronal channels is not influenced as markedly by phosphorylation as it is in heart ventricular myocytes. It is less likely that these channels are not substrates for phosphorylation. Indeed, it has been observed that in cultured rat DRGs, calcium channel currents are partially inhibited by adenosine imido-diphosphate, which reduces phosphorylation, and enhanced by cholera toxin and forskolin [107]. In addition, Gray and Johnston [111] have shown that β-adrenergic agonists enhance the activity of voltage-dependent calcium channels in hippocampal granule cells. In agreement with this, Lipscombe et al. [3] have reported a stimulation of single L-type Ca^{2+} channel activity in frog sympathetic neurones following noradrenaline application.

The evidence concerning the obligatory mediation of PKC in the inhibition of Ca^{2+} currents by neurotransmitters remains equivocal. It is clear that at least in some cell types PKC activators inhibit Ca^{2+} currents, since Rane and Dunlap [112] initially demonstrated an inhibition of the sustained Ca^{2+} current in chick DRGs by phorbol esters and OAG. In further studies from this group, it has been reported that this response and the inhibition by NA are inhibited by a specific peptide inhibitor of PKC [113]. One study that has provided evidence for an obligatory role of PKC in the inhibition of Ca^{2+} currents by a neurotransmitter in snail neurones is that of Hammond et al. [114]. The ability of cholecystokinin to inhibit Ca^{2+} currents was found to be mimicked by TPA and OAG, and enhanced by injection of PKC. In several other cell types (GH_3 and chromaffin cells), although OAG was found to inhibit whole cell Ca^{2+} currents, no evidence was presented for the possible hormone pathway involved or that the effect was prevented by PKC antagonists [115]. In another series of experiments, Ewald et al. [116] showed that down-regulation of PKC by chronic phorbol ester treatment, which results in complete PKC degradation following its translocation to the membrane, reduced the ability of NPY to inhibit the sustained but not the transient component of the Ca^{2+} current in neurones, whereas pertussis toxin completely prevented the response to NPY. This suggests that NPY is able to activate PKC by production of DAG via a pertussis-toxin-sensitive G-protein. The same group have provided evidence that this can indeed occur [117]. In contrast, Gross and MacDonald [118] have observed phorbol esters to inhibit the transient but not the sustained component of calcium currents in cultured mouse DRG neurones. This response is abolished in pertussis-toxin-treated cells, and therefore it may require a G-protein for its effect. Although it is known that PKC can phosphorylate inhibitory G-proteins [28], it is unclear whether phosphorylation of these G-proteins affects their ability to couple to effectors in systems other than adenylate cyclase. A further caveat is that non-PKC mediated inhibition of Ca^{2+} currents by phorbolesters has been demonstrated [118a].

Several other studies have failed to show either an effect on Ca^{2+} currents of PKC activators [92,105,107,119,120], or the ability of inhibitors of PKC to prevent the response of calcium currents to neurotransmitters or GTP analogues [107,119,120]. The ability of FMRF-amide to inhibit Ca^{2+} currents in *Aplysia* neurones [119], raises the possibility that it might use the same signal transduction pathway, involving the eicosanoids, as is involved in its modulation of K^+ channel activity [48]. However, we have not observed arachidonic acid to inhibit Ca^{2+} current in sensory neurones [107].

The role of Ca^{2+} influx in the inhibition of Ca^{2+} currents has also been examined. Although Ca^{2+} is well known to produce Ca^{2+}-dependent inactivation of Ca^{2+}

currents, most studies have found that inhibition by neurotransmitters still occurs if intracellular Ca^{2+} is buffered to very low levels [110,119,121]. An anomalous finding is that of Shimahara and Icard-Leipkalns [109] who showed that δ-opiate inhibition of transient Ca^{2+} currents in the neuroblastoma $\times$ glioma hybrid NG108–15 cells was prevented when internal Ca^{2+} levels were buffered to 10^{-9} M. Docherty and McFadzean [120] have also observed current-dependent inhibition of Ca^{2+} currents by NA in these cells.

Another means of determining whether a second messenger is involved in the inhibition of Ca^{2+} currents is to examine the effect of neurotransmitters on Ca^{2+} channels recorded in the cell-attached patch configuration, while applying the neurotransmitter either in the medium surrounding the cell or in the patch pipette (Figure 9.1). If the neurotransmitter is effective when exposed to the rest of the cell but not to the external face of the membrane under the pipette, then it must be producing its effect by a diffusible second messenger. The first study of this type was that of Forscher et al. [122], who measured Ca^{2+} currents in large patches using the cell-attached patch mode. Noradrenaline applied to the bath did not affect the Ca^{2+} currents recorded under the patch although it inhibited the whole cell Ca^{2+} current in the same cell. The authors thus concluded that no readily diffusible second messenger was involved. However, since very large diameter (2–4 μm) pipettes were used in this study, these may have been too large to observe effects of a second messenger such as diacylglycerol which is not readily diffusible into the cytoplasm, although it may diffuse in the plane of the membrane. It is unclear to what extent the patch pipette itself would form a barrier to diffusion of DAG into the patch. Another unknown factor is whether activated G-proteins might also be able to diffuse from the cytoplasmic surface of the cell membrane surrounding the patch pipette into the patch. Similar conclusions to those of Forscher et al. [122] have been reached by several other groups. Green and Cottrell [123] did not observe effects of baclofen on single channel currents in DRGs when it was applied externally to the patch electrode, in cells whose whole cell Ca^{2+} current had been shown to be baclofen-sensitive. Hirning et al. [124] also showed that whereas NPY inhibited the inactivating 'N' component of whole cell Ca^{2+} currents in myenteric neurones, it had no effect on N channels recorded in cell-attached patches when it was added to the medium outside the patch pipette. In a recent study Lipscombe et al. [124a] have observed that single N-type Ca^{2+} channels in sympathetic neurones are not inhibited by NA added outside the patch, but are inhibited by NA in the patch pipette; whereas 'L' channel activity is increased, presumably by a cAMP-dependent mechanism. In addition, the only effect of phorbol esters added outside the patch pipette was an increase in both 'N' and 'L' channel activity. In contrast, results have been obtained in chick DRGs show that noradrenaline added outside the patch inhibits L channels [125]. Thus, even this experimental protocol has not yet provided unequivocal evidence for, or against, direct G-protein coupling to high threshold N- or L-type Ca^{2+} channels.

A mechanism for the inhibition of calcium currents has been put forward by Bean [126]. He observed that the inhibitory effect of noradrenaline on Ca^{2+} currents in bullfrog DRG neurones is absent at high clamp potentials when the calcium current has reversed to an outward current. These results suggest that noradrenaline decreases Ca^{2+} currents by shifting the voltage dependence of channel opening to more positive potentials. This finding parallels previous work on β-adrenergic activation of Ca^{2+} currents, in which it was suggested that phosphorylation shifts the voltage dependence for activation to more hyperpolarized potentials [4].

INHIBITION OF LOW THRESHOLD CALCIUM CHANNELS

There are contradictory findings on the responsiveness of low voltage-activated or 'T' currents to inhibitory modulation by neurotransmitters or G-protein activation. Gross and MacDonald [127] showed no effect of dynorphin (acting at κ receptors) on low threshold currents in DRGs, although it produced a marked inhibition of the transient high threshold current. In agreement with this finding, somatostatin was observed to inhibit high but not low threshold Ca^{2+} currents in GH_3 cells [108]. In contrast, in the same cell type, Marchetti and Brown [115] have shown that PKC activators inhibit both 'T' and 'L' currents with no change in their activation kinetics. Marchettti *et al.* [128] have also shown that, whereas dopamine slows the rate of activation and reduces the amplitude of high threshold currents in DRG neurones, low threshold currents are inhibited without change in their kinetics. Low threshold single channel activity and ensemble averaged single channel 'T' currents recorded from inside-out patches were also inhibited by dopamine [128]. Similarly Scott and Dolphin [129] showed an adenosine agonist, 2-chloroadenosine, to inhibit both high and low threshold Ca^{2+} currents in rat DRGs.

Further evidence in these cells for the association of 'T' channels with a GTP binding protein comes from a study showing that photo-release of GTPγS, from an inactive 'caged' precursor, initially enhances 'T' currents (at a low concentration, 6 μM) and at higher concentrations (20 μM) inhibits 'T' currents [130]. From these studies it is clear that low threshold currents, responsible for the regulation of calcium influx and calcium-dependent events around the resting membrane potential, are also capable of modulation of neurotransmitters.

INTERACTION BETWEEN CALCIUM CHANNEL LIGAND BINDING SITES AND G-PROTEIN ACTIVATION

Recent work has suggested an interaction between G-protein activation and the effect of Ca^{2+} channel ligands on L-type channels in cultured rat DRGs and sympathetic neurones [131–132a]. Ca^{2+} channel antagonists, including nifedipine, $(-)$-202-791, diltiazem and D_{600} were observed to show only agonist properties in the presence of internal guanine nucleotide analogues (Figure 9.2). Further studies showed that agonist responses to both 'antagonist' and 'agonist' Ca^{2+} channel ligands were blocked by pertussis toxin in DRGs. This finding indicates that in the intact cells studied, G-protein activated by GTP or its non-hydrolysable analogues is an essential requirement for Ca^{2+} channel ligands to act as agonists [133] and may stabilize the channel in the resting state in which the calcium channel ligand binding site is its agonist conformation. Since purified DHP receptors have

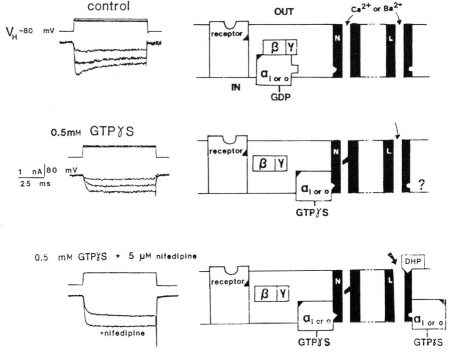

Figure 9.2 Modulation of N and L channels by GTP analogues.
Top panel: Whole cell calcium channel currents recorded from a cultured rat DRG have both a transient and a sustained component, which may be due to current flowing through N and L channels respectively [98]
Middle panel: GTPγS differentially inhibits transient current, and the activated G-protein may selectively inhibit N channels [98,99].
Bottom panel: Nifedipine in the presence of GTPγS acts as an agonist. This may be because an activated G-protein interacts with L channels and affects their conformation and gating kinetics so that the dihydropyridine binding site accepts only agonists. Nifedipine will bind to this site and have an agonist action [131,132]

Ca^{2+} channel activity when inserted in lipid bilayers, and show an agonist response to BaY K 8644 [134], it is possible that in intact neurones inactive G-proteins prevent the channel from existing in the conformation where the ligands can act as agonists. Under control conditions it would appear likely that there is partial tonic activation of these G-proteins [133], and this is responsible for the initial transient agonist response observed to many calcium channel ligands. There are parallels in this work with the effect of activated G_s on cardiac L channels [90] since the Ca^{2+} channel agonist Bay K 8644 appeared to stabilize these channels in a conformation in which they could interact with G_s, allowing activated G_s to increase the opening probability of the channel. In addition, although most binding studies have shown equivocal effects of guanine nucleotide analogues on calcium channel ligand binding, a recent report has shown that GTPγS enhanced the binding of Bay K 8644 to rat brain membranes in a pertussis toxin sensitive manner [135]. Evidence that this interaction may be physiologically relevant comes from the finding that pertussis toxin reduces the ability of the calcium channel agonist (+)-202–791 to enhance glutamate release from cultured cerebellar neurones (Huston and Dolphin, submitted). Another relevant piece of work by Horne *et al.* [136] shows that a protein fraction purified from skeletal muscle t tubules can be reconstituted to show calcium channel activity by the addition of G_o. However, this activity is then not sensitive to the antagonist verapamil.

These findings indicate that there is a complex interaction between calcium channels and G-proteins, which is likely to be unravelled by reconstitution experiments with pure constituent proteins.

Role of G-protein coupled ion channels in the regulation of synaptic transmission

The release of neurotransmitters from presynaptic terminals is dependent on an influx of Ca^{2+} through voltage-sensitive channels and can be modulated by several means. Many neurotransmitters have been shown to act on presynaptic receptors to inhibit neurotransmitter release, and these effects can be examined both at the biochemical and at the electrophysiological level. Classical presynaptic inhibition results from activation of $GABA_A$ receptors, and activation of a Cl^- conductance. However, presynaptic inhibition by other neurotransmitters may involve either an enhancement of a K^+ conductance, reducing the ability of action potentials to invade the terminal, or inhibition of Ca^{2+} currents, directly reducing the influx of Ca^{2+} reaching the active release zones. From the preceding sections it is clear that both these processes may involve pertussis-toxin-sensitive G-proteins.

In biochemical studies of neurotransmitter release, there are many examples of the ability of pertussis toxin to prevent presynaptic inhibition particularly in cultured cells. For example, inhibition of glutamate release by adenosine analogues in cultured cerebellar neurones [137], and by the $GABA_B$ agonist baclofen in the same system [138], and inhibition of substance P release by noradrenaline and GABA [139] in cultured DRGs, are all prevented by pertussis toxin. In intact tissues and brain slices the findings are more sparse. Kawata and Nomura [140] and Allgaier *et al.* [141] were able to prevent the α_2-mediated inhibition or noradrenaline release from cortical and hippocampal slices by incubating the isolated tissue with pertussis toxin, although this means of application has been unsuccessful in other systems [138]. In contrast, in vivo treatment with pertussis toxin did not attenuate α_2-adrenoceptor-mediated inhibition of noradrenaline release from mouse atria [142]. Another approach has been used by Fredholm [143], who incubated brain slices with the alkylating agent *N*-ethylmaleimide. At low concentrations this substance appears to selectively inactivate G_i-like G-proteins with no effect on G_s [144]. This treatment prevented adenosine-A_1-receptor-mediated inhibition of glutamate release from rat hippocampal slices [143].

There is less electrophysiological evidence that presynaptic inhibition of e.p.s.p. is prevented by pertussis toxin, although an early study showed that presynaptic inhibition of the twitch response in guinea pig ileum by opiates and the α_2 agonist clonidine is inhibited by pertussis toxin [145]. It has also recently been shown that adenosine inhibition of e.p.s.p. at the neuromuscular junction is sensitive to pertussis toxin [146]. In addition, Colmers and Williams [147] showed that the e.p.s.p. recorded in dorsal raphé neurones by focal stimulation of slices could be inhibited by about 20% by a $5-HT_1$ agonist, 5-carboxamidotryptamine, and this response was prevented by prior intraventricular administration of pertussis toxin.

There is contradictory data particularly with regard to $GABA_B$-mediated events. Whereas pertussis toxin has been shown to block the baclofen-mediated inhibition of noradrenaline, substance P and glutamate release [138,139], several studies have shown no effect of pertussis toxin on baclofen-mediated presynaptic inhibition of excitatory and inhibitory postsynaptic potentials. Harrison et al. [148–149] found 20 μM baclofen to inhibit the i.p.s.p. recorded between pairs of cultured hippocampal GABA-ergic interneurones by 92%. Similarly Colmers and Williams [147] showed baclofen to inhibit the e.p.s.p. in dorsal raphé neurones by 81%. Neither of these effects was pertussis-toxin-sensitive, although the latter study showed the postsynaptic K^+-dependent hyperpolarization by baclofen to be prevented by this treatment. Dutar and Nicoll have shown that in hippocampal pyramidal cells the presynaptic inhibition of the e.p.s.p. by baclofen differs from its postsynaptic effects, not only in its insensitivity to pertussis toxin, but also in the inability of the $GABA_B$ antagonist phaclofen to block the response [150]. This is in agreement with the results of Harrison [149].

These findings suggest that biochemical measurements of transmitter release evoked either by elevated K^+ depolarization or by electrical stimulation by pulse trains, do not exactly mimic the process of synaptic release of transmitter. There are several possible explanations for the discrepancies outlined above. It may be that there is a large excess of G-proteins in synaptic boutons, and only a small proportion of these are required to mediate a presynaptic inhibitory effect on the e.p.s.p. resulting from a single action potential. Since pertussis toxin treatment rarely results in complete ADP-ribosylation of all its substrate G-proteins, particularly when applied to intact tissues, it is possible that sufficient G-proteins remained unaffected. The observation of an effect of pertussis toxin on presynaptic inhibitory responses may be favoured in biochemical studies of neurotransmitter release because of the prolonged and repeated depolarization used. In addition, penetration of pertussis toxin into presynaptic terminals may be less efficient than into postsynaptic elements. Another reason for the ineffectiveness of pertussis toxin might be that a proportion of G-proteins in presynaptic terminals are tonically GTP-activated, and thus not substrates for pertussis toxin [151]. It is noteworthy that the pertussis-toxin-sensitive inhibition of transmitter release by the various agonists used in many different systems rarely reaches more than 50%, whereas inhibition of postsynaptic potentials by neuromodulators can be almost complete, particularly in cases where it is not pertussis-toxin-sensitive [147,148].

It is clearly also possible that presynaptic inhibition of release by these neuromodulators occurs by a mechanism involving neither inhibition of Ca^{2+} currents nor activation of potassium currents by pertussis-toxin-sensitive G-proteins. Changes in the mobilization of intracellular calcium or in the processes associated with vesicle fusion may also be involved, as they are in other secretory systems [24].

DIRECT INTERACTION OF SECOND MESSENGERS WITH ION CHANNELS

Since visual transduction is the subject of another chapter in this volume, the modulation of ion channels in the sensory transduction process will be mentioned only briefly here.

In the visual transduction system, the cytoplasmic level of cGMP in the visual receptor cells (rods and cones) is closely regulated both in its formation by guanylate cyclase, and in its breakdown by phosphodiesterase. The phosphodiesterase is stimulated by light via rhodopsin-induced activation of the G-protein transducin. In the plasma membrane of rod outer segments a non-specific cation channel is held open by the level of cytoplasmic cGMP which occurs in the dark [152]. It is rapidly closed by the light-induced reduction in cGMP (for review see [153]) and by the use of GTP analogues to activate transducin [154]. Thus in this system G-protein activation indirectly results in channel closure by causing a reduction in the concentration of the second messenger cGMP.

Most effects of cGMP have been thought to occur by activation of a cGMP-dependent protein kinase. However, it is clear from studies on isolated patches from rod outer segment membranes that this cyclic nucleotide response is a direct effect, either on the channel itself, or on a protein associated with the channel [152,155].

A similar signal transduction process is thought to occur in olfactory transduction, although in this case activation of receptors on the olfactory cilia results in stimulation of several G_s-like G-proteins and activation of adenylate cyclase [156]. Excised patches from olfactory cilia respond to both cAMP and cGMP, both of which gate a cation conductance in a phosphorylation-independent manner [157].

In contrast, although cAMP also mediates taste transduction, and depolarizes taste receptor cells, it appears to do so by a cAMP-dependent protein-kinase-mediated inhibition of a potassum conductance [158]. Thus the various sensory transduction processes function differently, both in terms of the species of cyclic nucleotide involved and in the requirement for a phosphorylation step.

CONCLUSIONS

G-proteins modulate ion channel activity in several ways: by altering the turnover of second messengers, by influencing intracellular calcium levels and by interacting directly with the membrane ion channels themselves. However, there remain many unanswered questions concerning ion channel modulation by G-proteins.

From the stoichiometry of G-proteins, receptors and ion channels, it is clear that the transducing G-proteins, particularly the pertussis toxin substrates, are present greatly in excess of either receptors or effectors. It remains unclear how specificity is achieved since some receptors couple to one G-protein, and some to several different G-proteins, and a number of receptors may affect the same ion channel. The same receptor subtype may couple to different effectors in different tissues. In addition, we must now grapple with the likelihood that released $\beta\gamma$ subunits also have physiological functions in many cell types. Activation of phospholipase A_2 [47a] and transduction of pheromone receptor signals [159], have already been shown to be mediated by $\beta\gamma$, and it is likely that other functions will be discovered. In this way two separate signal transduction pathways may be activated by the α and the $\beta\gamma$ subunits. It is very likely that there are more subclasses of G-proteins and receptors than have been identified to date. The differences may be very small, and may also involve post-translational modification. This may be sufficient to allow for specificity of coupling in the cell types in which they are expressed. It is also possible that the coupling is specified on insertion into the membrane. A fairly tight spatial association may be formed, which then prevents interaction of a particular receptor with other G-proteins. The arrays of receptors together with G-proteins may also be arranged around the ion channels whose activity they modulate, so that several receptor types can influence the same channel.

It is possible that G-proteins perform functions in the membrane or in the cytoplasm other than signal transduction. For example G_o comprises about 0.3% of neuronal membrane protein [160]. It does not appear to be concentrated at synaptic regions [160], and thus it must play more general roles in transduction throughout the plasma membrane. Its presence in the cytoplasm may also indicate an interaction with cytoskeletal elements or modulation of cytoplasmic enzyme activity.

It is likely that we can look forward to uncovering additional functions of G-proteins in the future.

REFERENCES

1. Cachelin, A. B., de Peyer, J. E., Kokubun, S., and Reuter, H. (1983). *Nature (London)*, **304**, 462–464.
2. Kameyama, M., Hofmann, F., and Trautwein, W. (1985) *Pflüger's Arch.*, **405**, 285–293.
3. Lipscombe, D., Bley, K., and Tsien, R. W. (1988) *Soc. Neurosci. Abstr.*, **14**, 64.12.
4. Bean, B. P., Nowycky, M. C., and Tsien, R. W. (1984) *Nature (London)*, **307**, 371–375.
5. Curtis, B. M., and Caterall, W. A. (1985) *Proc. Natl. Acad. Sci. USA*, **82**, 2528–2532.
6. Flockerzi, V., Oeken, H. J., Hofmann, F., Peltzer, D., Cavalié, A., and Trautwein, W. (1986) *Nature (London)*, **323**, 66–68.
7. Hartzell, H. C., and Fischmeister, R. (1987) *Mol. Pharmacol.*, **32**, 639–645.
8. Hescheler, J., Kameyama, M., and Trautwein, W. (1986) *Pflüger's Arch.*, **407**, 182–189.
9. Kameyama, M., Hescheler, J., Hofmann, F., and Trautwein, W. (1986) *Pflüger's Arch.*, **407**, 123–128.
10. Carbone, E., and Lux, H. D. (1987) *J. Physiol.*, **386**, 371–601.
11. Breitwieser, G. E., and Szabo, G. (1985) *Nature (London)*, **317**, 538–540.
12. Klein, M., Camardo, J. S., and Kandel, E. R. (1982) *Proc. Natl. Acad. Sci. USA*, **79**, 5713–5717.

13. Shuster, M. J., Camardo, J. S., Siegelbaum, S. A., and Kandel, E. R. (1985) *Nature (London)*, **313**, 392–395.
14. Volterra, A., Siegelbaum, S. A., Kandel, E. R., and Sweatt, J. D. (1988) *Soc. Neurosci. Abstr.*, **14**, 64.6.
14a. Kume, H., Takai, A., Tokuno, H., and Tomita, T. (1989) *Nature (London)*, **341**, 152–154.
15. Alger, B. E., and Nicoll, R. A. (1980) *Science*, **210**, 1122–1124.
16. Nicoll, R. A. (1988) *Science*, **241**, 545–551.
17. Madison, D. V., and Nicoll, R. A. (1984) *J. Physiol.*, **354**, 319–332.
18. Li, M., McCann, J. D., Liedtke, C. M., Nairn, A. C., Greengard, P., and Welsh, M. J. (1988) *Nature (London)*, **331**, 358–360.
19. Welsh, M. J. (1986) *Science*, **232**, 1648–1650.
20. Di Francesco, D., and Tromba, C. (1988) *J. Physiol.*, **405**, 493–510.
21. Huganir, R. L., Delcour, A. H., Greengard, P., and Hess G. P. (1986) *Nature (London)*, **321**, 774–776.
22. Nishizuka, Y. (1988) *Nature (London)*, **334**, 661–665.
23. Murakami, K., and Routtenberg, A. (1985) *FEBS Lett.*, **192**, 189–193.
24. Gomperts, B. D. (1983) *Nature (London)*, **306**, 64–66.
25. Burch, R. M., Luini, A., and Axelrod, J. (1986) *Proc. Natl. Acad. Sci. USA*, **83**, 7201–7205.
26. Cockcroft, S., and Gomperts, B. D. (1985) *Nature (London)*, **314**, 334–336.
27. Lapetina, E. G., Reep, B., Ganong, B. R., and Bell, R. M. (1985) *J. Biol. Chem.*, **260**, 1358–1361.
28. Jakobs, K. H., Bauer, S., and Watanabe, Y. (1985) *Eur. J. Biochem.*, **151**, 425–430.
29. De Riemer, S. A., Strong, J. A., Albert, K. A., Greengard, P., and Kaczmarek, L. K. (1985) *Nature (London)*, **313**, 313–316.
30. Strong, J. A., Fox, A. P., Tsien, R. W., and Kaczmarek, L. K. (1987) *Nature (London)*, **325**, 714–717.
31. Conn, P. J., Strong, J. S., and Kaczmarek, L. K. (1988) *Soc. Neurosci. Abstr.*, **14**, 64.11.
32. Braha, O., Klein, M., and Kandel, E. R. (1988) *Soc. Neurosci. Abstr.*, **14**, 262.3.
33. Dosemeci, A., Dhallan, R. S., Cohen, N. M., Lederer, W. J., and Rogers, T. B. (1988) *Circulation Res.*, **62**, 347–357.
34. Lacerda, A. E., Rampe, D., and Brown, A. M. (1988) *Nature (London)*, **335**, 249–251.
35. Young, S., Parker, P. J., Ullrich, A., and Stabel, S. (1987) *Biochem. J.*, **244**, 775–779.
36. Leonard, J. P., Nargeot, J., Snutch, T. P., Davidson, N., and Lester, H. A. (1987) *J. Neurosci.*, **7**, 875–881.
37. Sigel, E., and Baur, R. (1988) *Proc. Natl. Acad. Sci. USA*, **85**, 6192–6196.
38. Brown, D. A. (1988) *Trends Neurosci.*, **11**, 294–299.
39. Pfaffinger, P. (1988) *J. Neurosci.*, **8**, 3343–3353.
40. Brown, D. A., Marrion, N. V., and Smart, T. G., (1989) *J. Physiol.* (in press).
41. Fukuda, K., Higashida, H., Kubo, T., Maeda, A., Akiba, I., Bugo, H., Mishina, M., and Numa, S. (1988) *Nature (London)*, **335**, 355–358.
42. Clapp, L. H., Sims, S. M., Singer, J. J., and Walsh, J. V. (1988) *Soc. Neurosci. Abstr.*, **14**, 438.3.
43. Pfaffinger, P. J., Leibowitz, M., Subers, E. M., Nathanson, N. M., Almers, W., and Hille, B. (1984) *Neuron*, **1**, 477–484.
43a. Bosma, M., and Hille, B. (1988) *Soc. Neurosci. Abstr.*, **14**, 303.9.
44. Dutar, P., and Nicoll, R. A. (1988) *Neurosci. Letts.*, **85**, 89–94.
45. Malenka, R. C., Madison, D. V., Andrade, R., and Nicoll, R. A. (1986) *J. Neurosci.*, **6**, 475–480.
46. Stelzer, A., Kay, A. R., and Wong, R. K. S. (1988) *Science*, **241**, 339–341.
47. Muriyama, T., and Ui, M. (1985) *J. Biol. Chem.*, **260**, 7226–7233.
47a. Jelsema, C., and Axelrod, J. (1987) *Proc. Natl. Acad. Sci. USA*, **84**, 3623–3627.
48. Belardetti, F., and Siegelbaum, S. (1988) *Trends Neurosci.*, **11**, 232–238.
49. Volterra, A., and Siegelbaum, S. (1988) *Proc. Natl. Acad. Sci. USA*, **85**, 7810–7814.
49a. Kurachi, Y., Ito, H., Sugimota, T., Shimizu, T., Miki, I., and Ui, M. (1989) *Nature (London)*, **337**, 555–557.

49b. Kim, D., Lewis, D. L., Graziadei, L., Neer, E. J., Bar-Sagi, O., and Clapham, D. E. (1989) *Nature (London),* **337,** 557–560.
50. Evans, M. G., and Marty, A. (1986) *Proc. Natl. Acad. Sci. USA,* **83,** 4099–4103.
51. Scott, R. H., McGuirk, S. M., and Dolphin, A. C. (1988) *Brit. J. Pharmacol.,* **94,** 653–662.
52. Rogawski, M. A., Inoue, K., Suzuki, S., and Barker, J. L. (1988) *J. Neurophysiol.,* **59,** 1854–1870.
53. Chueh, S. H., Mullaney, J. M., Ghosh, T. K., Zachary, A. L., and Gill, D. L. (1987) *J. Biol. Chem.,* **262,** 13857–13864.
54. Dawson, A. P., Hills, G., and Comerford, J. G. (1987) *Biochem. J.,* **244,** 87–92.
55. Kaneko, S., Kato, K., Yamagishi, S., Sugiyama, H., and Nomura, Y. (1987) *Mol. Brain Res.,* **3,** 11–19.
56. Bolton, T. B., and Lim, S. P. (1989) *J. Physiol.,* **409,** 385–401.
57. Kobayashi, S., Somlyo, A. P., and Somlyo, A. V. (1988) *J. Physiol.,* **403,** 601–619.
58. Pfaffinger, P. J., Martin, J. M., Hunter, D. D., Nathanson, N. M., and Hille, B. (1985) *Nature (London),* **317,** 536–538.
59. Yatani, A. Codina, Y., Brown, A. M., and Birnbaumer, L. (1987) *Science,* **235,** 207–211.
60. Kurachi, Y., Nakajima, T., and Sugimoto, T. (1986) *Pflüger's Arch.,* **407,** 572–574.
61. Kurachi, Y., Nakajima, T., and Sugimoto, T. (1986) *Pflüger's Arch.,* **407,** 264–274.
62. Codina, J., Grenet, D., Yatani, A., Birnbaumer, L., and Brown, A. M. (1988) *FEBS. Lett.,* **216,** 104 106.
63. Kirsch, G. E., Yatani, A., Codina, J., Birnbaumer, L., and Brown, A. M. (1988) *Amer. J. Physiol.,* **254,** H1200–H1205.
64. Yatani, A., Hamm, H., Codina, Y., Mazzoni, M. R., Birnbaumer, L., and Brown, A. M. (1988) *Science,* **241,** 828–831.
65. Logothetis, D. E., Kurachi, Y., Galper, J., Neer, E. J., and Clapham, D. E. (1987) *Nature,* **325,** 321–326.
66. Katada, T., Northup, J. K., Bokoch, G. M., Ui, M., and Gilman, A. G. (1985) *J. Biol. Chem.,* **259,** 3578–3585.
67. Breitwieser, G. E., and Szabo, G. (1988) *J. Gen. Physiol.,* **91,** 469–493.
68. Kurachi, Y., Nakajima, T., and Sugimoto, T. (1987) *Pflüger's Arch.,* **410,** 227–233.
69. Trussell, L. O., and Jackson, M. B. (1987) *J. Neurosci.,* **7,** 3306–3316.
70. Andrade, R., Malenka, R. C., and Nicoll, R. A. (1986) *Science,* **234,** 1261–1265.
71. Thalmann, R. H. (1988) *J. Neurosci.,* **8,** 4589–4602.
72. Mihara, S., North, R. A., and Surprenant, A. (1987) *J. Physiol.,* **390,** 335–355.
73. Lacey, M. G., Mercuri, N. B., and North, R. A. (1988) *J. Physiol.,* **401,** 437–453.
74. Dolphin, A. C. (1987) *Trends Neurosci.,* **10,** 53–57.
75. Innis, R. B., and Aghajanian, G. K. (1987) *Brain Res.,* **411,** 139–143.
76. Sasaki, K., and Sato, M. (1987) *Nature (London),* **325,** 259–262.
77. North, R. A., Williams, Y. T., Surprenant, A.-M., and Christie, M. J. (1987) *Proc. Natl. Acad. Sci. USA,* **84,** 5487–5491.
78. Surprenant, A.-M., and North, R. A. (1988) *Proc. Roy. Soc. London B.,* **234,** 85–114.
79. Van Dongen, A. M. Y., Codina, J., Olate, Y., Mattera, R., Joho, R., Birnbaumer, L., and Brown, A. M. (1988) *Science,* **242,** 1433–1437.
80. Nakajima, Y., Nakajima, S., and Inoue, M. (1988) *Proc. Natl. Acad. Sci. USA,* **85,** 3643–3647.
81. Brown, L. D., Nakajima, S., and Nakajima, Y. (1988) *Soc. Neurosci. Abstr.,* **14,** 530.9.
82. Shah, A., Cohen, I. S., and Rosen, M. R. (1988) *Biophys. J.,* **84,** 219–225.
83. Carbone, E., and Lux, H. D. (1984) *Nature (London),* **310,** 501–502.
84. Hess, P., Lansman, J. B., and Tsien, R. W. (1984) *Nature (London),* **311,** 538–544.
85. Nowycky, M. C., Fox, A. P., and Tsien, R. W. (1985) *Nature (London),* **316,** 440–443.
86. Tsien, R. W., Lipscombe, D., Madison, D. V., Bley, K. R., and Fox, A. P. (1988) *Trends Neurosci.,* **11,** 431–437.
87. Llinas, R., and Yarom, Y. (1981) *J. Physiol.,* **315,** 569–584.

88. Swandulla, D., and Armstrong, C. M. (1988) *J. Gen. Physiol.*, **92**, 197–218.
89. Brum, G., Osterrieder, W., and Trautwein, W. (1984) *Pflüger's Arch.*, **401**, 111–118.
90. Yatani, A., Codina, J., Reeves, J. P., Birnbaumer, L., and Brown, A. M. (1987) *Science*, **238**, 1288–1292.
91. Yatani, A., Imoto, Y., Codina, J., Hamilton, S. L., Brown, A. M., and Birnbaumer, L. (1988) *J. Biol. Chem.*, **263**, 9887–9895.
92. Hescheler, Y., Rosenthal, W., Hinsch, K. D., Wulfern, M., Trautwein, W., and Schultz, G. (1988) *EMBO J.*, **7**, 619–624.
93. Irving, H. R., and Exton, J. H. (1987) *J. Biol. Chem.*, **262**, 3440–3443.
94. Dunlap, K., and Fischbach, G. (1981) *J. Physiol.*, **317**, 519–535.
95. Scott, R. H., and Dolphin, A. C. (1986) *Neurosci. Letts.*, **56**, 59–64.
96. Holz, G. G., Rane, S. G., and Dunlap, K. (1986) *Nature (London)*, **319**, 670–672.
97. Lewis, D. L., Weight, F. F., and Luini, A. (1986) *Proc. Natl. Acad. Sci.*, **83**, 9035–9039.
98. Dolphin, A. C., and Scott, R. H. (1987) *J. Physiol.*, **386**, 1–17.
99. Dolphin, A. C., Wootton, J. F., Scott, R. H., and Trentham, D. R. (1988) *Pflüger's Arch.*, **411**, 628–636.
100. Walker, M. W., Ewald, D. A., Perney, T. M., and Miller, R. J. (1988) *J. Neurosci.*, **8**, 2438–2446.
101. Hescheler, J., Rosenthal, W., Trautwein, W., and Schultz, G. (1987) *Nature (London)*, **325**, 445–447.
102. Ewald, D. A., Sternweis, P. C., and Miller, R. J. (1988) *Proc. Natl. Acad. Sci USA*, **85**, 3633–3637.
102a. McFadzean, I., Mullaney, I., Brown, D. A., and Milligan, G. (1989) *Neuron*, **3**, 177–182.
103. Ewald, D. A., Miller, R. J., and Sternweis, P. C. (1989) *Neuron*, **2**, 1185–1193.
104. Harris-Warwick, R. M., Hammond, C., Paupardin-Tritsch, D. Homburger, V., Rouot, B., Bockaert, J., and Gerschenfeld, H. M. (1988) *Nueron*, **1**, 27–32.
105. Wanke, E., Ferroni, A., Malgaroli, A., Ambrosinsi, A., Pozzan, T., and Meldolesi, Y. (1987) *Proc. Natl. Acad. Sci USA*, **84**, 4313–4317.
106. Dolphin, A. C., Forda, S. R., and Scott, R. H. (1986) *J. Physiol.*, **386**, 1–17.
107. Dolphin, A. C., McGuirk, S. M., and Scott, R. H. (1989) *Brit. J. Pharmacol.*, **97**, 263–273.
108. Rosenthal, W., Hescheler, J., Hinsch, K. D., Spicher, K., Trautwein, W., and Schultz, G. (1988) *EMBO J.*, **7**, 1627–1633.
109. Shimahara, T., and Icard-Liepkalns, C. (1987) *Brain Res.*, **415**, 357–361.
110. Akopayan, A. R., Chemeris, N. K., and Iljin, V. I. (1985) *Brain Res.*, **326**, 145–148.
111. Gray, R., and Johnston, D. (1987) *Nature (London)*, **327**, 620–622.
112. Rane, S. G., and Dunlap, K. (1986) *Proc. Natl. Acad. Sci. USA*, **83**, 184–188.
113. Rane, S. G., Walsh, M. P., and Dunlap, K. (1987) *Soc. Neurosci. Abstr.*, **13**, 222.4
114. Hammond, C., Paupardin-Tritsch, D., Nairn, A. C., Greengard, P., and Gerschenfeld, H. M. (1987) *Nature (London)*, **325**, 809–811.
115. Marchetti, C., and Brown, A. M. (1988) *Amer. J. Physiol.*, **254**, C206–210.
116. Ewald, D. A., Matthies, H. J. G., Perney, T. M., Walker, M. W., and Muller, R. J. (1988) *J. Neurosci*, **8**, 2447–2451.
117. Perney, T. M., and Miller, R. J. (1988) *Soc. Neurosci. Abstr.*, **14**, 485.1
118. Gross, R. A., and MacDonald, R. L. (1988) *Neurosci. Letts.*, **88**, 50–56.
118a. Hockberger, P., Toselli, M., Swandulla, D., and Lux, H. D. (1989) *Nature*, **338**, 340–342.
119. Brezina, V., Eckert, R., and Erxleben, C. (1987) *J. Physiol.*, **388**, 565–595.
120. McFadzean, I., and Docherty, R. J. *Euro. J. Neurosci.*, **1**, 141–150.
121. Forscher, P., and Oxford, G. (1985) *J. Gen. Physiol.*, **85**, 743–763.
122. Forscher, P., Oxford, G. S., and Schultz, D. (1986) *J. Physiol.*, **379**, 131–144.
123. Green, K. A., and Cottrell, G. A. (1987) *Brit. J. Pharmacol.*, **94**, 235–246.
124. Hirning, L. D., Fox, A. P., and Miller, R. J. (1988) *Soc. Neurosci. Abstr.*, **14**, 363–367.
124a. Lipscombe, D., Kongsamut, S., and Tsien, R. W. (1989) *Nature*, **340**, 639–642.

125. Anderson, C. S., and Dunlap, K. (1988) *Soc. Neurosci. Abstr.*, **14**, 262.5.
126. Bean, B. P. (1989) *Nature*, **340**, 153–156.
127. Gross, R. A., and MacDonald, R. L. (1987) *Proc. Natl. Acad. Sci. USA*, **84**, 5469–5473.
128. Marchetti, C., Carbonne, E., and Lux, H. D. (1986) *Pflüger's Arch.*, **406**, 104–111.
129. Scott, R. H., and Dolphin, A. C. (1987) In *Topics and Perspective in Adenosine Research* (Gerlach E., and Becker, B. F., eds). Springer-Verlag, Berlin.
130. Dolphin, A. C., Scott, R. H., and Wootton, J. F. (1989) *J. Physiol.*, **410**, 16 pp.
131. Scott, R. H., and Dolphin, A. C. (1987) *Nature (London)*, **330**, 760–762.
132. Scott, R. H., and Dolphin, A. C. (1988) *Neurosci. Letts.*, **89**, 170–175.
132a. Dolphin, A. C., and Scott, R. H. (1989) *J. Physiol.*, **413**, 271–288.
133. Dolphin, A. C., and Scott, R. H. (1988) *Trends Pharmacol. Sci.* **9**, 394–395.
134. Hymel, L., Streissnig, J., Glossman, H., and Schindler, H. (1988) *Proc. Natl. Acad. Sci. USA*, **85**, 4290–4294.
135. Bergamaschi, S., Govoni, S., Cominetti, P., Parenti, M., and Trabucchi, M. (1988) *Biochem. Biophys. Res. Comm.*, **56**, 1279–1286.
136. Horne, W. A., Abdel-Ghany, M., Racker, E., Wieland, G. A., Oswald, R. E., and Cerione, R. A. (1988) *Proc. Natl. Acad. Sci. USA*, **85**, 3718–3722.
137. Dolphin, A. C., and Prestwich, S. A. (1985) *Nature*, **316**, 148–150.
138. Ulivi, M., Wojcik, W., and Costa, E. (1988) *Soc. Neurosci. Abstr.*, **14**, 352.2.
139. Holz, G. G., Kream, R. M., Spiegel, A., and Dunlap, K. (1988) *J. Neurosci* (in press).
140. Kawata, K., and Normura, Y. (1987) *Neurosci. Res.*, **4**, 236–240.
141. Allgaier, C., Feuerstein, T. J., Jakisch, R., and Hertting, G. (1985) *Naunyn-Schmiedeberg's Arch. Pharmacol.*, **331**, 235–239.
142. Musgrave, I., Marley, P., and Majewski, H. (1987) *Naunyn-Schmiedeberg's Arch. Pharmacol.*, **336**, 280–286.
143. Fredholm, B. B., Fastbom, J., and Lindgren, E. (1986) *Acta Physiol. Scand.*, **127**, 381–386.
144. Ukena, D., Poeschla, E., Hüttemann, E., and Schwabe, U. (1984) *Naunyn-Schmiedeberg's Arch. Pharmacol.*, **327**, 247–253.
145. Tucker, J. F. (1984) *Brit J. Pharmacol.*, **83**, 326–328
146. Silinsky, E. M., Solsona, C., and Hirsh, J. K. (1989) *Brit. J. Pharmacol.*, **97**, 16–18.
147. Colmers, W. F., and Williams, J. T. (1988) *Neurosci. Letts.*, **93**, 300–306.
148. Harrison, N. L., Lange, G. D., and Barker, J. L. (1988) *Neurosci. Letts.*, **85**, 105–109.
149. Harrison, N. (1988) *Soc. Neurosci. Abstr.*, **14**, 439.9.
150. Dutar, P., and Nicoll, R. A. (1988) *Neuron*, **1**, 585–591.
151. Mattera, R., Codina, J., Sekura, R. D., and Birnbaumer, L. (1987). *J. Biol. Chem.*, **262**, 11247–11251.
152. Fesenko, E. E., Kolesnikov, S. S., and Lyubarsky, A. L. (1985) *Nature (London)*, **313**, 310–313.
153. Stryer, L. (1986) *Ann. Rev. Neurosci.*, **9**, 87–119.
154. Lamb, T. D., and Matthews, H. R. (1988) *J. Physiol.*, **407**, 463–487.
155. Haynes, L. W., and Yau, K-W (1986) *Nature*, **317**, 61–64.
156. Pace, U., Hanski, E., Salomon, Y., and Lancet, D. (1985) *Nature*, **316**, 255–258.
157. Nakamura, T., and Gold, G. H. (1987) *Nature*, **325**, 442–444.
158. Avenet, P., Hofmann, F., and Lindemann, B. (1988) *Nature (London)*, **331**, 351–354.
159. Whitway, M., Hougan, L., Dignard, D., Thomas, D. Y., Bell, L., Saari, G. C., Grant, F. J., O'Hara, P., and MacKay, V. L. (1989) *Cell*, **56**, 467–477.
160. Brabet, P., Dumuis, A., Sebben, M., Pantaloni, C., Bockaert, J., and Homburger, V. (1988) *J. Neurosci.*, **8**, 701–708.

10

INOSITOL PHOSPHATE METABOLISM AND G-PROTEINS

Irene Litosch

Department of Pharmacology, University of Miami School of Medicine, Miami, Florida 33101, USA

INTRODUCTION

A primary receptor-mediated event invoked upon stimulation of cells by a variety of hormones, neurotransmitters and growth factors is the activation of a phosphoinositide-specific phospholipase C. Hydrolysis of phosphoinositides by phospholipase C results in the generation of intracellular mediators which promote an increase in levels of cytosolic Ca^{2+} and an activation of protein kinase C. Phospholipase C, like the hormone-regulated adenylyl cyclase, appears to be subject to G-protein regulation. Two G-proteins regulate adenylyl cyclase activity: a stimulatory (G_s) and an inhibitory (G_i) protein. The identity of the G-protein (G_{PLC}) involved in phospholipase C regulation has not yet been established.

G-proteins are a subclass of cellular GTP binding proteins which include proteins involved in protein synthesis (EF-Tu, IF-2, eIF-2) and regulation of cytoskeletal assembly (tubulin). A major criterion which distinguishes G-proteins from other cellular GTP binding proteins is that the activity of G-proteins is directly regulated by cell surface receptors which respond to extracellular signals. G-proteins are oligomeric, consisting of α, β and γ subunits. The α subunits of G-proteins are unique and range in apparent molecular weight from 39 000 to 52 000, as determined by SDS–polyacrylamide gel electrophoresis (SDS–PAGE). The β subunits migrate as a doublet at 35/36 kDa on SDS–PAGE and are functionally interchangeable among the different G-proteins. Less is known about the γ subunits which have an apparent molecular weight of approximately 8000. The current status of the role of G-proteins and their mechanism of interaction with multiple systems has been recently reviewed [1].

The primary specificity for both the receptor and the target system resides in the α subunit. Receptor activation promotes a GTP and Mg^{2+}-dependent dissociation of the oligomer into a $\beta\gamma$ subunit complex and an activated α subunit. It is the α subunit that

---

G-Proteins as Mediators of Cellular Signalling Processes
Edited by Miles D. Houslay and Graeme Milligan. © 1990 John Wiley & Sons Ltd

interacts with specific target systems. Hydrolysis of GTP by the α subunit results in a deactivation of the α subunit, dissociation of α from the target enzyme, re-association of the α subunit with $\beta\gamma$ and a return to the inactive state. The $\beta\gamma$ subunits have a role in the anchoring of the α subunit [2], promotion of GDP exchange on the α subunit [3,4] and attenuation of α-mediated events [5-7] thereby conferring an additional level of control to the α-regulated system.

In addition to oligomeric G-proteins, smaller molecular weight GTP binding proteins are also present in cells. First isolated from placenta, these GTP binding proteins have an apparent molecular weight of 21 000 [8]. Studies have shown that cells contain many GTP binding proteins of this class [9]. Interest in these proteins arises from their similarity in size to ras proteins which are thought to be involved in the regulation of cellular transformation [10]. There is little evidence, however, that this class of low molecular weight GTP binding proteins is linked to transmembrane signalling processes or that they interact with $\beta\gamma$ subunits.

G-proteins of the $\alpha\beta\gamma$ subunit structure mediate agonist regulation of cGMP phosphodiesterase of rod outer segments [1,11] as well as Ca^{2+} and K^+ channels [12,13]. The major evidence in support of the involvement of a G-protein in the phospholipase C system has come from studies in membranes and permeabilized cells which have demonstrated a guanine nucleotide dependency for agonist stimulation of phospholipase C activity [14]. However, unlike the situation with adenylyl cyclase and cGMP phosphodiesterase, the identity and properties of the phospholipase C involved in transmembrane signalling are not known. It is little wonder, therefore, that the identity of G_{PLC} remains elusive. In addition, as will be discussed in the following sections, the phospholipase C system is complex. There appear to be multiple levels of regulation exerted on the system that may hinder identification of primary receptor-mediated events and thus identification of the key components of the system.

CHARACTERISTICS OF THE PHOSPHOINOSITIDE RESPONSE

Phosphoinositide hydrolysis in cells

The participation of phosphoinositides in receptor mechanism(s) linked to the elevation of cytosolic Ca^{2+} was first outlined by Michell's review in 1975 [15]. Michell noted that agonists which promoted an increase in cytosolic Ca^{2+} levels also stimulated a phospholipase C-mediated hydrolysis of phosphatidylinositol. The agonist-stimulated loss of phosphatidylinositol, in many cells, could not be induced by an increase in cytosolic Ca^{2+} and was relatively insensitive to omission of Ca^{2+} from the medium. These observations suggest that phosphatidylinositol hydrolysis was independent of the agonist-induced changes in Ca^{2+} levels. Phosphatidylinositol hydrolysis could therefore constitute part of a primary receptor mechanism involved in transmembrane signalling which led to an increase in Ca^{2+} levels. It was subsequently demonstrated that the initial event which occurred upon receptor activation was a rapid, phospholipase-C-mediated hydrolysis of phosphatidylinositol-4,5-bisphosphate, a phosphorylated derivative of phosphatidylinositol. Hydrolysis of phosphatidylinositol-4,5-bisphosphate and accumulation of water-soluble products could be detected within 15 s of agonist addition [15–22]. The connection between the increase in cytosolic Ca^{2+} and phosphoinositide hydrolysis was consolidated with the demonstration that inositol-1,4,5-trisphosphate, derived from phosphatidylinositol-4,5-bisphosphate hydrolysis, induced the release of Ca^{2+} from the endoplasmic reticulum, an intracellular store of cellular Ca^{2+} [23].

These findings are consistent with a scheme linking phospholipase C activation to a primary receptor event. However, the recent realization that many forms of phospholipase C are present in cells, together with the observed complexity of the inositol phosphate response observed in many cells, suggests that net measured cellular phospholipase C activity reflects multiple levels of regulation.

In many cells, a biphasic inositol phosphate response is observed in response to agonist stimulation. In guinea pig longitudinal smooth muscle, carbachol stimulated a rapid production of inositol-1,4,5-trisphosphate within 2 s [24]. The initial, agonist-stimulated increase in inositol-1,4,5-trisphosphate was followed by a decline in inositol-1,4,5-trisphosphate levels and then a rapid secondary sustained increase in inositol-1,4,5-trisphosphate production. A biphasic inositol phosphate response also occurs in response to thyrotropin releasing hormone (TRH) in GH_3 cells [25] and pituitary lactotroph cells [26], gonadotropin releasing hormone in pituitary cells [27], EGF in A431 cells [28], EGF and angiotensin II in hepatocytes [29] and α_1-adrenergic amine stimulation in rat aorta [30] to cite a few examples.

The initial, agonist-stimulated increase in inositol-1,4,5-trisphosphate is of sufficient rapidity to be consistent with this event being directly linked to receptor activation and thus functioning in the mechanism involved in the elevation of cytosolic Ca^{2+}. Analysis of the water-soluble products indicates that the major product is the Ca^{2+}-mobilizing isomer, inositol-1,4,5-trisphosphate [24,26,29]. The mechanism and physiological significance of the secondary increase in inositol-1,4,5-trisphosphate levels is not known. In rat pituitary lactotroph cells, omission of Ca^{2+} from the medium markedly reduced the secondary phase but had little effect on the initial increase in inositol phosphate production due to TRH [26]. In A431 cells, Ca^{2+} omission attenuated the secondary increase in inositol phosphates but not the initial increase due to EGF [28]. These results suggest that the magnitude of the secondary increase in inositol phosphates may be regulated by Ca^{2+}. However, in GH_3 cells, both the initial and secondary increase in inositol trisphosphate was relatively insensitive to Ca^{2+} omission from the medium [25]. The basis for the secondary increase in inositol phosphate production is thus unresolved. Ascertaining whether both phases of inositol phosphate production reflect activation of the same phospholipase C and/or hydrolysis of the same pool of phosphatidylinositol-4,5-bisphosphate may provide needed insight into this phenomenon.

In some cells, agonist-stimulated inositol-1,4,5-trisphosphate production lags behind the increase in cytosolic Ca^{2+}. An analysis of the time course for the production of inositol-1,4,5-trisphosphate in GH_4C_1 cells [31], BC3H-1 and 1321N1 cells [32] has shown that increases in inositol trisphosphate levels occurred in 30 s. However, increases in cytosolic Ca^{2+} levels were observed within 2 ms of agonist stimulation [31]. This disparity in the time course for inositol trisphosphate production and elevation in cytosolic Ca^{2+} suggests that cytosolic Ca^{2+} levels may increase through a mechanism independent of inositol-1,4,5-trisphosphate production, i.e. influx through receptor-regulated Ca^{2+} channels. The mechanism underlying the late activation of phospholipase C and the significance of this late increase in phospholipase C activity is therefore unclear.

A difficulty with these studies is that they rely on measurement of labelled inositol trisphosphate. The ability to detect inositol trisphosphate is critically dependent on the specific activity of the labelled phosphoinositide pool. Variability in the specific activity of inositol trisphosphate may partially account for the observed differences in the time course for production of this compound in various cells. Clearly, resolution of this issue will be of importance in furthering our understanding of the regulation of phospholipase C as well as the temporal relationship between phospholipase C activation and cellular activation.

An even more complex response occurs in rat pancreatic islet cells. Glucose and carbamylcholine stimulated a rapid increase in inositol-1,4,5-trisphosphate levels [33]. Removal of extracellular Ca^{2+} or treatment with verapamil, a blocker of voltage-gated Ca^{2+} channels, reduced the glucose-stimulated inositol-1,4,5-trisphosphate production but had little effect on the inositol-1,4,5-trisphosphate response due to carbamycholine. Depolarization by high K^+ also evoked a rapid increase in inositol-1,4,5-trisphosphate production suggesting that an increase in cellular Ca^{2+} may be sufficient to activate phospholipase C activity. These studies highlight the existence of multiple pathways for stimulation of phospholipase C activity. A key question which needs to be resolved concerns the identity and mechanism of activation of phospholipase C by these different agents.

Phospholipase C activity is also regulated by agonists which are not linked to the activation of phospholipase C. In rat anterior pituitary cells, dopamine inhibited the increase in inositol phosphate production in response to TRH stimulation [34,35]. Inhibition by dopamine was antagonized by incubation of cells in Ca^{2+}-free medium or treatment with verapamil [26]. These results suggested that dopamine inhibited TRH-stimulated inositol phosphate production through an inhibition of the Ca^{2+} influx which was necessary for induction or maintenance of the secondary phase of inositol phosphate production. Dopamine effects were inhibited by pertussis toxin treatment suggesting a G-protein involvement in dopamine action [26,34,35]. Interestingly, pertussis toxin did not inhibit TRH-stimulated inositol phosphate production suggesting that the mechanism involved in the activation of phospholipase C was insensitive to pertussis toxin. Similarly, excitatory amino acids inhibited carbachol and K^+-stimulated inositol phosphate production in hippocampus slices [36]. In these studies, inositol phosphate production was measured after a 30 min incubation and thus the time course for onset of inhibition is not known. In GH_3 cells, adenosine inhibited basal and TRH-stimulated inositol phosphate production [25]. Effects of adenosine were detected within 15 s and appeared to be insensitive to omission of Ca^{2+} from the medium. While these results are consistent with the existence of a primary receptor mechanism involved in the inhibition of phospholipase C activity, the effects of adenosine may be complex as evidenced by the stimulated synthesis of phosphoinositides which was observed in the presence of adenosine. Cumulatively, however, these studies indicate that net phospholipase C activity is inhibited by dopamine, excitatory amino acids and adenosine. However, the mechanism of inhibition remains to be determined.

Growth factors, EGF, PDGF, insulin and FGF potentiated the effects of thrombin in fibroblasts [37] while having little effect on inositol phosphate production per se. In Swiss 3T3 cells, a similar synergistic effect of insulin on bombesin action was shown [38]. The synergistic effect of growth factors on the agonist-stimulated production of inositol phosphates is of obvious interest. However, the mechanism underlying their action has not been ascertained. Growth factors may directly regulate phospholipase C activity, facilitate receptor and G-protein activation of phospholipase C or induce other cellular changes which amplify the initial agonist-stimulated increase in phospholipase C activity. The measured synergistic effect responses are relatively late (2 min) and the major effects are an increase in the rate and extent of production of inositol phosphates [37].

Involvement of G-proteins in agonist-stimulated phosphoinositide hydrolysis

Major evidence for the involvement of G-proteins in the regulation of phospholipase C activity in cells has come from studies with pertussis toxin. Pertussis toxin mediates the ADP-ribosylation and inactivation of G_i-like proteins. The site for pertussis toxin action is a cysteine, positioned four amino acids from the carboxyl terminus of the α subunit. ADP-ribosylation at this site inhibits agonist-promoted dissociation of the $\alpha\beta\gamma$ subunits [1]. Pertussis toxin has been a useful tool for the identification of G-protein-mediated events in studies on the regulation of adenylyl cyclase, cGMP phosphodiesterase and ion channels [1]. A summary of the results from numerous studies which have examined the effects of pertussis toxin treatment on agonist-stimulated inositol phosphate production is shown in Table 10.1. Pertussis toxin treatment attenuated fMet–Leu–Phe-stimulated inositol phosphate production in neutrophils [39–41], mast cells [42], human HL-60 leukaemic cells [43], EGF-stimulation in hepatocytes [29], thrombin stimulation in platelets [44], PAF stimulation in macrophages [45], bradykinin stimulation in Madin–Darby canine kidney cells [46] and α_1-adrenergic amine stimulation in adipocytes [47]. Pertussis toxin treatment, however, did not affect angiotensin II or vasopressin-stimulated inositol phosphate production in hepatocytes [29,48], α_1-adrenergic amine stimulation in brown fat cells [49], muscarinic stimulation in neuroblastoma N1E-115 cells [50], cholinergic stimulation in anterior pituitary cells [51], bradykinin stimulation in neuroblastoma × glioma hybrid NG108-15 cells [52], gonadotropin releasing hormone stimulation in pituitary gonadotrophs [53], EGF-stimulation in A431 cells [54], TRH stimulation in GH_3 cells [55] and muscarinic stimulation in 1321NI cells [56] to cite a few examples. The observation that pertussis toxin treatment inactivated the agonist-stimulated inositol phosphate response in some cells but not in others suggested that pertussis toxin effects might be cell specific. Thus, the success or failure of pertussis toxin

Table 10.1 Effect of pertussin toxin treatment on agonist-stimulated inositol phosphate production

Cell type	Pertussis toxin sensitivity	
	Sensitive	Insensitive
	Agonists	
neutrophil	fMet–Leu–Phe	
mast cell	fMet–Leu–Phe	
HL-60	fMet–Leu–Phe	
hepatocytes	EGF	angiotensin II, vasopressin
platelets	thrombin	
macrophage	platelet-activating factor	
MDCK cells	bradykinin	
adipocytes	α_1-adrenergic amine	
brown fat cells		α_1-adrenergic amine
NIE-115 cells		muscarinic
anterior pituitary cells		muscarinic
NG108-15 cells		bradykinin
pituitary gonadotrophs		GRH
A431 cells		EGF
Swiss 3T3 cells	bombesin, vasopressin	PDGF, FGF, PGF$_\alpha$
NIH 3T3 cells	bombesin	PDGF
GH$_3$ cells		TRH
132INI cells		muscarinic

Abbreviations: EGF, epidermal growth factor; PDGF platelet-derived growth factor; THR, thyrotropin releasing hormone; GRH, gonadotrophin releasing hormone; FGF, fibroblast growth factor.

treatment might reflect a cell-dependent phenomenon rather than the existence of two G-proteins [14]. However, in hepatocytes, pertussis toxin treatment inhibited EGF but not angiotensin II or vasopressin-stimulated inositol phosphate production [29,48]. In NIH 3T3 cells, bombesin-stimulated inositol phosphate production was unaffected by pertussis toxin treatment while that due to PDGF was inhibited [57]. In Swiss 3T3 cells, bombesin and vasopressin action was attenuated by pertussis toxin treatment while that due to PDGF, and PGF$_{2\alpha}$ was unaffected [58]. These studies thus demonstrated that in the same cell, pertussis toxin treatment inhibited the response to one agonist but not to another. Thus, differences in the ability of pertussis toxin to completely ADP-ribosylate G-proteins in different cells does not appear to explain all the negative findings.

These observations have led to the view that at least two G-proteins function in the phospholipase C system, i.e. a pertussis-toxin sensitive and a pertussis-toxin insensitive G-protein. There is little reason to exclude the possible involvement of two G-proteins in the stimulation of phospholipase C activity. However, G-protein utilization is clearly not receptor specific. Thus, EGF stimulation is pertussis-toxin sensitive in hepatocytes but pertussis-toxin insensitive in A431 cells. Similar considerations apply to bradykinin and vasopressin for which both pertussis toxin sensitivity and insensitivity have been reported. These data suggest that other factors may be responsible for some of the inconsistencies observed with pertussis toxin.

There are many potential difficulties with the use of pertussis toxin in cellular studies. Adequate documentation is needed to prove that all G-proteins were fully ADP-ribosylated. This has not been done in all studies. Pertussis toxin may not be transported across cell membranes with a similar efficiency in all cell types. The cellular levels of NAD may not be sufficient to support complete ADP-ribosylation of the G-proteins in question. Pertussis toxin may have effects other than G-protein inactivation. In Swiss 3T3 cells, pertussis toxin inhibited mitogenic stimulation of DNA synthesis without producing a corresponding inhibition in the ability of mitogens to activate phospholipase C [58]. In adrenal glomerulosa cells, pertussis toxin inhibited the angiotensin-II-induced increase in Ca^{2+} but had little detectable effect on the formation of inositol trisphosphate due to angiotensin II [59]. These results suggest that pertussis toxin has multiple sites of action within the cell and/or that pertussis toxin has multiple effects on G-proteins. Finally, pertussis toxin may affect the functioning of other GTP binding proteins which are indirectly linked to the regulation of phospholipase C activity. The recent realization that G-proteins are present in microsomal fractions may indicate a potential role of G-proteins in the regulation of Ca^{2+} release from intracellular stores [60,61]. In this context, studies on permeabilized rabbit main pulmonary artery have shown that hydrolysis-resistant guanine nucleotides stimulated the release of Ca^{2+} from intracellular storage sites, presumably sarcoplasmic reticulum [62]. These effects were antagonized by GDPβS and exhibited a high sensitivity for GTPγS. These results suggested that GTPγS may induce its effects through a regulatory GTP binding protein linked to the release of Ca^{2+} from intracellular sites. However, the presence of G-proteins in these subcellular fractions does not necessarily indicate that they have a physiological function and their involvement in cellular activation remains to be demonstrated.

In summary, the studies with pertussis toxin indicate that some agonists mediate their effects through a pertussis-toxin sensitive mechanism while other agonists utilize a pertussis-toxin insensitive mechanism. Whether these confusing results reflect the complexity of the phospholipase C system or the pertussis toxin remains to be determined.

STUDIES IN CELL-FREE SYSTEMS

Prelabelled membranes and permeabilized cells

Insight into the mechanism(s) underlying agonist stimulation of phospholipase C activity was derived from studies in cell-free systems in which it was demonstrated that guanine nucleotides were required for agonist stimulation of phospholipase C activity. The general approach for the study of phospholipase C regulation has involved the use of prelabelled membranes or permeabilized cells. Activation of phospholipase C by agonists is directly monitored by measurement of inositol phosphate production. Studies in permeabilized platelets first showed that GTPγS stimulated phospholipase C activity [63]. Similarly in neutrophil membranes, it was found that GTPγS and Gpp(NH)p stimulated phospholipase C activity. Half-maximal activation of phospholipase C activity occurred at 10 μM GTPγS and corresponded to a degradation of 10% of phosphatidylinositol-4,5-bisphosphate after a 10 min incubation at 37 °C in the presence of 100 nM Ca^{2+} [64].

Agonist-stimulated phospholipase C activity was first demonstrated in blowfly salivary gland membranes [65]. GTP, Gpp(NH)p and GTPγS promoted a dose-dependent stimulation of phospholipase C activity by 5-methyltryptamine, an analogue of 5-hydroxytryptamine which acts selectively on receptors coupled to the phospholipase C system rather than adenylyl cyclase. The effects of guanine nucleotides and agonist were synergistic. The rank order potency of guanine nucleotides in promoting agonist stimulation of phospholipase C activity was GTPγS > Gpp(NH)p ≫ GTP. Half-maximal activation of phospholipase C activity occurred at 0.1 μM GTPγS. The specificity for the guanine nucleotide effect, in conjunction with the demonstrated rank order of potency of guanine nucleotides in promoting agonist stimulation of phospholipase C activity, suggested that a regulatory G-protein functioned in the regulation of phospholipase C.

Studies by other laboratories using different membrane systems have subsequently also demonstrated that receptor activation of phospholipase C activity required the presence of guanine nucleotides [66–76]. Several general features of the response which are apparent from these studies are outlined below:

(1) The rank order of potency in promoting basal and agonist stimulation of phospholipase C has been GTPγS > Gpp(NH)p ≫ GTP. Other nucleotides are ineffective.
(2) Synergistic activation of phospholipase C is observed in the presence of agonist and guanine nucleotides.
(3) NaF activates basal phospholipase C activity.
(4) Optimum effects are observed in the presence of nanomolar concentrations of Ca^{2+} and at physiological pH.
(5) Activation by guanine nucleotides appears without an apparent lag in onset.
(6) The primary substrate hydrolysed is phosphatidylinositol-4,5-bisphosphate although some hydrolysis of phosphatidylinositol-4-phosphate is also observed.

The observed rank order of potency for guanine nucleotides to promote phospholipase C stimulation is similar to that observed in the adenylyl cyclase system [1] suggesting that a similar class of G-proteins is involved in the regulation of phospholipase C. The observed effects of guanine nucleotides with agonists are synergistic rather than additive. This is an important criterion since it suggests that both agonist and guanine nucleotides act through a shared transduction pathway rather than through separate mechanisms. NaF which has been used to study G_s regulation of adenylyl cyclase also activates phospholipase C, presumably through G_{PLC} [10,73,76]. The assay conditions which allow expression of guanine nucleotide stimulation have generally included Ca^{2+} which is needed to support the catalytic activity of the enzyme. A survey of the Ca^{2+} requirement for phospholipase C, however, indicates a considerable variability in various membrane preparations. In neutrophil membranes, Ca^{2+} concentrations less than 10 μM had little effect on basal phospholipase C activity. Incubation in the presence of fMet–Leu–Phe plus GTP lowered the Ca^{2+} requirement for phospholipase C activation to 0.1 μM [77]. In liver membranes, basal phospholipase C activity was activated at 0.1 μM Ca^{2+} [44]. Ca^{2+} activated basal phospholipase C at 33 nM in GH_3 cells [69,70] and 10 μM in cerebral cortical membranes [71]. In each case, Gpp(NH)p or GTPγS increased phospholipase C activity at a given Ca^{2+} concentration [69–71]. The range in Ca^{2+} concentrations required to activate phospholipase C in different membrane preparations may reflect the presence of different phospholipase C activities in these membranes and the loss of activators or inhibitors of phospholipase C. The apparent increase in Ca^{2+} sensitivity observed in the presence of guanine nucleotides has important regulatory consequences. It provides a mechanism for activation of phospholipase C at resting Ca^{2+} concentrations, an event which should occur if activation of phospholipase C activity preceded the increase in cytosolic Ca^{2+}. It remains to be established, however, whether the increased sensitivity to Ca^{2+} in the presence of guanine nucleotides actually reflects an increase in affinity for Ca^{2+}.

The role of Mg^{2+} in guanine nucleotide activation of phospholipase C is controversial. A fundamental property of the functionally characterized G-proteins is that Mg^{2+} is required for the activation process. This reflects the critical role of Mg^{2+} in promoting both the dissociation of the $\alpha\beta\gamma$ complex and the maintenance of the dissociated subunits [1]. Although Mg^{2+} has been included in membrane studies, conclusive studies on the role of Mg^{2+} in the activation of phospholipase C have not been carried out. In neutrophil membranes, Mg^{2+} was not required for GTPγS activation of phospholipase C activity [64]. A difficulty with assessing the role of Mg^{2+} in membrane preparations is that Mg^{2+} activates phosphatidylinositol-4,5-bisphosphate and phosphatidylinositol-4-phosphate phosphatase activity [78,79]. Activation of these lipid phosphatases results in the depletion of lipid substrate resulting in a corresponding decrease in apparent measured phospholipase C activity. Thus, the activation of a mutually conflicting process may preclude a thorough evaluation of the role of Mg^{2+} in the guanine nucleotide activation of phospholipase C in membranes. In cerebral cortical membranes [80] and liver membranes [81], Mg^{2+} was clearly required to promote guanine nucleotide stimulation of phospholipase C. Whether this Mg^{2+} effect was solely exerted on the activation of G_{PLC} or at some other site remains to be established.

Analysis of the time course for GTPγS and agonist-stimulated increases in phospholipase C activity have not, in general, demonstrated an appreciable lag in the onset of activation of phospholipase C as is characteristic of the adenylyl cyclase system. However, a lag of approximately 2 min in the onset of stimulation by 10 μM GTPγS in liver membranes has been demonstrated [81].

The primary substrate hydrolysed is phosphatidylinositol-4,5-bisphosphate although hydrolysis of phosphatidylinositol-4-phosphate may also occur. Little detectable hydrolysis of phosphatidylinositol is observed in response to agonist stimulation.

Exogenously added substrate

Studies with prelabelled membranes have provided considerable information on the mechanism of phospholipase C activation. One difficulty with the endogenous substrate approach is that some cells and tissues do not readily incorporate label into the phosphoinositides, precluding an analysis of the regulation of phospholipase C in these systems. An alternative method which has been used to measure phospholipase C activity utilizes exogenously added substrate. In this protocol, unlabelled membranes which contain receptors, G-proteins and phospholipase C are incubated with radiolabelled phosphatidylinositol-4,5-bisphosphate. Phospholipase C activity is directly measured by analysis of the production of water-soluble product in response to agonist stimulation. Stimulation of phospholipase C activity on exogenously added substrate has been observed in response to serotonin in blowfly salivary glands [82], carbachol in porcine coronary artery membranes [83], vasopressin in liver membranes [81], bombesin in MDCK membranes [46], thrombin in fibroblast membranes [84], guanine nucleotides in rat aorta membranes [85], guanine nucleotides in cerebral cortical membranes [80], and norepinephrine in DDT MF-2 membranes [86]. Guanine nucleotides were required for agonist activation of phospholipase C activity in these studies although guanine nucleotide selectivity was not demonstrated in all cases [46,82–84,86]. Ca^{2+} activated basal phospholipase C activity and guanine nucleotides lowered the Ca^{2+} required for a given degree of phospholipase C activation. The similarity of these results with those of prelabelled membranes suggested that exogenously added substrate might provide a complementary approach to the study of membrane phospholipase C regulation.

The nature of the interaction of substrate with membrane, however, has not been addressed in many cases. In some studies, the substrate was presented either as pure phosphatidylinositol-4,5-bisphosphate [80,82,83,85,86], in a mixed micelle [46,81,84] and in the absence or presence of trace amounts of detergent [81,86]. At least two possible mechanisms of interaction of membranes with substrate were possible. The membranes could hydrolyse both the endogenous unlabelled substrate and soluble exogenous substrate. Alternatively, the exogenous substrate might become incorporated into the membrane and serve as a native substrate for the enzyme. Studies on mitochondrial phospholipase A_2 have demonstrated that the exogenous substrate is incorporated into the membrane and that it is the incorporated substrate that is hydrolysed by phospholipase A_2 [87,88]. Similarly, hydrolysis of exogenously added ganglioside by a membrane sialidase occurs after the ganglioside micelle substrate has been associated into the membrane [89,90]. In the assay with cerebral cortical membranes, a major fraction of the added phosphatidylinositol-4,5-bisphosphate was rapidly associated with membranes. Measured phospholipase C activity was dependent on the degree of association with substrate indicating that the degree of substrate association with membrane could markedly affect the measured phospholipase C activity [91]. While these results did not rule out a possible concurrent hydrolysis of soluble substrate by the membrane, it was clear that a major fraction of the substrate became membrane associated. Thus, the use of exogenous substrate to assay regulation of membrane phospholipase C activity needs to be done with careful consideration as to the possible effects of incubation conditions on the partitioning of substrate to the membrane. Increases in phospholipase C activity in response to an agent could simply reflect an enhanced accessibility of membrane phospholipase C to substrate if substrate binding is not at steady state at the time of the assay.

Guanine nucleotide activation of soluble phospholipase C

Phospholipase C activity, present in the supernatant of disrupted cells is activated by guanine nucleotides [92,93]. In platelets, Gpp(NH)p, GTPγS and GTP stimulated soluble phospholipase C activity [93]. The potency of Gpp(NH)p, GTPγS and GTP in promoting stimulation of phospholipase C was similar. The pH optimum for guanine nucleotide effects was 5.5. These results were taken to indicate the involvement of a cytosolic G-protein in the regulation of cytosolic phospholipase C activity. However, there are important differences between the studies on supernatant and membranes. In cerebral cortical membranes, Gpp(NH)p activated phospholipase C activity while GTP was ineffective indicating a selectivity for Gpp(NH)p versus GTP [80]. Gpp(NH)p stimulation was optimum at 6.75 with little effect at pH 5.5. Studies in prelabelled membranes have also demonstrated that the pH optimum for guanine nucleotide activation of phospholipase C activity is approximately 7.0 [55,70]. These results suggest that enzyme activity in membranes is different from the supernatant activity. Therefore, it remains to be established whether the stimulatory effect of guanine nucleotides observed in supernatant fractions is indicative of a G-protein-mediated regulation. However, these studies have led to the view that the cytosolic phospholipase C is the enzyme which is regulated by receptor activation. This has yet to be proved.

PROPERTIES OF PURIFIED PHOSPHOLIPASE C

The intense interest in the mechanism underlying agonist stimulation of phospholipase C activity has prompted a massive effort by many laboratories to purify phospholipase C. Soluble phospholipase C has been purified to apparent homogeneity from several tissues. In general, the enzyme(s) are activated at micromolar Ca^{2+} concentrations, hydrolyse all the phosphoinositides with a preference for phosphatidylinositol-4,5-bisphosphate and exhibit a pH optimum at neutral pH in the presence of deoxycholate. The use of deoxycholate in most assay conditions is unfortunate since deoxycholate will shift the apparent pH optimum of the enzyme from pH 5 to pH 7 [94]. As summarized in Table 10.2, a number of phospholipase C activities are present in the cytosol of cells. The specific activities of the purified enzymes are generally similar despite the use of different assay conditions.

Table 10.2 Properties of phospholipase C purified to homogeneity

Source	Designation	Apparent molecular weight (SDS–PAGE)	Specific activity	Assay condition
Soluble or salt extracts of membranes				
sheep seminal vesicles	PLC-I PLC-II	65 000	28 µmol/min/mg	250 µM PI 1 mg/ml DOC 1 mM $CaCl_2$ 0.5 mg/ml BSA 100 mM NaCl 50 mM HEPES (pH 7.0)
bovine brain	PLC-I	150 000	13 µmol/min/mg	300 µM soybean PI 0.1% DOC
	PLC-II	145 000	24 µmol/min/mg	3 mM EGTA, 1 mM $CaCl_2$ 50 mM HEPES (pH 7.0)
	PLC-III	85 000		
		88 000	57 µmol/min/mg	100 µM PIP_2 1 mg/ml DOC 0.1 mM DTT 50 mM sodium phosphate (pH 6.8)

Source	Enzyme			Conditions
rat brain	PLC-II	85 000	15 μmol/min/mg	100–500 μM PIP_2, 50 μM PE
	PLC-III	85 000	13 μmol/min/mg	200 μM KCl
				1 mg/ml BSA
				50 mM MES buffer (pH 6.8)
guinea pig	PI-PLC-I	62 000	.7 μmol/min/mg	10 μM PI
uterus	PI-PLC-II			2.4 mM DOC
				1 mM $CaCl_2$
				50 mM Tris (pH 7.0)
liver		87 000	407 μmol/min/mg	17 μM PIP_2
				2.4 mM DOC
				300 μM $CaCl_2$, 100 μM EGTA
				180 mM NaCl
				50 mM HEPES (pH 7.0)
Detergent-solubilized membranes				
platelet	mPLC-I			200 μM PIP_2
membrane	mPLC-II	61 000	6 μmol/min/mg	1 mg/ml DOC
				80 mM KCl, 10 μM Ca^{2+}
				unspecified cholate
				50 mM Tris-maleate (pH 6.75)

Abbreviations: DOC, deoxycholate; PI, phosphatidylinositol; PIP_2, phosphatidylinositol-4,5-bisphosphate

Studies on sheep seminal vesicles were the first to document the existence of at least two forms of soluble phospholipase C. PLC-I has an apparent molecular weight of 65 000 [94]. PLC-I appears immunologically distinct from PLC-II as determined by immunoprecipitation studies with antibodies directed to each enzyme. However, distinct biochemical differences between the two forms have not been documented.

Soluble phospholipase C has been purified from bovine brain. At least three activities are present with apparent molecular weights of 150 000, 145 000 [95,96] 85 000 [92,96] and 88 000 [97]. The activity of the 145 kDa protein was increased in the presence of 1 mM GTP, ATP and GTPγS by approximately 200% whereas Gpp(NH)p was ineffective. The significance of the stimulation by these high concentrations of guanine nucleotides is not readily apparent. Stimulatory effects of nucleotides have not been reported in other phospholipase C preparations. The enzyme has been sequenced and cloned. An amino acid sequence similar to the products of tyrosine kinase-related oncogenes was noted. The hydropathic profile of the amino acid sequence did not show appreciable hydrophobic segments which are shared by other transmembrane proteins suggesting that this enzyme was not an integral component of the membrane [98]. A 148 kDa phospholipase C has been independently isolated and cloned [99]. It appears to correspond to the 145 kDa protein. The 148 kDa phospholipase C has significant homology with non-receptor protein kinases. This led to considerable speculation that the enzyme was regulated by phosphorylation. The clone was expressed in transfected COS-1 cells [100]. Activity against phosphatidylinositol was detected but there has been little information as to the Ca^{2+} requirement of the expressed enzyme or its substrate specificity.

Two forms of soluble phospholipase C (PLC-II and PLC-III) are present in rat brain cytosol [101]. Both enzyme activities were purified to apparent homogeneity and have apparent molecular weights of 85 000. PLC-II and PLC-III appeared to be immunologically distinct from each other and exhibited different substrate specificities. PLC-II hydrolysed phosphatidylinositol, phosphatidylinositol-4-phosphate and phosphatidylinositol-4,5-bisphosphate with similar efficiency while PLC-III had little activity on phosphatidylinositol. The mechanism conferring this substrate selectivity is not known.

Guinea pig uterine cytosol contains two forms of phospholipase C. One form (PI-PLC I) was purified to apparent homogeneity and has an apparent molecular weight of 62 000 [102]. Subcellular distribution studies with polyclonal antibodies raised to the purified enzyme indicated that 25% of the total immunoreactive material was present in unfractionated membranes obtained from this tissue, suggesting that this enzyme was membrane associated and possibly involved in transmembrane signalling. These studies with antibodies, however, did not exclude the possible presence of an immunologically distinct membrane phospholipase C activity or the possibility that antibody reacted with cytosolic enzyme adsorbed to the membrane. An interesting aspect of this study was the observation that stimulation of RBL cells with 1 μM phorbol myristic acid promoted the phosphorylation of a 62 kDa protein. Antibody to purified phospholipase C recognized the phosphoprotein on Western blots suggesting that phospholipase C was phosphorylated. However, there is little evidence that the activity of the phospholipase C was affected by phosphorylation. The 62 kDa enzyme has been cloned and the amino acid sequence determined. The amino acid sequence revealed potential serine and threonine phosphorylation sites but no consensus sequence for tyrosine phosphorylation or N-linked glycosylation. The protein appeared predominantly hydrophilic although two hydrophobic domains were evident. The sequence of PLC-II (62 kDa) was dissimilar to that of PLC-II (145 kDa) suggesting that these enzymes are distinct. This is consistent with lack of immunocrossreactivity between the two enzymes [103].

Phospholipase C in liver cytosol has an apparent molecular weight of 87 000 [104]. Basic proteins activate the enzyme. The significance of this activation is unclear.

Phospholipase C (mPLC-I and mPLC-II) is present in cholate extracts obtained from KCl-washed platelet membranes [105,106]. mPLC-II has an apparent molecular weight of 61 000. The detergent-extracted enzyme was activated at micromolar Ca^{2+} concentrations. Chelators such as EDTA and EGTA inhibited phospholipase C activity. Mg^{2+} in the micromolar range did not stimulate phospholipase C activity. Nucleotides such as GTP, GDP, App(NH)p did not affect phospholipase C activity. Whether mPLC-II represents a true integral protein remains to be determined.

This brief discussion of the status of current phospholipase C research highlights the complexity of this enzyme system. At this time, it is not known which activity, if any, is involved in the initial receptor-activation process. This awaits the successful demonstration of regulation by a G-protein in a reconstituted system.

G-PROTEIN INVOLVEMENT IN REGULATION OF PHOSPHOLIPASE C: STUDIES IN RECONSTITUTED SYSTEMS

The identity of the putative G-protein(s) involved in the regulation of phospholipase C has not been established. Reconstitution studies have not, so far, resulted in tremendous insight into the regulation of this system. The first published reconstitution study was done on differentiated human leukaemic HL-60 cells. In these cells, pertussis toxin treatment markedly reduced fMet–Leu–Phe-stimulated inositol phosphate production [107]. Addition of purified G_i and G_o to membranes derived from pertussis toxin-treated cells restored fMet–Leu–Phe sensitivity. G_i and G_o were equipotent in their effects. Prior ADP-ribosylation of G_i/G_o with pertussis toxin resulted in a loss in the ability of G_i/G_o to restore fMet–Leu–Phe sensitivity. While these studies suggested that G_i and G_o might function in the regulation of phospholipase C activity, other G proteins and resolved α subunits were not tested. Thus it is unclear as to which component in the G_i/G_o preparation was responsible for restoration of the fMet–Leu–Phe response.

In *Xenopus* oocytes, injection of $\beta\gamma$ subunits blocked muscarinic stimulation of Cl^- current [108]. Since Cl^- channel activation is an inositol-1,4,5-trisphosphate-mediated event, it was concluded that $\beta\gamma$ subunits inhibited the receptor-regulated G-protein activation of phospholipase C. Although these results are intriguing, the measured effects were indirect. Thus, the status of $\beta\gamma$ subunits in the regulation of phospholipase C remains unclear.

G_i and G_o stimulated phospholipase C activity in a partially purified phospholipase C preparation from platelet membranes [105]. A similar reconstitution experiment with the enzyme which was purified to homogeneity was unsuccessful [106]. This suggests that loss of a crucial component conferring guanine nucleotide sensitivity to phospholipase C may have occurred during the purification.

Reconstitution of GTPγS-stimulated activity in supernatant-derived phospholipase C has been reported [109]. In these studies a soluble phospholipase C obtained from thymocytes was partially purified. The addition of column fractions containing a 27 kDa GTPγS binding protein conferred GTPγS sensitivity to phospholipase C. The effects of other nucleotides were not examined to ascertain guanine nucleotide specificity or the dose dependency of the reconstitution effect. Thus it remains to be demonstrated as to whether the 27 kDa protein confers the appropriate sensitivity and rank order potency for guanine nucleotide activation of phospholipase C. In a similar type of study, GTPγS regulation of platelet-derived soluble phospholipase C could be restored upon addition of fractions containing GTPγS binding activity [110]. A GTP binding protein of 29 kDa was identified as being responsible for conferring GTPγS sensitivity to the phospholipase C. Similar considerations apply here as in the thymocyte studies. Guanine nucleotide specificity was not demonstrated. Furthermore, it is unclear as to how these low molecular weight GTP binding proteins, which have not yet been demonstrated to interact with receptors, could mediate transmembrane activation of phospholipase C.

CONCLUSIONS

It should be evident from this brief discussion that there are still many unexplored and poorly defined areas in the area of phosphoinositide metabolism. The phosphoinositide response is clearly a complex signal transduction pathway. The characteristics of the response, as measured by net inositol phosphate production, vary in different cell types and in response to different activators within the same cell type.

A summary of the regulation of this system is outlined in Figure 10.1. The primary receptor-mediated event involves stimulation of phospholipase C by the putative regulatory G-protein (G_{PLC}). Activation of phospholipase C may occur as a consequence of a conformational change in the enzyme that lowers the Ca^{2+} concentration required for enzyme activation. The activated membrane-associated phospholipase C hydrolyses phosphatidylinositol-4,5-bisphosphate resulting in the generation of Ins-1,4,5-P_3 and diacylglycerol. Ins-1,4,5-P_3 stimulates the release of Ca^{2+} from the endoplasmic reticulum. This increase in cytosolic Ca^{2+} levels may increase net inositol phosphate production through either activation of cytosolic phospholipase C activities or activation of the membrane-associated phospholipase C activity or both. Influx of Ca^{2+} also occurs during agonist stimulation. Inositol-1,3,4,5-tetrakisphosphate, a phosphorylated derivative of inositol-1,4,5-trisphosphate appears to have a role in mediating Ca^{2+} influx [111]. The increase in Ca^{2+} levels resulting from influx may further stimulate phospholipase C

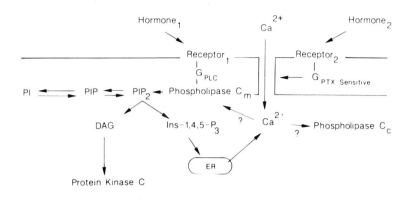

Abbreviations:

PI	phosphatidylinositol
PIP	phosphatidylinositol-4-phosphate
PIP_2	phosphatidylinositol-4,5-bisphosphate
G_{PLC}	The putative G-protein mediating receptor activation of phospholipase C
phospholipase C_m	membrane associated phospholipase C
phospholipase C_c	cytosolic phospholipase C
Ins-1,4,5-P_3	inositol-1,4,5-trisphosphate
DAG	diacylglycerol
ER	endoplasmic reticulum

Figure 10.1 Events involved in the activation of the phosphoinositide-specific phospholipase C system

activity. Protein kinase C may also regulate phospholipase C activity through phosphorylation, however direct evidence for this is lacking.

Activation of other receptors, dopamine-like, may inhibit net phospholipase C activation through a pertussis-toxin-sensitive G-protein linked to the inhibition of Ca^{2+} influx. In addition, phospholipase C activity may be subject to an inhibitory G-protein-mediated regulation [112]. As has been described for adenylyl cyclase, growth factors may potentiate net measured phospholipase C activity through an as yet undetermined mechanism. Clearly, the more complicated the cell, the more difficult it may be to identify the primary receptor event.

The potential involvement of G-proteins in the receptor-mediated activation of phospholipase C is intriguing. The identity of this protein remains elusive despite the intense research effort in this area. The lack of definitive reconstitution studies may reflect several possibilities. The studies are not being conducted on the appropriate phospholipase C, i.e. it is clear that cells contain many phospholipase C activities. One would presume that a specific phospholipase C would be activated upon receptor activation. However a priori, there is little reason to make this assumption. Identification and isolation of the appropriate enzyme may be difficult if it is present in low quantities or unstable.

It is possible that G_{PLC} is not similar to the functionally characterized G-proteins which regulate adenylyl cyclase and thus may not be of the $\alpha\beta\gamma$ subunit structure. Other proteins may be essential for activation of G_{PLC} in reconstituted systems. The participation of protein factors in the regulation of G-proteins and ras proteins has been demonstrated. Efficient ADP-ribosylation of G_s by cholera toxin requires ARF, ADP-ribosylation factor which is present in cytosolic and membrane fractions [113,114]. GAP protein regulates the GTPase activity of ras proteins [115]. Since the GTPase activity determines the degree of activation of GTP binding proteins, GAP may have an important regulatory function. Whether a similar protein functions in the phospholipase C system remains to be determined.

Given the complexity of both the phospholipase C system and G-proteins, it is not surprising that a considerable amount of basic information may still be required to achieve a successful reconstitution. Achievement of this goal will not only increase our understanding of the phospholipase C system but will also expand our appreciation of the mechanism of G-protein target system interactions.

ACKNOWLEDGEMENTS

Supported by an Established Investigatorship of The American Heart Association, funds contributed by the American Heart Association, Florida Affiliate and NIH DK 37007.

I acknowledge the excellent secretarial skill of Mrs Gerry Trebilcock in the preparation of this manuscript.

REFERENCES

1. Gilman, A. G. (1987) *Ann. Rev. Biochem.*, **56**, 615–649.
2. Sternweis, P. C. (1986) *J. Biol. Chem.*, **261**, 631–637.
3. Fung, B. K.-K (1983) *J. Biol. Chem.*, **258**, 10495–10503.
4. Correze, C., d'Alayer, J., Coussen, F., Berthiller, G., Deterre, P., and Monneron, A. (1987) *J. Biol. Chem.*, **262**, 15182–15187.
5. Katada, T., Northup, J. K., Bokoch, G. M., Ui, M., and Gilman, A. G. (1984) *J. Biol. Chem.*, **259**, 3578–3585.
6. Katada, T., Bokoch, G. M., Northup, J. K., Ui, M., and Gilman, A. G. (1984) *J. Biol. Chem.*, **259**, 3568–3577.

7. Cerione, R. A., Codina, J., Kilpatrick, B. F., Staniszewski, C., Gierschik, P., Somers, R. L., Spiegel, A. M., Birnbaumer, L., Caron, M. G., and Lefkowtiz, R. J. (1985) *Biochemistry*, **24**, 4499–4503.
8. Evans, T., Brown, M. L., Fraser, E. D., and Northup, J. K. (1986) *J. Biol. Chem.*, **261**, 7052–7059.
9. Kikuchi, A., Yamashita, T., Kawata, M., Yamamoto, K., Ikeda, K., Tanimoto, T., and Takai, Y. (1988) *J. Biol. Chem.*, **263**, 2897–2904.
10. Gibbs, J. B., Sigal, I., and Scolnick, E. M. (1985) *Trends Biochem. Sci.*, **10**, 350–353.
11. Stryer, L., Hurley, J. B., and Fung, B. K.-K. (1981) *Curr. Topics Mem. Transport*, **15**, 93–108.
12. Brown, A. M., and Birnbaumer, L. (1988) *Amer. J. Physiol.*, **254**, H401–H410.
13. Rosenthal, W., Hescheler, J., Trautwein, W., and Shultz, G. (1988) *FASEB J.*, **2**, 2784–2790.
14. Litosch, I. (1987) *Life Sci.*, **41**, 251–258.
15. Michell, R. H. (1975) *Biochim. Biophys. Acta*, **415**, 81–147.
16. Kirk, C. J., Creba, J. A., Downes, C. P., and Michell, R. H. (1981) *Biochem. Soc. Trans.*, **9**, 377–379.
17. Rhodes, D., Prpic, V., Exton, J. H., and Blackmore, P. F. (1983) *J. Biol. Chem.*, **258**, 2770–2773.
18. Litosch, I., Lin, S.-H., and Fain, J. N. (1983) *J. Biol. Chem.*, **258**, 13727–13732.
19. Weiss, S. J., McKinney, J. S., and Putney, J. W. (1982) *Biochem. J.*, **206**, 555–560.
20. Rebecchi, M. J., and Gerschengorn, M. C. (1983) *Biochem. J.*, **216**, 287–294.
21. Berridge, M. J. (1983) *Biochem. J.*, **212**, 849–858.
22. Litosch, I., Lee, H. S., and Fain, J. N. (1984) *Amer. J. Physiol.*, **246**, C141–C147.
23. Streb, H., Irvine, R. F., Berridge, M. J., and Schultz, I. (1983) *Nature (London)*, **306**, 67–69.
24. Salmon, D. M., and Bolton, T. B. (1988) *Biochem. J.*, **254**, 553–557.
25. Delahunty, T. M., Cronin, M. J., and Linden, J. L. (1988) *Biochem. J.*, **255**, 69–77.
26. Vallar, L., Vicentini, L. M., and Meldolesi, J. (1988) *J. Biol. Chem.*, **263**, 10127–10134.
27. Balla, T., Baukal, A. J., Guillemette, G., and Catt, K. J. (1988) *J. Biol. Chem.*, **263**, 4083–4091.
28. Wahl, M. and Carpenter, G. (1988) *J. Biol. Chem.*, **263**, 7581–7590.
29. Johnson, R. M., and Garrison, J. C. (1987) *J. Biol. Chem.*, **262**, 17285–17293.
30. Pijuan, V., and Litosch, I. (1988) *Biochem. Biophys. Res. Commun.*, **156**, 240–245.
31. Tashjian, A. H., Jr, Heslop, J. P., and Berridge, M. J. (1987) *Biochem. J.*, **243**, 305–308.
32. Ambler, S. K., Thompson, B., Solski, P. A., Brown, J. H., and Taylor, P. (1987) *Mol. Pharmacol.*, **32**, 376–383.
33. Biden, T. J., Peter-Reisch, B., Schlegel, W., and Wollheim, C. B. (1987) *J. Biol. Chem.*, **262**, 3567–3571.
34. Journot, L., Homburger, V., Pantaloni. C., Priam, M., Bockaert, J., and Enjalbert, A. (1987) *J. Biol. Chem.*, **262**, 15106–15110.
35. Enjalbert, A., Sladeczek, F., Guillon. G., Bertrand, P., Shu, C., Epelbaum, J., Garcia-Sainz, A., Jard, S., Lombard, C., Kordon, C., and Bockaert, J. (1986) *J. Biol. Chem.*, **261**, 4071–4075.
36. Baudry, M., Evans, J., and Lynch, G. (1986) *Nature (London)*, **319**, 329–331.
37. Paris, S., Chambard, J. C., and Pouyssegur, J. (1988) *J. Biol. Chem.*, **263**, 12893–12900.
38. Heslop, J. P., Blakeley, D. M., Brown, K. D., Irvine, R. F., and Berridge, M. J. (1986) *Cell*, **47**, 703–709.
39. Bradford, P. G., and Rubin, R. P. (1985) *FEBS Lett.*, **183**, 317–320.
40. Shefcyk, J., Yassin, R., Volpi, T. F. P., Molski, P. H., Naccache, J. J., Munoz, E. L., Becker, M. B., Feinstein, M. B., and Shaafi, R. I. (1985) *Biochem. Biophys. Res. Commun.*, **126**, 1174–1181.
41. Verghese, M. W., Smith, C. D., and Synderman, R. (1985) *Biochem. Biophys. Res. Commun.*, **127**, 450–457.
42. Ohta, H., Okajima, F., and Ui, M. (1985) *J. Biol. Chem.*, **260**, 15771–15780.
43. Brandt, S. J., Dougherty, R. W., Lapetina, E. G., and Niedel, J. E. (1985) *Proc. Natl. Acad. Sci. USA*, **82**, 3277–3280.

44. Brass, L. F., Laposata, M., Banga, H. S., and Rittenhouse, S. E. (1986) *J. Biol. Chem.*, **261**, 16838–16847.
45. Huang, S. J., Monk, P. N., Downes, C. P., and Whetton, A. D. (1988) *Biochem. J.*, **249**, 839–845.
46. Portilla, D., Morrissey, J., and Morrison, A. R. (1988) *J. Clin. Invest.*, **81**, 1896–1902.
47. Rapiejko, P. J., Northup, J. K., Evans, T., Brown, J., and Malbon, C. (1986) *Biochem. J.*, **240**, 35–40.
48. Uhing, R. J., Prpic, V., Jiang, H., and Exton, J. H. (1986) *J. Biol. Chem.*, **261**, 2140–2146.
49. Schimmel, R. J., and Elliott, M. E. (1986) *Biochem. Biophys. Res. Commun.*, **135**, 823–829.
50. Martin, J. M., Liles, W. C., and Nathanson, N. M. (1988) *Brain Res.*, **455**, 370–376.
51. Schrey, M. P., and Read, A. M. (1988) *Mol. Cell. Endocrinol.*, **57**, 107–113.
52. Osugi, T., Imaizumi, T., Mizushima, A., Uchida, S., and Yoshida, H. (1987) *Eur. J. Pharmacol.*, **137**, 207–218.
53. Naor, Z., Azrad, A., Limor, R., Zakut, H., and Lotan, M. (1986) *J. Biol. Chem.*, **261**, 12506–12512.
54. Pike, L. J., and Eakes, A. T. (1987) *J. Biol. Chem*, **262**, 1644–1651.
55. Martin, T. F. J., Lucas, D. O., Bajjalieh, S. M., and Kowalchuk, J. A. (1986) *J. Biol. Chem.*, **261**, 2918–2927.
56. Masters, S. B., Martin, M. W., Harden, T. K., and Brown, J. H. (1985) *Biochem. J.*, **227**, 933–937.
57. Hoshijima, M., Ueda, T., Hamamori, Y., Ohmori, T., and Takai, Y. (1988) *Biochem. Biophys. Res. Commun.*, **152**, 285–293.
58. Taylor, C. W., Blakeley, D. M., Corps, A. N., Berridge, M. J., and Brown, K. D. (1988) *Biochem. J.*, **249**, 917–920.
59. Kojima, I., Shibata, H., and Ogata, E. (1986) *FEBS Lett.*, **204**, 347–351.
60. Scherer, N. M., Toro, M.-J., Entman, M. L., and Birnbaumer, L. (1987) *Arch. Biochem. Biophys.*, **259**, 431–440.
61. Codina, J., and Kraus-Friedman, N. (1988) *Biochem. Biophys. Res. Commun.*, **150**, 848–852.
62. Kobayashi, S., Somylo, A. P., and Somylo, A. V. (1988) *J. Physiol.*, **403**, 601–619.
63. Haslam, R. J., and Davidson, M. M. L. (1984) *J. Receptor Res.*, **4**, 605–629.
64. Cockroft, S., and Gomperts, B. D. (1985) *Nature (London)*, **314**, 534–536.
65. Litosch, I., Wallis, C. W., and Fain, J. N. (1985) *J. Biol. Chem.*, **260**, 5464–5471.
66. Wallace, M. A., and Fain, J. N. (1985) *J. Biol. Chem.*, **260**, 9527–9530.
67. Smith, C. D., Lane, B. C., Kusaka, I., Verghese, M. W., and Snyderman, R. (1985) *J. Biol. Chem.*, **260**, 5875–5878.
68. Uhing, R. J., Prpic, V., Jiang, H., and Exton, J. H. (1986) *J. Biol. Chem.*, **261**, 2140–2146.
69. Lucas, D. O., Bajjalieh, S. M., Kowalchyk, J. A., and Martin, T. F. J. (1985) *Biochem. Biophys. Res. Commun.*, **132**, 721–720.
70. Straub, R. E., and Gershengorn, M. C. (1986) *J. Biol. Chem.*, **261**, 2712–2717.
71. Gonzales, R. A., and Crews, F. T. (1985) *Biochem. J.*, **232**, 799–804.
72. Kikuchi, A. K., Kozo, K., Katada, T., Ui, M., and Takai, Y. (1986) *J. Biol. Chem.*, **261**, 11558–11562.
73. Taylor, C. W., Merritt, J. E., Putney, J. W. Jr, and Rubin, R. P. (1986) *Biochem. Biophys. Res. Commun.*, **136**, 362–368.
74. Guillon, G., Balestre, M.-N., Mouillac, B., and Devilliers, G. (1986) *FEBS Lett.*, **196**, 155–159.
75. Hrbolich, J. K., Culty, M., and Haslam, R. J. (1987) *Biochem. J.*, **243**, 457–465.
76. Harden, T. K., Stephens, L., Hawkins, P. T., and Downes, C. P. (1987) *J. Biol. Chem.*, **262**, 9057–9061.
77. Smith, C. D., Cox, C. C., and Snyderman, R. (1986) *Science*, **231**, 79–99.
78. Dawson, R. M. C., and Thompson, W. (1964) *Biochem. J.*, **91**, 244–250.
79. Irvine, R. F., Letcher, A. J., and Dawson, R. M. C. (1984) *Biochem. J.*, **218**, 177–185.

80. Litosch, I. (1987) *Biochem. J.*, **244**, 35–40.
81. Taylor, S. J., and Exton, J. H. (1987) *Biochem. J.*, **248**, 791–799.
82. Litosch, I., and Fain, J. N. (1985) *J. Biol. Chem.*, **260**, 16052–16055.
83. Sasaguri, T., Hirata, M., and Kuriyama, H. (1985) *Biochem J.*, **231**, 497–503.
84. Magnaldo, D., Morrissey, J., and Morrison, A. R. (1988) *J. Clin. Invest.*, **81**, 1896–1902.
85. Roth, B. L. (1987) *Life Sci.*, **41**, 629–634.
86. Fulle, H.-J., Hoer, R. D., Lache, W., Rosenthal, W., Schultz, G., and Oberdisse, E. (1987) *Biochem. Biophys. Res. Commun.*, **145**, 673—679.
87. de Winter, J. N., Lenting, H. B. M., Neys, F. W., and van den Bosch, H. (1987) *Biochim. Biophys. Acta*, **917**, 169–177.
88. Lenting, H. B. M., Neys, F. W., and van den Bosch, H. (1987) *Biochim. Biophys. Acta*, **917**, 178–185.
89. Scheel, G., Acevedo, E., Conzelmann, E., Nehrkorn, H., and Sandhoff, K. (1982) *Eur. J. Biochem.*, **127**, 245–253.
90. Scheel, G., Schwarzmann, G., Hoffman-Bleihauer, P., and Sandhoff, K. (1985) *Eur. J. Biochem.*, **153**, 29–35.
91. Litosch, I. (1989) *Biochem. J.*, **261**, 325–331.
92. Baldassare, J. J., and Fisher, G. J. (1986) *Biochem. Biophys. Res. Commun.*, **137**, 801–805.
93. Deckmyn, H., Tu, S-M., and Majerus, P. W. (1986) *J. Biol. Chem.*, **261**, 16553–16558.
94. Hofmann, S. L., and Majerus, P. W. (1982) *J. Biol. Chem.*, **257**, 6461–6469.
95. Ryu, S. H., Cho, K. S., Lee, K. Y., Suh, P. G., and Rhee, S. G. (1987) *J. Biol. Chem.*, **262**, 12511–12518.
96. Ryu, S. H., Suh, P. G., Cho, K. S., Lee, K.-Y., Rhee, S. G. (1987) *Proc. Natl. Acad. Sci. USA*, **84**, 6649–6653.
97. Rebecchi, M. J., and Rosen, O. M. (1987) *J. Biol. Chem.*, **262**, 12526–12532.
98. Suh, P. G., Ryu, S. H., Moon, K. H., Suh, W. S., and Rhee, S. G. (1988) *Proc. Natl. Acad. Sci. USA*, **85**, 5419–5423.
99. Katan, M., and Parker, P. J. (1987) *Eur. J. Biochem.*, **168**, 413–418.
100. Stahl, M. L., Ferenz, C. R., Kelleher, K. L., Kriz, R. W., and Knopf, J. L. (1988) *Nature (London)*, **322**, 269–275.
101. Homma, Y., Imaki, J., Nakanishi, O., and Takenawa, T. (1988) *J. Biol. Chem.*, **263**, 6592–6598.
102. Bennett, C. F., and Crooke, S. T. (1987) *J. Biol. Chem.*, **262**, 13789–13797.
103. Bennett, C. F., Balcarek, J. M., Varrichio, A., and Crooke, S. T. (1988) *Nature (London)*, **334**, 268–270.
104. Fukui, T., Lutz, R. J., and Lowenstein, J. M. (1988) *J. Biol. Chem.*, **263**, 17730–17737.
105. Banno, Y., Nagao, S., Katada, T., Nagata, K., Ui, M., and Nozawa, Y. (1987) *Biochem. Biophys. Res. Commun.*, **146**, 861–869.
106. Banno, Y., Yada, Y., and Nozawa, Y. (1988) *J. Biol. Chem.*, **263**, 11459–11477.
107. Kikuchi, A., Kozawa, O., Kaibuchi, K., Katada, T., Ui, M., and Takai, Y. (1986) *J. Biol. Chem.*, **261**, 11558–11562.
108. Moriarty, T. M., Gillio, B., Carty, D. J., Premont, R. T., Landau, E. M., and Iyengar, R. (1988) *Proc. Natl. Acad. Sci. USA*, **85**, 8865–8870.
109. Baldassare, J., Knipp, M. A., Henderson, P. A., and Fisher, G. J. (1988) *Biochem. Biophys. Res. Commun.*, **154**, 351–357.
110. Wang, P., Toyoshima, S., and Osawa, T. (1987) *J. Biochem.*, **102**, 1275–1287.
111. Irvine, R. F., and Moor, R. M. (1986) *Biochem. J.*, **240**, 917–920.
112. Litosch, I. (1989) *Biochem. J.*, **261**, 245–251.
113. Enomoto, K., and Gill, M. (1979) *J. Supramol. Structure*, **10**, 51–60.
114. Schleifer, L. S., Kahn, R. A., Hanski, E., Northup, J. K., Sternweis, P. C., and Gilman, A. G. (1982) *J. Biol. Chem.*, **257**, 20–23.
115. Trahey, M., and McCormick, F. (1987) *Science*, **238**, 542–547.

11
RAS AND RAS-RELATED GUANINE NUCLEOTIDE BINDING PROTEINS

A. Hall

Institute of Cancer Research, Chester Beatty Laboratories, Fulham Road, London SW3 6JB, UK

INTRODUCTION

The control of mammalian adenylate cyclase has provided a paradigm for cellular regulation by guanine nucleotide binding proteins. Heterotrimeric G-proteins shuttle between an agonist-stimulated receptor (the detector) and adenylate cyclase (the effector) to transduce a signal from the external environment into the cytoplasm. The specific biochemical details of this and related processes are described elsewhere in this book. In the last few years a novel class of small (21–24 kDa) monomeric guanine nucleotide binding proteins has been described of which the prototype is ras. These proteins have some similarities with the classical G-proteins; they exist in a resting GDP form and an active GTP form and though they are capable of hydrolysing GTP this is not required for activity. However, the precise function of none of these proteins is yet known and it is not clear how far any analogy with classical G-protein systems can be taken.

This chapter will deal with the structure and biochemical function of this family. The ras proteins were the first examples of this class of regulatory molecule to be studied in detail and a description of ras in both mammalian and yeast systems will form the bulk of the chapter. The rest of the chapter will deal with the growing number of ras-related proteins that have been identified through screening cDNA and genomic libraries and from direct purification of GTP binding proteins. There is little doubt that this is currently a very active area of research and that an understanding of this group of proteins will shed light on the regulation of a multitude of biochemical processes.

RAS PROTEINS

Ras genes were first characterized as viral oncogenes carried by two closely related transforming retroviruses [1,2], Harvey Murine Sarcoma Virus (v-Ha-*ras*) and Kirsten

Murine Sarcoma Virus (v-Ki-*ras*). Since then a great deal of attention has been focused on the mammalian cellular homologues of these genes, namely the three proto-oncogenes, c-Ha-*ras*, c-Ki-*ras* and N-*ras* [3]. A significant proportion of human malignancies contain an activated (oncogenic) version of these cellular genes, the mechanism of activation being a point mutation in the coding sequence of the gene [4]. This leads to the production of an oncoprotein carrying a single amino acid substitution either at position 12, 13 or 61 [3–7]. An understanding of the role of the three ras proteins in the control of normal cell growth and of the biochemical consequences of amino acid substitutions that ultimately lead to malignant growth has become a major goal of cancer research.

Structure

PRIMARY STRUCTURE

Each of the three mammalian *ras* proto-oncogenes encodes a 21 kDa protein, $p21^{ras}$ and in the case of N- and Ha-, the protein is 189 amino acids [3]. $p21^{Ki\text{-}ras}$ differs from the other ras proteins in that the cardoxy terminus is encoded within two alternative fourth exons, 4A and 4B [8]. In all mammalian cells so far analysed, splicing ensures that only exon 4B is used generating a protein of 188 amino acids (v-Ki-ras protein contains 4A sequences and is 189 amino acids). The three ras proteins are highly similar with a maximum of 15 differences between any two proteins within the first 164 amino acids (see Figure 11.1). Eight of these differences are clustered at 121–135 though most are conservative substitutions. Between 165 and 185 there is a highly variable region with almost no homology between the three ras proteins and its functional significance, if any, is unknown. The C-terminal amino acids 186–189 form a conserved CAAX box (C, cysteine; A, an aliphatic residue; X, any residue), required for correct post-translational processing of p21 (see later) [9].

Ras genes have been cloned from a variety of other organisms including yeast. *Saccharomyces cerevisiae* contains two *ras* genes *RAS1* and *RAS2* that encode proteins of 309 and 322 amino acids respectively [10,11]. The first 120 amino acids of these two proteins have a striking (70%) homology to mammalian ras. This is followed by a highly diverged region including around one hundred additional amino acids. Like mammalian ras, the yeast proteins have a conserved C-terminal CAAX motif. Interestingly, the single *ras* gene so far identified in *Schizosaccharomyces pombe* (SP*RAS*) encodes a 219 amino acid protein with a structure much closer to mammalian than to the *cerevisiae ras* genes [12].

THREE-DIMENSIONAL STRUCTURE

The 3-D structure at 2.7 Å has been solved for the GDP form of $p21^{Ha\text{-}ras}$ [13] and shows many common features with the *Escherichia coli* GTP binding protein EF-Tu [14,15]. The protein consists of six β sheets, four α helices and nine connecting loops and the GDP molecule is bound in a pocket formed by four of the loops. Loop one consists of residues 10–16 and is close to the β phosphate. It is highly likely that this loop affects the ability

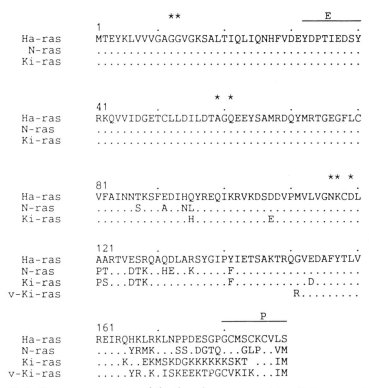

Figure 11.1 Sequence comparison of the three human ras proteins. Dots represent identical residues. Asterisks show positions of oncogenic mutations and the effector region (E) (aa32–40) and the carboxy terminal post-translational processing signals (P) are indicated. The alternative fourth exon (exon 4A) of Ki-ras (present in viral Ki-ras) is also shown for comparison of processing signals

to hydrolyse an attached γ phosphate explaining why oncogenic mutations in this region often affect the GTPase activity of ras. Residues in loops 7 and 9 (116–120 and 145–157) form a pocket for the guanine moiety and it is known that mutations in this region decrease dramatically the affinity and specificity for guanine nucleotide binding. The second site for oncogenic activation (59, 61 and 63) is found in loop 4 which is in close contact with the 10–16 loop. Presumably these mutations affect the configuration of this neighbouring loop leading indirectly to an effect on GTPase activity. The C-terminal 18 residues were not present in the protein used for crystallographic analysis and it has been proposed that they protrude from the globular bulk of the molecule in an α-helical-like stem.

EXPRESSION AND PROCESSING

All three ras proteins appear to be expressed in most cell types and at all stages of development [16]. There are few examples where ras levels can be significantly altered as a consequence of proliferation or differentiation signals but PHA stimulation of lymphocytes is known to give a significant increase in N-*ras* transcription [17]. The proteins have been localized to the inner surface of the plasma membrane [18] and mutational analysis has revealed that cys186 is essential for a post-translational modification step [19] permitting membrane localization. The mobility of membrane-bound p21^{ras} on SDS-polyacrylamide gels is faster than the cytosolic pro p21 form, and it had originally been speculated that amino acids might be clipped from the C-terminus of the protein [20,21]. Later it was shown that [^{3}H]palmitic acid could be incorporated into p21$^{Ha\text{-}ras}$ in a thioester linkage and this was tentatively localized on cys186 [21,22]. More recent experiments have shown that this acylation step is a dynamic process with the palmitic acid attached to ras turning over with a half-life of 20 min [23], whereas the protein itself has a half-life of 20 h in vivo [23,24]. It appears, therefore, that ras undergoes a post-translational acylation/deacylation cycle and this presumably explains why non-acylated p21, purified from *E. coli* expression systems, is biologically active after microinjection into mammalian cells.

It is now known that C-terminal processing of p21^{ras} is much more complicated and still not fully characterized (see Figure 11.2). Work in our laboratory has shown that the three amino acids C-terminal to cys186 are removed during processing, while others have shown that carboxy terminal methylation occurs [25,26]. If p21^{ras} is engineered to terminate at cys186, correct processing and membrane localization do not occur showing that the removal of these C-terminal amino acids is an obligatory step for correct processing in vivo [27]. Deletion analysis has revealed that amino acids 181–185 are also

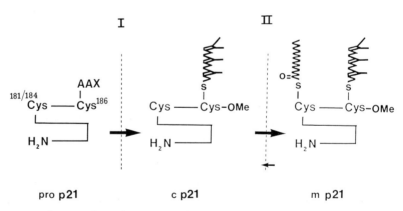

Figure 11.2 Post-translational processing of p21*ras*. Cytoplasmic pro p21 is the substrate for a carboxy peptidase, a methyltransferase and polyisoprenylation (a mutation at cys186 blocks all these steps). The protein then appears in a detergent (Triton X114)-soluble form (c p21) but is not yet stably associated with plasma membranes. Isoprenylation is detected by incorporation of label from mevalonic acid, but the exact structure of the modification on c p21 has not yet been determined. In the case of Ha- and N-ras this is followed by palmitoylation at cys181 or 184 and mature (m) p21 becomes tightly associated with plasma membrane (Step II). Step I is essentially irreversible but palmitic acid on p21 turns over with a half-life of ~20 min. Step II and palmitoylation do not occur with Ki-ras

essential for correct processing of p21$^{Ha\text{-}ras}$ and removal of these blocks addition of palmitic acid [27]. This is somewhat surprising since there is little homology between the three ras proteins in this region (see Figure 11.1). However, a comparison of these sequences in cellular Ha- and N-ras proteins as well as in viral Ha- and viral Ki-ras (exon 4A carboxy terminus) shows that each has cysteine residues at either 181, 184 or both. All of these proteins have been shown to be palmitoylated [20–23]. Cellular Ki-ras protein (exon 4B carboxy terminus) does not have cysteine in the region 181–185 and recent results from our laboratory have shown that cellular p21$^{Ki\text{-}ras}$ is not palmitoylated [27].

The carboxy terminal conserved sequence CAAX is also found in yeast mating factor proteins. As with p21^{ras}, the residues AAX are removed and in the case of the mating factor an isoprenoid farnesyl group is added to the now C-terminal cysteine [28]. We have recently shown that all ras proteins, including cellular p21$^{Ki\text{-}ras}$ contain a covalently linked isoprenoid derivative, presumably at cys186 [27]. This form of ras (c p21, see Figure 11.2) is soluble in the detergent Triton X114 but it not yet stably associated with plasma membranes. In the case of Ha-ras and N-ras, palmitoylation then occurs at cys181 or 184 leading to a tight association with plasma membrane (Step II, Figure 11.2). The significance of this dual lipid attachment is unclear but it is an intriguing possibility that the rapid turnover of the palmitic acid modification provides some regulatory function. Palmitoylation of Ki-ras does not occur and unlike Ha- and N-ras, Ki-ras can be dissociated from plasma membranes in high salt [27]. It is possible that the highly positively charged carboxy terminus is important for plasma membrane localization of this protein.

Biochemical properties

PROTEIN PURIFICATION

Purification of large quantities of p21^{ras} has been achieved by many groups using a wide variety of *E. coli* expression systems [29–32]. The protein can be obtained in a soluble form in high yields (1–10% of total protein) and a two-column purification procedure using DEAE ion exchange chromatography followed by gel filtration is usually sufficient to give 95% purity. Purification in the presence of Mg^{2+} and in the absence of detergent prevents nucleotide release, and in this case purified normal p21 contains one mole of GDP bound per mole of protein [31,33]. Since the concentration of GTP in *E. coli* is much higher than that of GDP, it is likely that the intrinsic GTPase activity of ras accounts for the final nucleotide composition. Indeed we have found that Val12 mutated p21^{ras} (which has a ten-fold reduced GTPase activity, see later) is 60% in the GTP form after purification [34]. In our hands between 10% and 30% of p21$^{Ha\text{-}ras}$ and p21$^{N\text{-}ras}$ purified from *E. coli* moves as a faster migrating species on SDS gels, and is clipped at the carboxy terminus by an endogenous protease. However, when p21$^{Ki\text{-}ras}$ is expressed, more than 90% is clipped at the carboxy terminus, even when extracted in the presence of protease inhibitors.

P21^{ras} proteins purified from *E. coli* do not contain the modifications found on mammalian ras, e.g. palmitoylation. Nevertheless, protein microinjected into mammalian cells is active and is modified after entry. Attempts have been made to obtain high level expression of modified p21^{ras} and human Ha-ras expressed in yeast appears to be correctly modified and acylated, though no attempts have been reported to purify and characterize this protein further [35]. High level expression using baculovirus insect vectors has also been achieved; in the case of Ha-ras 50% of the total cellular protein is p21 and around 10% of this appears to be correctly modified and is membrane bound [36].

NUCLEOTIDE EXCHANGE

The normal ras proteins when isolated from *E. coli* expression systems in the absence of denaturants have one mole of GDP bound per mole of protein and indeed the proteins are unstable in the absence of bound nucleotide [31,33,37]. The dissociation constants for GDP and GTP are $\sim 2 \times 10^{-11}$ M [37,38] and the rate determining step for nucleotide exchange is the nucleotide off-rate, i.e. exchange is pseudo first order [31]. Guanine nucleotide exchange rates on ras are critically dependent on Mg^{2+} concentration [31]; in high free Mg^{2+} exchange rates are very slow and the half-life for nucleotide dissociation is around 60 min. In low free Mg^{2+} (0.1 μM) exchange is fast and the half-life is less than 1 min. Expression of *ras* gene constructs in mammalian cells has allowed the isolation of plasma membranes containing high levels of p21 [39] and, using an immunoprecipitation technique, it is possible to measure the nucleotide exchange rate of membrane-bound p21^{ras} [40]. These rates are almost identical to those of soluble p21.

GTP HYDROLYSIS AND INTERACTION WITH GTPASE-ACTIVATING PROTEIN

The ras proteins have an intrinsic Mg^{2+}-dependent GTPase activity, which can be measured by pre-equilibrating ras with radioactively [α-^{32}P]-labelled GTP in low Mg^{2+}. Nucleotide exchange is stopped by increasing the Mg^{2+} concentration to 5 mM and GTP hydrolysis begins. The conversion of GTP to GDP can then be measured using thin layer chromatography. Alternatively, ras can be pre-equilibrated with [γ^{32}P]GTP and the loss of counts from the protein (as ^{32}P$_i$) measured by a filter binding assay. The half-life for this first-order process is ~ 30 min in vitro for the normal protein [29,38,40].

A major breakthrough in looking at ras function came in 1987 when Trahey and McCormick measured the in vivo exchange and GTP hydrolysis rates of p21^{ras} [41]. They pre-equilibrated ras with [α^{32}P]GTP, microinjected the protein into *Xenopus* oocytes, and re-extracted and analysed the nucleotide content of ras after various intervals. Although the in vivo exchange rate was as predicted by in vitro measurements, hydrolysis of GTP to GDP occurred very much faster than in vitro. This has led to the identification of a GTPase-activating protein, GAP, which has subsequently been found in the cytoplasm of all vertebrate cells, though not in yeast. This 116 kDa protein has now been purified and cloned and in vitro at least appears to work enzymatically to convert p21^{ras}GTP → p21^{ras} GDP [42–44].

EFFECT OF ONCOGENIC MUTATIONS

Although mutations at only codons 12, 13 and 61 have been found in human tumours [4], alterations at 59, 63, 116, 117 and 119 have also been shown to lead to an oncogenically activated protein [45–48]. Furthermore, a complete analysis of all possible substitutions at codon 12 has revealed that Gly can be replaced by any other amino acid except Pro [49] and that Gln at codon 61 can be replaced by all but Pro and Glu [50] to generate a transforming gene.

A comparison of the biochemical properties of normal and oncogenic ras proteins produced in *E. coli* expression systems has revealed that alterations at codon 12, 13 or 61 have only small effects on nucleotide binding and exchange rates [31,38,51]. However, many of the alterations do lead to a drop in intrinsic GTPase activity (e.g. Val12 ten-fold reduced) and this has long been viewed as an explanation for the oncogenic properties of the protein [3,29,49]. Several groups have questioned this, however, and shown that there is no good correlation between the extent of GTPase reduction and oncogenic activity [30,50]. This situation has now been clarified with the discovery that the GAP protein is unable to stimulate the GTPase activity of oncogenically activated proteins [41,43]. The difference in the in vivo GTPase rates of normal and all oncogenic proteins is therefore several hundred-fold. It has been shown, however, that mutant ras proteins still bind to GAP, indeed some oncogenic proteins such as Leu61 have a higher affinity than the normal protein [43]. It appears that oncogenic mutations attenuate the ability of ras proteins to respond to a GAP-induced increase in GTPase activity and this is the biochemical basis of oncogenic activation.

Although activating alterations at codons 116, 117 and 119 have not been detected in human malignancies, in vitro produced protein is oncogenic when introduced into cells [46–48]. These mutations have a dramatic effect on the nucleotide exchange rate which can be increased by more than two orders of magnitude (e.g. Asn119 has a half-life for nucleotide exchange of approx. 30 s), and this may be the basis for their oncogenic activity [46,47,52].

Function

YEAST *RAS*

Deletion of both *S. cerevisiae* ras genes *RAS1* and *RAS2* is a lethal event and the cells are incapable of vegetative growth, whereas introduction of a valine mutation at codon 19 (equivalent to codon 12 in mammalian ras) results in the inability to arrest growth in response to nutritional deprivation [53]. The use of the Val19 substitution has allowed detailed analysis of the pathway regulated by ras in yeast. The mutant behaves very similarly to *bcy*1, a mutation in the regulatory subunit of adenylate cyclase which suppresses the phenotypic consequences of adenylate cyclase deficiency. Furthermore, *bcy*1 can suppress the lethal phenotype *ras*1$^-$ *ras*2$^-$, suggesting that *ras* is intimately involved in controlling adenylate cyclase [53]. In agreement with this, cells expressing a Val19 mutant protein have four-fold higher levels of cAMP [54]. It has proved possible to reconstitute purified ras into yeast membrane systems; thus the *RAS*2 protein, when in the GTP form but not in the GDP form, can activate adenylate cyclase in membranes made from *ras*1$^-$ *ras*2$^-$ *bcy*1 cells [54]. The Val19 mutant is even more efficient at producing cAMP in this reconstituted system but the normal protein can also be 'activated' by use of the non-hydrolysable analogue GTPγS.

More recent work has focused on defining the exact biochemical relationship between the *RAS* protein and adenylate cyclase. In an attempt to define biochemically the downstream target for *RAS*1 a leaky mutation was constructed in its effector domain and introduced into a *ras*2$^-$ background [55]; these cells were viable but grew poorly. Second-site suppressor mutants were then isolated and it was found that a single amino acid change in the adenylate cyclase protein (*CYR*-1) was capable of overcoming the effector site mutation. Although this does not prove there is direct protein/protein contact between *RAS* and *CYR*-1, it does suggest a close interaction. More recently yeast adenylate cyclase has been partially purified from a yeast strain over-expressing this protein [56]. Addition of purified yeast ras (or mammalian ras) is capable of stimulating cyclase activity and as expected the GTP and not the GDP form of the protein is required. Interestingly, the GDP form of oncogenic ras is not active, confirming that mutations do not constitutively activate the protein per se; it still requires GTP to induce an activating conformational change. The adenylate cyclase purified from yeast had a 70 kDa protein associated with the 200 kDa cyclase protein and it is still not clear therefore whether *RAS* interacts directly with cyclase or through this other protein.

The role of the *cdc*25 gene product in *RAS* function in *S. cerevisiae* is of great interest. Mutations in *CDC*25 result in a dramatic decrease in cAMP levels and it seems that both *CDC*25 and *RAS* control adenylate cyclase activity [57]. *CDC*25 mutations can be suppressed by Val19 *RAS*2 but not by over-expression of normal *RAS*2, suggesting that normal *RAS* function is dependent on *CDC*25. It has been postulated that *CDC*25 may be an exchange factor for yeast *RAS* but as yet there is no direct biochemical evidence for this [57–59]. In fact, analysis has shown that even in growing yeast normal *RAS* is predominantly in the GDP form [59], while mutant (Val19) *RAS*1 expressed in yeast is, as expected, predominantly in the GTP form. It is still not clear why significant amounts of GTP cannot be found in growing yeast on *RAS*1 or *RAS*2 despite the fact that these proteins are essential for growth and are only active in the GTP form. It seems probable that normal *RAS* is rapidly cycling between a GDP and GTP form and that at steady state only a small percentage of the protein is in the active form.

Genetic analysis has shown that *ras* in the fission yeast *S. pombe* has a different role. SP*RAS* disruption does not interfere with growth rates or adenylate cyclase activity but instead completely blocks mating [60]. The biochemical basis of this phenomenon is unknown.

ASSAYS FOR MAMMALIAN p21

The classical biological phenomenon associated with oncogenic ras protein is cellular transformation. Introduction of an oncogenic *ras* gene into established rodent fibroblast cell lines causes complete malignant conversion [3]. This can be observed as changes in cell shape and morphology, stimulation of DNA synthesis and formation of tumours when cells are injected into nude mice. The concentration of oncogenic *ras* required to transform established fibroblast cell lines is equivalent to the endogenous $p21^{ras}$ levels (around 50 000 molecules/cell) and over-expression is not required. Over-expression of normal ras in some, though apparently not in all, cell types [61], also leads to transformation and levels of around 20–50-fold higher than endogenous levels seem to be required [39].

Not all cells respond similarly to oncogenic ras, however. In established fibroblasts stimulation of cell profileration clearly occurs, whereas in the PC12 cell line, ras blocks proliferation and induces differentiation to neurone-like cells mimicking the action of nerve growth factor [62]. It is assumed that ras generates the same signal in the two cases but that this is 'read' differently by different cells.

Microinjection of purified protein into established fibroblasts also results in the induction of DNA synthesis and morphological transformation though the lifetime of the effect is determined by the half-life of the protein, around 1–2 days [63,64]. In addition time lapse video analysis shows a dramatic stimulation of cell motility [30]. Microinjection of p21 into *Xenopus* oocytes has proved a very sensitive assay for ras; oncogenic protein induces progression from prophase to metaphase of meiosis and this maturation process is accompanied by the appearance of a white spot in the oocyte and by germinal vesicle breakdown [65].

An assay for the function of normal mammalian $p21^{ras}$ has been developed making use of a neutralizing rat monoclonal antibody, Y-259 [66]. Microinjection of Y-259 into ras-transformed fibroblasts causes phenotypic reversion lasting for the lifetime of the antibody [67]. Perhaps more interestingly, microinjection of Y-259 into quiescent fibroblasts blocks the stimulatory effect of added growth factors [68,69]. Time course experiments show that p21 is essential for 8 h during growth factor stimulation but is not required afterwards during S phase. The antibody blocked stimulation of DNA synthesis by PDGF, serum, phosphatidic acid and $PGF2\alpha$ [69]. Assuming this antibody is completely specific for ras as has been suggested [70] and does not cross-react with other cell components, then these results would show that ras has a key common regulatory role in controlling cell proliferation. In agreement with this, antibody 259 also reverted the transformed phenotype induced by growth factor receptor or plasma membrane oncogenes such as *fms, fes* or *src*, but not by cytoplasmic oncogenes such as *raf* [71].

MUTATIONAL ANALYSIS

The effects of oncogenic mutations on the biochemistry of $p21^{ras}$ and the biological consequences of introducing such mutated genes into cells have already been covered. To recap, it seems likely that these mutations result in ras being constitutively in the active GTP form because the intrinsic GTPase is unable to be stimulated by GAP. The biological consequences of this are uncontrolled cell growth and morphological transformation. It has been reported that deletions in the variable region of oncogenic ras (aa165–184) have no effect on its biological activity [72]. Because this is the only region of significant divergence between the three p21s, and because it is believed that oncogenic ras no longer needs to interact with an upstream exchange factor, it has been speculated that the detector domain of ras might be localized in this variable region. In other words, this would be the region of $p21^{ras}$ responsible for binding to a GTP/GDP exchange factor. However, there is no direct evidence for such a factor in mammalian cells and until such a molecule is identified any ras sequences required for its activity remain speculative.

The region corresponding to amino acids 32–40 of ras has received a lot of attention. This region is located in an external loop (loop 2) of the molecule [13] and is totally conserved in the three proteins. Originally it was reported that some mutations in this area completely destroyed the biological activity of oncogenic ras as assayed by transformation but had no effect on nucleotide binding, intrinsic GTPase rates or post-translational processing [73,74]. On this basis the region is presumed to interact with the downstream target for ras and has been called the effector binding domain.

Mutational analysis has been used to map the binding site on ras for the GAP protein [40,75]. Mutations in a wide variety of positions throughout the protein have been shown to have little effect on GAP activity but mutations in the effector domain were found to block the GAP-induced increase in GTPase activity of normal ras. In fact there is a very good correlation between those mutations in this region that block GAP interaction and those that inactivate the biological activity. It appears then that GAP binds directly to the effector domain implicating it as the target for regulation by ras. More recently it has been shown by direct binding studies that ras protein containing an effector site mutation in either the GDP or GTP form no longer binds to GAP [43]. It was also demonstrated that the antibody Y-259 blocks binding of GAP to $p21^{ras}$ [75] but since the epitope recognized by 259 is localized to a loop on p21 (around amino acids 63–73) that is very close to the effector loop [76], it is likely that 259 blocks GAP interaction by steric interference.

ANALYSIS OF EFFECTS ON SECOND MESSENGER SYSTEMS

One approach taken to look for the function of mammalian ras has been to assay directly for changes in known second messenger systems. All attempts to find an effect of ras on mammalian or *Xenopus* oocyte adenylate cyclase activity [65,77] have failed. Thus microinjection of oncogenic ras protein into *Xenopus* oocytes mimics progesterone or insulin in inducing cell maturation [65]. However, whereas progesterone addition decreases cAMP production, neither $p21^{ras}$ nor insulin has any effect on cAMP levels. Interestingly, antibodies to ras when microinjected into *Xenopus* oocytes block insulin but not progesterone-stimulated maturation [78] suggesting that ras either synergizes with, or is part of, the insulin stimulated signalling pathway.

A lot of effort has gone into looking for effects of ras on the phosphatidyl inositol (PI) pathway in mammalian cells. Comparison of ras-transformed cell lines with parental cells has revealed increases of between two- and four-fold in the basal rate of production of inositol phosphates, leading several groups to conclude that oncogenic ras leads to stimulation of a phosphatidylinositol-specific phospholipase C [79–81]. We have reported the use of a cell line containing oncogenic ras attached to the mouse mammary tumour virus (MMTV) inducible promotor [80]. Addition of inducer leads to the production of Lys-61, $p21^{ras}$ and we observe a four-fold stimulation in the basal rate of inositol phosphate production. Another cell line containing an inducible normal $p21^{ras}$ showed no effect on basal turnover rates when fully induced [82].

Microinjection of $p21^{ras}$ into *Xenopus* oocytes, was found to stimulate some inositol phosphate release, although the largest increases were detected in the levels of 1,2-diacylglycerol (DAG) [81]. Other groups have not found significant changes in inositol phosphate levels in transformed cell lines but have found increased levels of DAG, measured either directly or via its effect through protein kinase C (PKC) on phosphorylation of an 80 kDa substrate [83,84]. This has led to the idea that DAG is generated primarily from a non-inositol-containing phospholipid, perhaps phosphatidylcholine or phosphatidylethanolamine.

A major problem with analysing cell lines is that it is impossible to distinguish primary induced ras events from long-term effects of ras expresssion. To try to overcome this limitation, we have used 'scrape-loading' to introduce ras protein into a large number of Swiss 3T3 cells so that biochemical changes can be measured [34]. Under these conditions the ras protein behaves identically to microinjected protein and leads to morphological transformation and stimulation of DNA synthesis. Unlike microinjection experiments, however, scrape-loading allows biochemical analyses to be undertaken and at short times after introducing the protein. Using this technique we find that Val12 $p21^{ras}$ has no effect on basal rates of inositol phosphate production or an adenylate cyclase activity after scrape-loading. However, 5–10 min after introducing the protein we can see phosphorylation of the 80 kDa substrate of protein kinase C implying increased levels of DAG. It is still not clear where this early source of DAG is coming from but so far we have been unable to detect significant changes in phosphatidylcholine or phosphatidylethanolamine at these early times.

Several groups have shown that protein kinase C is essential for the biological activity of Val12 $p21^{ras}$ [34,85]. In our scrape-loading experiments the only exogenous factor required for Val12 $p21^{ras}$ to give maximal DNA synthesis is insulin [34]. DNA synthesis is, however, completely blocked if protein kinase C is first removed by prolonged treatment with TPA, suggesting that this kinase is essential for stimulation of DNA synthesis by oncogenic ras. It is striking to note that growth-factor-stimulated DNA synthesis in Swiss 3T3 cells is not blocked by removal of protein kinase C [34]. Since experiments using the 259 ras antibody have shown that ras is essential for the activity of many growth factors [69] the simplest rationalization of these apparently contradictory results is to suggest that protein kinase C is either not the first or not the only target of control by ras. In support of this we have found oncogenic-ras-induced events that are independent of protein kinase C. In particular, morphological transformation occurs even in the absence of protein kinase C, as does stimulation of c-*myc* transcription [86]. We believe this suggests that the increase in DAG is not the first event triggered by ras but that an earlier signal is generated which in turn triggers several events, one of which is the generation of DAG. The activation of a kinase would be an obvious candidate for this early event though we have no evidence for this as yet.

The effect of normal $p21^{ras}$ introduced into cells has been more controversial. We and others have shown that normal $p21^{ras}$, even when expressed at high levels, has no effect on basal PI turnover [80,82]. However, in one cell line we constructed containing normal N-*ras* attached to the MMTV inducible promoter, we found an enhanced stimulation by bombesin when, and only when, the normal ras protein was being expressed [82]. We could detect no change in bombesin receptor numbers in these cells and concluded that $p21^{ras}$ was capable of synergism with the activated bombesin receptor to stimulate a phospholipase C. We have not, however, been able to produce this effect in other cell lines containing constitutively expressed N-*ras*. It seems likely that there is something special about the original clone isolated and the real basis for the observed synergism with bombesin on PI turnover in this cell line remains obscure. An increase in bradykinin response has been detected in Ki-*ras*-transformed cells but in this case the mechanism is through a ras-mediated increase in bradykinin receptor numbers [87].

THE RAS GTPASE ACTIVATING PROTEIN, GAP

The original identification of this protein has already been discussed [41]. It appears to be predominantly a cytoplasmic protein though since ras is active at the plasma membrane it seems likely that GAP translocates to the membrane to interact with ras GTP. Indeed it has been speculated that this translocation event could be the primary effect of active ras [40,75]. The 116 kDa GAP protein has been purified and cloned from bovine brain and human placenta (42–44), but sequence analysis reveals no striking homology to other proteins. There is a weak (20%) homology to the regulatory subunit of yeast adenylate cyclase [43] and some limited homology to the non-catalytic domain of phospholipase C and non-receptor tyrosine kinases. However, the function of GAP is unknown and it may itself be a regulatory component of a larger catalytic complex.

The binding constants for normal or oncogenic ras and GAP are of the order of 0.1 mM in the GTP bound state and > 1 mM in the GDP form, and if this is similar in vivo then a GAP–ras–GTP complex will be very short lived. It is of course possible that in the plasma membrane this complex could be stabilized by other components. If induced GTP hydrolysis is faster than dissociation then one molecule of active ras will cause the translocation and activation of one GAP molecule. As suggested by others, this scenario would be similar to the elongation factor EF-Tu which on delivering the correct aminoacyl tRNA to the ribosome is very rapidly converted to the inactive GDP form, by a ribosome-induced increase in EF-Tu GTPase activity [88,89].

Although it appears highly likely that GAP lies downstream and is the target of $p21^{ras}$, an alternative model is possible. In this model GAP lies upstream of ras and ensures that ras–GTP is continually converted to the resting GDP form. Activation of ras through spontaneous or catalysed GTP exchange would require inactivation of GAP allowing ras to interact with its target, presumably through the effector site. At the moment the exact role of GAP in the ras pathway is unclear but the availability of recombinant GAP protein and GAP antibodies should allow a much more detailed examination of the physiological role of the ras/GAP interaction.

RAS-RELATED PROTEINS

Structure

A growing number of ras-related proteins have been identified in mammalian cells. The use of oligonucleotide probes has been very successful in pulling out sequences from cDNA libraries to reveal ral [90], the *rap* family (*rap*1A, *rap*1B, *rap*2) [91] and the *rab* genes (1–4) [92]. Rho genes were first identified in the mollusc *Aplysia* but later three *rho* genes, A, B and C, were identified in mammalian genomes [93,94]. R-*ras* was identified in a genomic library using a *ras* cDNA probe [95], and BRL-*ras* was identified in a rat liver cDNA library [96]. The *SEC*4 gene was identified from a genetic analysis of a yeast strain defective in secretion [97].

The sequence of the *ras*-related genes reveals predicted proteins of ~ 21–23 kDa and though the amino acids important for guanine nucleotide binding in $p21^{ras}$ are highly conserved in the ras-related genes, the rest of the molecules show divergence to varying extents. Figure 11.3 shows proteins with around 50% overall homology to ras; the closest homology being found in rap1 and R-ras each with around 55% identical amino acids [91,95]. In particular the effector sites of R-ras and the three rap proteins are either identical or have only one amino acid difference from ras, suggesting a common function (see later). Surprisingly, however, rap contains Thr instead of Gln at codon 61 (see Table 11.1), equivalent to an oncogenic mutation at codon 61 of ras. Ral has around 50% homology to ras but has three amino acid differences from ras in the effector region. Ral and R-ras seem to be expressed in most tissues and at levels similar to ras (see Table 11.1) [98].

The rab group of proteins are around 30% homologous to ras and around 40% homologous to each other as shown in Figure 11.4 [92]. They have a widely diverged effector site and do not have Gly at the equivalent of codon 12 in ras (see Table 11.1). The reason for classifying these proteins as a group is that they all have a mutation at the equivalent to codon 12 in ras and they have more homology to each other than they do to ras. The recently identified BRL-*ras* appears to be related to this rab group [96] as does the *SEC*4 protein from yeast [97]. More recently rab3 has been purified independently from brain, and subsequently cloned [99,100]. This has revealed that rab3 is itself a member of a very closely related subfamily A, B, C (also called smg25A, B, C) which are around 80% homologous to each other and are exclusively expressed in brain [98]. As

```
                                                      **
Ha-ras    1                            MTEYKLVVVGAGGVGKSALTI
rap1A     1                            .R......L.S........V
rap2      1                            .R...V..L.S........V
R-ras     1           MSSGAASGTGRGRPRGGGPGPGDPPPSETH......G..........
ral       1                       MAANKPGQNSLALH.VIM..S.........L
                           E                      * *
Ha-ras   22           QLIQNHFVDEYDPTIEDSYRKQVVIDGETCLLDILDTAGQEEYSAMR
rap1A    22           .FV.GI..EK.............EV.CQQ.M.E......T.QFT...
rap2     22           .FVTGT.IEK.......F...EIEV.SSPSV.E......T.QFAS..
R-ras    48           .F..SY..SD........T.ICSV..IPAR..........FG...
ral      33           .FMYDE..ED.E..KA.....K..L...EVQI.........D.A.I.

H-ras    69           DQYMRTGEGFLCVFAINNTKSFEDIHQYREQIKRVKDSDDVPMVLVG
rap1A    69           .L..KN.Q..AL.YS.TAQST.N.LQDL....L.....TE....I...
rap2     69           .L.IKN.Q..IL.YSLV.QQ..Q..KPM.D..I...RYEK..VI...
R-ras    95           E....A.H...L.....DRQ..NEVGKLFT..L....R..F.V....
ral      80           .N.F.S........S.TEME..AATADF....L...EDEN..FL...
                      ** *
Ha-ras  116           NKCDLAA RTVESRQAQDLARSY GIPYIETSAKTRQGVEDAFYTLV
rap1A   116           ...EDE.V.V.GKE.G.N...QWCNAFL.S...SKIN.NEI..D..
rap2    116           ..V..ESE.E.S.SEGRA..EEW .C.FM.....SKTM.DEL.AEI.
R-ras   142           ..A..ESQ.Q.PRSE.SAFGA.H HVA.F.A...L.LN.DE..EQ..
ral     127           ..S..EDK.Q.SVEE.KNR.DQW NVN.V.......AN.DKV.FD.M

Ha-ras  161           REIRQHKLRKLNPPDESGPGC         MSCKCVLS
rap1A   163           .Q.NRKTPVEKKKPKKKS            .L.L
rap2    162           .QMNYAAQPDKDDPCCSA            .NIQ
R-ras   188           .AV.KYQEQE.P.SPP.A.RKK        GGG.P...L
ral     173           ....AR.MEDSKEKNGKKKRKSLAKRIRER.CIL
```

Figure 11.3 Sequence comparison of ras-related proteins with 50% or more amino acid homology to ras. Asterisks show amino acids in ras that are sites of oncogenic mutation. The effector domain of ras (aa32–40) is shown

with the ras family most sequence divergence is concentrated at the carboxyl terminal end of the proteins. Rab1 expression is ubiquitous [98] and Rab1 appears to be the mammalian equivalent of the yeast YPT-1 protein originally identified in a yeast cDNA library using a mammalian *ras* gene probe [101,102]. *Rab2* and 4 and BRL-*ras* also seemed to be expressed in most cells [96,98]. A mammalian equivalent to *SEC*4 has not yet been identified.

Finally, the rho family consists of three members, A, B, C, each with around 35% homology to ras (see Figure 11.4) As can be seen in Figure 11.5, the rho family are around 85% homologous to each other and most of the divergence is located in a variable region towards the C-terminus in an analogous fashion to ras. The rho effector domain is significantly diverged from ras and, in addition, rho contains Ala instead of Gly at the equivalent of codon 13 in ras (see Table 11.1). However, we have recently shown that although a Val mutation at codon 13 of Ha-ras is oncogenic, an Ala substitution is not,

Table 11.1 Ras and ras-related proteins. The numbers shown in the sequence panel refer to codon positions in ras. The amino acids corresponding to the sites of oncogenic mutation (12,13,61) and the effector domain (32–40) in ras are shown

Protein	Expression	12	13	61	32								40	Properties
ras (Ha, Ki, N)	Most cells	G	G	Q	Y	D	P	T	I	E	D	S	Y	Involved in growth control. Interacts with 116 kDa GAP. Plasma membrane
R-ras	Most cells	G	G	Q	.	.	.	.	.	.	.	F	.	Interacts with 116 kDa GAP. Reverts ras-transformed cells
rap1 (A, B)		G	G	T	.	.	.	.	.	.	.	.	.	
rap2		G	G	T	.	.	.	.	.	.	.	.	.	
ral (A, B)	Most cells	G	G	Q	.	E	.	.	K	A	.	.	.	
rho (A, B, C)	Most cells	G	A	Q	.	V	.	.	V	F	E	N	.	Interacts with 30 kDa GAP
rab1	Most cells	S	G	Q	.	I	S	.	.	G	V	D	F	May be involved in vesicle trafficking and localized in Golgi
rab2	Most cells	T	G	Q	H	.	L	.	M	G	V	E	F	
rab3 (A, B, C)	Brain	S	S	Q	F	V	S	.	V	G	I	D	F	
rab4	Most cells	A	G	Q	S	N	H	.	.	G	V	E	F	
BRL-ras	Most cells	S	G	Q	.	K	A	.	.	G	A	D	F	

and indeed the Ala13 p21^{ras} still interacts with and is affected by GAP [103]. The yeast *S. cerevisiae* contains two *rho* genes *RHO1* and *RHO2*. *RHO2* is somewhat different from *RHO1* and its three mammalian rho protein homologues [104], and it is likely that a mammalian equivalent to *RHO2* is yet to be identified.

It is possible that more ras-related proteins await discovery. In addition, a number of more distantly related 21–24 kDa guanine nucleotide binding proteins have been purified. For example, the ADP ribosylation factor (ARF) required for ADP ribosylation of Gs by cholera toxin is itself a GTP binding protein and has some similarity to ras [105]. A protein from *E. coli*, Era, has 316 amino acids and 91 out of the first 197 are similar or identical to ras [106].

```
                                  **
Ha-ras    1                MTEYKLVVVGAGGVGKSALTIQLIQNHFVDE
rhoA      1                MAAIRK...I..D.AC..TC.L.VFSKDQ.PEV
rab1      1                MSSMNPEYDYLF..LLI.DS.....C.LLRFADDTYTES
rab2      1                MAYAYLF.YIII.DT.....C.LL.FTDKR.QPV
rab3A     1    MASATDARYGQKESSDQNFDYMF.ILII.NSS...TSFLFRYADDS.TPA
BRL-ras   1                MLL.VIIL.DS....TS.MN.YVNKK.SNQ
rab4      -
                                 *  *
Ha-ras    32   YDPTIEDSY  RKQVVIDGETCLLDILDTAGQEEYSAMRD QYMRTGEGFL
rhoA      34   .V..VFEN.  VADIEV..KQVE.ALW......D.DRL.PLS.PD.DVILM
rab1      40   .IS..GVDFKIRTIEL..K.IK.Q.W......RFRTITS S.Y.GAH.II
rab2      35   H.L.MGVEFGARMITI..KQIK.Q.W......SFRSITR S.Y.GAA.A.
rab3A     51   FVS.VGIDFKV.TIYRNDKRIK.Q.W......R.RTITT A.Y.GAM..I
BRL-ras   31   .KA..GADFLT.E.MV.DRLVTMQ.W......RFQSLGV AFY.GADCCV
rab4      -              EFGQ.IINVG.KYVK.Q.W......RFRSVTT S.Y.GAA.A.

Ha-ras    80   CVFAINNTKSFEDIHQYREQIKRV   KDSDDVPMVLVGNKCDL
rhoA      83   ..S.DSPD.L.N.PELWTPEVK     HFCPN..II.....K..RNDEHT
rab1      89   V.YDVTDQE..NNVK.WLQE.D.    YA.EN.NKL........
rab2      84   L.YD.TRRDT.NHLTTWL.DARQ    HSNSNMVIM.I...S..
rab3A     100  LMYD.T.EE..NAVQDWST...T    YSW.NAQVL.......M
BRL-ras   80   L..DVTAPNT.KTLDSW.DEFLIQASPR.PENF.F.VL...I..
rab4      -    L.YD.TSRETYNALTNWLTDARM    LA.QNIVII.C...K..

Ha-ras    127         AA RTVESRQAQDLARSYGIPYIETSAKTRQGVEDAFYTLVRE
rhoA      128  RRELAKMKQEPVKPEEGRDMAN.IGAFG.M.C....KD..REV.EMAT.A
rab1      135          TTKKV.DYTT.KEF.D.L...FL.....NEKN..QS.MTMAA.
rab2      130          ESR.E.KKEEGEAF..EH.LIFM......ASN..E..INTAK.
rab3A     146          EDE.V.S.ERGRQ..DHL.FEFF.A....DNIN.KQT.ER..DV
BRL-ras   130          ENRQVATK.AQAWCYSKNN...F.....EAIN..Q..Q.IA.N
rab4      -            D.D.E.TFLE.SRF.QENELMFL....L.GEN..E..MQCARK

Ha-ras    163  IRQHKLRKLNPPDESGP          GCMSCKCVLS
rhoA      178  AL.ARRG.KKSG                    .LVL
rab1      172  .KKRMGPGATAGGAEKSNVKIQSTPVKES   GGGCC
rab2      167  .YERIQEGVFDINNEANGIKIGPQHAATNASHGGNQGGQQAGGGCC
rab3A     183  .CEKMSESLDTADPAVTGAKQGPQLTDQQAPPHQD    CAC
BRL-ras   167  ALKQETEVELYNEFPEPIKLDKNERAKASAES       CSC
rab4      -    .LNKIESGELDPERMGSGIQYGDAALRQLRSPRRTQAPSAQE CGC
```

Figure 11.4 Sequence comparison of ras with ras-related genes having 30–40% overall amino acid homology. The rab genes and BRL-ras are rat protein sequences, ras and rho are human

```
                                                           E
                        **                .         .
rhoA    1    MAAIRKKLVIVGDGACGKTCLLIVFSKDQFPEVYVPTVFE
rhoB         .......V.................E..........
rhoC         ........................................


             _          .         .   *     .          .
rhoA   41    NYVADIEVDGKQVELALWDTAGQEDYDRLRPLSYPDTDVI
rhoB         ........................................
rhoC         ..I.....................................


                   .         .         .         .
rhoA   81    LMCFSIDSPDSLENIPELWTPEVKHFCPNVPIILVGNKKD
rhoB         .....V............K.V..............A....
rhoC         ................K.......................


                   .         .         .         .
rhoA  121    LRNDEHTRRELAKMKQEPVKPEEGRDMANRIGAFGYMECS
rhoB         ..S...V.T...R......RTDD..A..V..Q.YD.L...
rhoC         ..Q..............RS..........S....L...


                        .         .         .
rhoA  161    AKTKDGVREVFEMATRAALQARRGKKKSG    CLVL
rhoB         ....E.......T.......K.Y.SQNGCINC.K..
rhoC         ....E............G..V.KN.RRR.    .PI.
```

Figure 11.5 Sequence comparison of the three mammalian rho proteins. Asterisks indicate amino acids corresponding to codons 12, 13 and 61 of ras and the putative effector domain (E) is shown

Biochemical properties and function

The guanine nucleotide exchange rates have been determined for the yeast ypt protein [107], mammalian R-ras and mammalian rho [103]. They each show similar kinetics to ras; a very slow nucleotide exchange rate in the presence of Mg^{2+} but in low free Mg^{2+} this is increased several hundred-fold. The in vitro GTPase rates of R-ras and rho are very similar to ras with half-lives of 30 min and 15 min respectively, whereas the ypt protein has a low GTPase activity, presumably as a consequence of not having Gly at codon 17 (equivalent to codon 12 of ras) [107]. A mutation at codon 14 of rho decreases its intrinsic GTPase activity around ten-fold in a similar fashion to codon 12 changes in ras [103]. Interestingly, however, it appears that rab3 (smgA) has if anything a higher GTPase activity than ras despite having alterations at positions equivalent to codon 12 and 13 in ras [100]. Clearly with proteins having only 30% overall homology to ras it is dangerous to predict the effect of mutations based on their effect in ras.

The function of YPT in yeast has been extensively investigated making use of the genetic systems available in this organism and deletion of YPT-1 is lethal [108]. The phenotype of *ypt* mutations is similar to that of secretion-defective mutants with cells accumulating membranes and vesicles, and it has been suggested that *YPT-1* and its mammalian homologue, Rab1, are localized in the Golgi and function in intracellular vesicle trafficking [109]. Another group has arrived at a different conclusion, namely that YPT-1 is involved in regulating cytoplasmic Ca^{2+} concentration [110]. The SEC4 protein is certainly involved in secretion but probably at a late stage in the pathway during the fusion of Golgi with plasma membrane [97].

The function of none of the other mammalian ras-related proteins is clear. We have recently shown that $p23^{R-ras}$ interacts with the same 116 kDa GAP protein as $p21^{ras}$ [103], which increases the GTPase activity of R-ras in a similar fashion to that of ras. A comparison of effector site sequences (Table 11.1) would suggest that at least rap1A and 1B and possibly rap2 will interact with GAP, though since rap proteins have a mutation at the equivalent to codon 61 in ras, it may be that a rap/GAP interaction would not be followed by an increase in GTPase activity. If GAP is the target for ras then it is tempting to speculate that rap and R-ras are also involved in controlling cell growth. However, the *rap* gene itself is not oncogenic and when a mutation, equivalent to Val12 in ras, is introduced into R-*ras* this is also not oncogenic [111]. It has recently been shown that over-expression of rap can revert ras-transformed cells raising the possibility that rap and perhaps R-ras are negative regulators of GAP activity and of cell growth [91,112]. More experiments are required to delineate the exact inter-relationship of the four proteins known to interact with the 116 kDa GAP protein and the three others that are predicted to interact. Based on the effector site sequence (Table 11.1), it is likely that the ral protein will not interact with ras–GAP and will probably have its own GAP. However, since ral has around 50% overall homology to ras, it is tempting to speculate that it too might somehow be involved in growth control.

We have recently found a GAP-like activity towards $p21^{rho}$ in mammalian and *Xenopus* oocyte cytoplasmic extracts but not in yeast [103]. Mutations in rho that are equivalent to oncogenic changes in ras also block the ability of this rho–GAP to increase the GTPase activity of rho. The protein is approximately 30 kDa as judged by gel filtration and we are currently trying to purify rho–GAP with the hope that this will shed light on the function of rho.

The rab proteins and BRL-ras each have different effector sites and are probably involved with distinct biochemical processes. Interaction with their targets is not predicted to give an increased GTPase activity since rab proteins contain mutations known to block the GAP-induced increase in GTPase of ras. However, it may be possible to assay for and identify GAP-like proteins for the rab molecules if these amino acids are first replaced.

Recently there have been numerous reports of small molecular weight substrates for ADP-ribosylation by a transferase (exoenzyme C3) from *Clostridium botulinum* types C and D [113–118]. Many cells contain a 22 kDa substrate, and this has been purified and shown to bind GTP and GDP [114] but the protein is not $p21^{ras}$ [115,116,118]. One substrate has now been purified from brain and shown to be rho protein [117]. We have recently shown that $p21^{rho}$ purified from *E. coli* expression vectors is a substrate for the *botulinum* transferase, whereas ras, R-ras, ral, ypt and rap1A proteins are not substrates [118]. It remains to be seen what effects ribosylation has on the biochemical properties of $p21^{rho}$ and whether other ras-related proteins are substrates for this transferase.

DISCUSSION

Many details of the function of ras proteins in *S. cerevisiae* have been elucidated using the elegant genetics available in this organism but unfortunately such approaches are not yet possible in mammalian cells. In view of the fact that mammalian ras can function in yeast and that yeast ras (after removal of the long carboxy terminus) will function in mammalian cells, it was disappointing to find that the biochemical functions have diverged significantly in the two organisms. Yeast ras controls adenylate cyclase in yeast but mammalian ras has no effect on mammalian adenylate cyclase.

A large effort has been put into looking for changes in other second messengers in mammalian cells and it seems clear now that $p21^{ras}$ does not directly control the PIP_2-specific phospholipase C. $p21^{ras}$ does lead to the activation of protein kinase C and this seems to be an obligatory step for at least part of its oncogenic effect, though whether this is a primary effect of ras still needs to be resolved. Other ras-mediated effects such as morphological transformation are independent of protein kinase C and the signals generated to elicit these events are unknown.

The identification of the GAP protein has given new hope that an alternative and more direct approach to identifying ras function in mammalian cells is at hand. It is clear that GAP binds to the effector domain of ras though it is still not proven whether GAP functions as an upstream regulator of ras or a downstream target for ras. It is possible it could do both. Now that GAP is available from recombinant sources it should be possible to determine the function of GAP.

Ras has turned out to be just one branch of a large superfamily of small GTP binding regulatory proteins. Around 20–30 are now known and for none of these do we yet know the function. Some (*rho*, *rab*1) are also present in yeast (*RHO1*, *YPT*) and perhaps a genetic analysis of the yeast counterparts will be more fruitful than was the case with ras. At least one ras-related protein (R-ras) and probably three more (rap1A, 1B and 2) interact with the same 116 kDa GAP as the three ras proteins and the way they are involved in regulating cell proliferation is intriguing but unknown. There is clearly a wide range of biochemical activities that are regulated by these small molecular weight guanine nucleotide binding proteins and of which we know very little at present; however, with the intense investigation now being undertaken the nature of these should soon become apparent.

REFERENCES

1. Harvey, J. J. (1964) *Nature (London)* **204**, 1104–1105.
2. Kirsten, W. H., and Mayer, L. A. (1967) *J. Natl. Cancer Inst.*, **39**, 311–335.
3. Barbacid, M. (1987) *Ann. Rev. Biochem.*, **56**, 779–827.
4. Bos, J. L. (1988) *Mutation Res.*, **195**, 255–271.
5. Tabin, C. J., Bradley, S. M., Bargmann, C. I., Weinberg, R. A., Papageorge, A. G., Scolnick, E. M., Dhar, R., Lowy, D. R., and Chang, E. H. (1982) *Nature (London)*, **300**, 143–149.

6. Bos, J. L., Toksoz, D., Marshall, C. J., Vries, M. V., Veeneman, G. H., Van der Eb, A. T., van Bloom, J. H., Janssen, J. W. G., and Steenvoorden, A. C. M. (1986) *Nature (London)*, **315**, 726–730.
7. Brown, R., Marshall, C. J., Pennie, S. G., and Hall, A. (1984) *EMBO J.*, **3**, 1321–1326.
8. McGrath, J. P., Capon, D. J., Smith, D. H., Chen, E. Y., Seeburg, P. H., Goeddel, D. V., and Levinson, A. D. (1983) *Nature (London)*, **304**, 501–506.
9. Magee, A. I., and Hanley, M. (1988) *Nature*, **335**, 114–115.
10. DeFeo-Jones, D., Scolnick, E. M., Koller, R., and Dhar, R. (1983) *Nature (London)*, **306**, 707–709.
11. Powers, S., Kataoka, T., Fasano, O., Goldfarb, M., Strathern, J., and Wigler, M. (1984) *Cell*, **36**, 607–612.
12. Fukui, Y., and Kaziro, Y. (1985) *EMBO J.* **4**, 687–691.
13. DeVos, A. M., Tong, L., Milburn, M. V., Matias, P. M., Jancarik, J., Noguchi, S., Nishimura, S., Miura, K., Ohtsuka, E., and Kim, S. H. (1988) *Science*, **239**, 888–893.
14. Jurnak, F. (1985) *Science*, **230**, 32–36.
15. McCormick, F., Clark, B. F. C., La Cour, T. F. M., Kjeldgaard, M., Lauritsen, L. N., and Nyborg, J. (1985) *Science*, **230**, 78–82.
16. Leon, J., Guerrero, I., and Pellicer, A. (1987) *Mol. Cell. Biol.*, **7**, 1535–1540.
17. Reed, J. C., Alpers, J. D., Nowell, P. C., and Hoover, R. G. (1986) *Proc. Natl. Acad. Sci. USA*, **83**, 3982–3986.
18. Willingham, M. C., Pastan, I., Shih, T. Y., and Scolnick, E. M. (1980) *Cell*, **19**, 1005–1014.
19. Willumsen, B. M., Norris, K., Papageorge, A. G., Hubert, N. L., and Lowy, D. R. (1984) *EMBO J.*, **3**, 2581–2585.
20. Fijiyama, A., and Tamanoi, F. (1986) *Proc. Natl. Acad. Sci. USA*, **83**, 1266–1270.
21. Chen, Z. Q., Ulsh, L. S., DuBois, G., and Shih, T. Y. (1985) *J. Virol.*, **56**, 607–612.
22. Buss, J. E., and Sefton, B. M. (1986) *Mol. Cell. Biol.*, **6**, 116–122.
23. Magee, A. I., Gutierrez, L., McKay, I. A., Marshall, C. J., and Hall, A. (1987) *EMBO J.*, **6**, 3353–3357.
24. Ulsh, L. S., and Shih, T. Y. (1984) *Mol. Cell. Biol.*, **4**, 1647–1652.
25. Gutierrez, L., Magee, A. I., Marshall, C. J., and Hancock, J. (1989) *EMBO J.*, **8**, 1093–1098.
26. Clarke, S., Vogel, J. P., Descheres, R. J., and Stock, J. (1988) *Proc. Natl. Acad. Sci. USA*, **85**, 4643–4647.
27. Hancock, J. F., Magee, A. I., Childs, J. E., and Marshall, C. J. (1989) *Cell*, **57**, 1167–1177.
28. Ishibashi, Y., Sakagami, Y., Isogai, A., and Suzuki, A. (1984) *Biochemistry*, **23**, 1399–1404.
29. McGrath, J. P., Capon, D. J., Goeddel, D. V., and Levinson, A. D. (1984) *Nature (London)*, **310**, 644–649.
30. Trahey, M., Mulley, R. J., Cole, G. E., Innis, M., Paterson, M., Marshall, C. J., Hall, A., and McCormick, F. (1987) *Mol. Cell. Biol.*, **7**, 541–544.
31. Hall, A., and Self, A. (1986) *J. Biol. Chem.*, **261**, 10963–10965.
32. Sweet, R. W., Yokayama, S., Kamata, T., Feramisco, J. R., Rosenberg, M., and Gross, M. (1984) *Nature (London)*, **311**, 273–275.
33. Poe, M., Scolnick, E. M., and Stein, R. B. (1985) *J. Biol. Chem.*, **260**, 3906–3909.
34. Morris, J. D. H., Price, B., Lloyd, A. C., Self, A. J., Marshall, C. J., and Hall, A. (1989) *Oncogene*, **4**, 27–31.
35. Clark, S. G., McGrath, J. P., and Levinson, A. D. (1985) *Mol. Cell. Biol.*, **5**, 2746–2752.
36. Lowe, P. N., Bradley, S. A., Hall, A., Murphy, M., Rhodes, S., Skinner, R. H., and Page, M. (unpublished results).
37. Feuersten, J., Goody, R. S., and Wittinghofer, A. (1987) *J. Biol. Chem.*, **262**, 8455–8460.
38. Neal, S., Eccleston, J. F., Hall, A., and Webb, M. R. (1989) *J. Biol. Chem.*, **263**, 19718–19722.
39. McKay, I. A., Marshall, C. J., Calés, C., and Hall, A. (1986) *EMBO J.*, **5**, 2617–2621.
40. Calés, C., Hancock, J. F., Marshall, C. J., and Hall, A. (1988) *Nature (London)*, **332**, 548–551.
41. Trahey, M., and McCormick, F. (1987) *Science*, **238**, 542–545.

42. Gibbs, J. B., Shaber, M. D., Allard, W. J., Sigal, I. S., and Scolnick, E. M. (1988) *Proc. Natl. Acad. Sci. USA*, **85**, 5026–5030.
43. Vogel, U., Dixon, R. A. F., Schaber, M. D., Diehl, R. E., Marshall, M. S., Scolnick, G. M., Sigal, I. S., and Gibbs, J. B. (1988) *Nature (London)*, **335**, 90–93.
44. Trahey, M., Wong, G., Halenbeck, R., Rubinfield, B., Martin, G. A. Ladner, M., Long, C. M., Crosier, W. J., Watt, K., Koths, K., and McCormick, F. (1988) *Science*, **242**, 1697–1699.
45. Fasano, O., Aldrich, T., Tamonoi, F., Taparowsky, E., Furth, M., and Wigler, M. (1984) *Proc. Natl. Acad. Sci. USA*, **81**, 4008–4012.
46. Walter, M., Clark, S. G., and Levinson, A. D. (1986) *Science*, **233**, 649–652.
47. Sigal, F. S., Gibbs, J. B., D'Aonzo, J. S., Temeles, G. L., Wolanski, B. S., Socher, S. H., and Scolnick, E. M. (1986) *Proc. Natl. Acad. Sci. USA*, **83**, 952–956.
48. Reynolds, S. M., Stowers, S. J., Patterson, R. M., Maronpot, R. R., Aaronson, S. A., and Anderson, M. W. (1987) *Science*, **237**, 1309–1316.
49. Seeburg, P. H., Colby, W. W., Capon, D. J., Goeddel, D. V., and Levinson, A. D. (1984) *Nature (London)*, **312**, 71–75.
50. Der, C. J., Finkel, T., and Cooper, G. M. (1986) *Cell*, **44**, 167–176.
51. John, J., Frech, M., and Wittinghofer, A. (1988) *J. Biol. Chem.*, **263**, 11792–11799.
52. Feig, L. A., Pan, B. T., Roberts, T. M., and Cooper, G. M. (1986) *Proc. Natl. Acad. Sci. USA*, **83**, 4607–4611.
53. Toda, T., Uno, I., Ishikawa, T., Powers, S., Kataoka, T., Broek, D., Cameran, S., Broach, J., Matsumoto, K., and Wigler, M. (1985) *Cell*, **40**, 27–36.
54. Broek, D., Samiy, N., Fasano, O., Fujiyama, A., Tamanoi, F., Northrup, J., and Wigler, M. (1985) *Cell*, **41**, 763–769.
55. Marshall, M. S., Gibbs, J. B., Scolnick, G. M., and Sigal, I. S. (1988) *Mol. Cell Biol.* **8**, 52–61.
56. Field, J., Nikawa, J. I., Broek, D., MacDonald, B., Rodgers, L., Wilson, I. A., Lerner, R. A., and Wigler, M. (1988) *Mol. Cell. Biol.*, **8**, 2159–2165.
57. Broek, D., Toda, T., Michaeli, T., Levin, L., Birchmeier, C., Zoller, M., Powers, S., and Wigler, M. (1987) *Cell*, **48**, 789–799.
58. Daniel J., Becker, J. M., Enari, E., and Levitzki, A (1987) *Mol. Cell. Biol.*, **7**, 3857–3861.
59. Gibbs, J. B., Schaber, M. D., Marshall, M. S., Scolnick, E. M., and Sigal, I. S. (1987) *J. Biol. Chem.*, **262**, 10426–10429.
60. Fukui, Y., Kozasa, T., Kaxiro, Y., Takeda, T., Yamamoto, M. (1986) *Cell*, **44**, 329–336.
61. Rickets, M. M., and Levinson, A. D. (1988) *Mol. Cell. Biol.*, **8**, 1460–1468.
62. Bar-Sagi, D., and Feramisco, J. R. (1985) *Cell*, **42**, 841–848.
63. Stacey, D. N., and Kung, H. F. (1984) *Nature (London)*, **310**, 508–511.
64. Feramisco, J. R., Gross, M., Kamata, T., Rosenberg, M., and Sweet, R. W. (1984) *Cell*, **38**, 109–117.
65. Birchmeier, C., Brock, D., and Wigler, M. (1985) *Cell*, **43**, 615–621.
66. Furth, M. E., Davis, L. J., Fleuredelys, B., and Scolnick E. M. (1982) *J. Virol.*, **43**, 294–304.
67. Feramisco, J. R., Clark, R., Wong, G., Arnheim, N., Milley, R., and McCormick, F. (1985) *Nature (London)*, **314**, 639–642.
68. Mulcahy, L. S., Smith, M. P., and Stacey, D. W. (1985) *Nature (London)*, **313**, 241–243.
69. Yu, C. L., Tsai, M. H., and Stacy, D. W. (1988) *Cell*, **52**, 63–71.
70. Papageorge, A. G., Willumsen, B. M., Johnson, M., Kung, H. F., and Stacey, D. W. (1986) *Mol. Cell. Biol.*, **6**, 1843–1846.
71. Smith, M. R., DeGudicibus, S. J., and Stacey, D. W. (1988) *Nature*, **320**, 540–543.
72. Willumsen, B. M., Papageorge, A. G., Hubber, N., Bekesi, E., Kung, H. F., and Lowy, D. R. (1985) *EMBO J.* **4**, 2893–2896.
73. Sigal, I. S., Gibbs, J. B., D'Alonzo, J. S., and Scolnick, E. M. (1986) *Proc. Natl. Acad. Sci. USA*, **83**, 4725–4729.

74. Willumsen, B. M., Papageorge, A. G., Kung, M. F., Bekesi, E., Robins, T., and Lowy, D. R. (1986) *Mol. Cell. Biol.*, **6**, 2646–2654.
75. Adari, H., Lowy, D. R., Willumsen, B. M., Der, C. J., and McCormick, F. (1988) *Science*, **240**, 518–521.
76. Sigal, I. S., Gibbs, J. B., D'Alonzo, J. S., and Scolnick, E. M. (1986) *Proc. Natl. Acad. Sci. USA*, **83**, 4725–4729.
77. Beckner, S. K., Hattori, S., and Shih, T. H. (1985) *Nature (London)*, **317**, 71–72.
78. Korn, L. J., Siebel, C. W., McCormick, F., and Roth, R. A. (1987) *Science*, **236**, 840–843.
79. Fleischman, L. F., Chawala, S. B., and Cantley, L. (1986) *Science*, **231**, 407–410.
80. Hancock, J. F., Marshall, C. J., McKay, I. A., Gardner, S., Houslay, M., Hall, A., and Wakelam, M. J. O. (1988) *Oncogene*, **3**, 187–193.
81. Lacal, J. C., De La Pena, P., Moscat, J., Garcia-Barreno, P., Anderson, P. S., and Aaronson, S. A. (1987) *Science*, **238**, 533–536.
82. Wakelam, M. J. O., Davies, S. A., Houslay, M. D., McKay, I., Marshall, C. J., and Hall, A. (1988) *Nature (London)*, **323**, 173–176.
83. Lacal, J. C., Moscat, J., and Aaronson, S. A. (1987) *Nature (London)*, **330**, 269–271.
84. Seuwen, K., Lagarde, A., and Pouyssegur, J. (1988) *EMBO J.*, **7**, 161–168.
85. Lacal, J. C., Fleming, T. P., Warren, B. S., Blumberg, P. M., and Aaronson, S. A. (1987) *Mol. Cell. Biol.* **7**, 4146–4149.
86. Lloyd, A. C., Paterson, H. F., Morris, J. D. H., Hall, A., and Marshall, C. J. (1989) *EMBO J.*, **8**, 1099–1104.
87. Downward, J., DeGunzburg, J., Riehl, R., and Weinberg, R. A. (1988) *Proc. Natl. Acad. Sci. USA*, **85**, 5774–5778.
88. Bourne, M. (1988) *Cell*, **83**, 669–671.
89. Kaziro, Y. (1978) *Biochem. Biophys. Acta*, **505**, 95–127.
90. Chardin, P., and Tavitian, A. (1986) *EMBO J.*, **5**, 2203–2208.
91. Pizon, V., Chardin, P., Leroscy, I., Olofsson, B., and Tavitian, A. (1988) *Oncogene*, **3**, 201–204.
92. Touchot, N., Chardin, P., and Tavitian, A. (1987) *Proc. Natl. Acad. Sci. USA*, **84**, 8210–8214.
93. Madaule, P., and Axel, R. (1985) *Cell*, **41**, 31–40.
94. Yeramian, P., Chardin, P., Madaule, P., and Tavitian, A. (1987) *Nucl. Acids Res.*, **15**, 1869.
95. Lowe, D. G., Capon, D. J., Delwart, E., Sakaguchi, A. Y., Naylor, S. L., and Goeddel, D. V. (1987) *Cell*, **48**, 137–146.
96. Bucci, C., Franzio, R., Chiariotti, L., Brown, A. L., Rechler, M. M., and Bruni, C. B. (1988) *Nucl. Acids Res.*, **16**, 9979–9993.
97. Salminen, A., and Novick, P. J. (1987) *Cell*, **49**, 527–538.
98. Olofsson, B., Chardin, P., Touchot, N., Zahraoni, A., and Tavitian, A. (1988) *Oncogene*, **3**, 231–234.
99. Yamamoto, K., Kim, S., Kikuchi, A., and Takai, Y. (1988) *Biochem. Biophys. Res. Comm.*, **3**, 1284–1292.
100. Matsui, Y., Kikuchi, A., Kondo, J., Hishida, T., Teranishi, Y., and Takai, Y. (1988) *J. Biol. Chem.*, **263**, 11071–11074.
101. Gallwitz, D., Donath, C., and Sander, C. (1983) *Nature (London)*, **306**, 704–707.
102. Haubruck, H., Disela, G., Wagner, P., and Gallwitz, D. (1987) *EMBO J.*, **6**, 4049–4053.
103. Garrett, M. D., Self, A. J., Van Oers, C., and Hall, A. (1989) *J. Biol. Chem.*, **264**, 10–13.
104. Madaule, P., Axel, R., and Myers, A. M. (1987) *Proc. Natl. Acad. Sci. USA*, **84**, 779–783.
105. Sewell, J. L., and Kahn, R. A. (1988) *Proc. Natl. Acad. Sci. USA*, **85**, 4620–4624.
106. March, P. E., Lerner, R. G., Ahan, J., Cui, X., and Inouye, M. (1988) *Oncogene*, **2**, 539–544.
107. Wagner, P., Molenaar, C. M. T., Rauh, A. J. G., Brökel, R., Schmitt, H. D., and Gallwitz, D. (1987) *EMBO J.*, **6**, 2373–2379.
108. Schmitt, H. D., Wagner, P., Pfaff, E., and Gallwitz, D. (1986) *Cell.*, **47**, 401–412.
109. Segev, N., Mulholland, J., and Botstein, D. (1988) *Cell*, **52**, 915–924.

110. Schmitt, H. O., and Gallwitz, D. (1988) *Cell*, **53**, 635–647.
111. Lowe, D. G., and Goeddel, D. V. (1987) *Mol. Cell. Biol.*, **7**, 2845–2856.
112. Kitayama, H., Sugimoto, Y., Matsuzaki, T., Ikawa, Y., and Noda, M. (1989) *Cell*, **56**, 77–84.
113. Quillan, L. A., Brown, J. H., and Buss, J. E. (1988) *FEBS Lett.*, **238**, 22–26.
114. Morii, N., Sekue, A., Ohashi, Y., Nakao, K., Imura, H., Fujiwara, M., and Narumija, S. (1988) *J. Biol. Chem.*, **263**, 12420–12426.
115. Rubin, E. J., Gill, M. D., Boquet, P., and Popoff, M. R. (91988) *Mol. Cell. Biol.*, **8**, 418–426.
116. Adam-Vizi, V., Knight, D., and Hall, A. (1987) *Nature (London)*, **328**, 581.
117. Narumiya, S., Sekine, A., and Fujiwara, M. (1988) *J. Biol. Chem.*, **263**, 17255–17257.
118. Aktories, K., Braun, U., Rösener, S., and Hall, A. (1989) *Biochem. Biophys. Res. Comm.*, **158**, 209–213.

12
ALTERED EXPRESSION AND FUNCTIONING OF GUANINE NUCLEOTIDE BINDING (REGULATORY) PROTEINS DURING GROWTH, TRANSFORMATION, DIFFERENTIATION AND IN PATHOLOGICAL STATES

Miles D. Houslay

Molecular Pharmacology Group, Institute of Biochemistry, University of Glasgow, Glasgow G12 8QQ, Scotland, UK

INTRODUCTION

Types of G-proteins

Cell surface receptors may broadly be classified into two groups. In the first group the receptor molecule possesses the ability both to recognize a ligand and generate, of itself, an intracellular signal, e.g. the nicotinic acetylcholine receptor. In the second group the receptor can recognize a ligand but relies upon another plasma membrane protein, the signal generator, to produce an intracellular signal. In the second instance it appears that the linkage between the receptor and signal generator is mediated by a class of proteins which act as 'go-betweens'. These are the guanine nucleotide regulatory (binding) proteins or G-proteins (Figure 12.1).

As reviewed elsewhere in this book and in refs [1–3], the discovery of guanine nucleotide regulatory proteins (G-proteins) centred around research performed on adenylate cyclase, the enzyme which produces the intracellular second messenger cAMP. This enzyme is commonly found to be under dual control, being regulated through

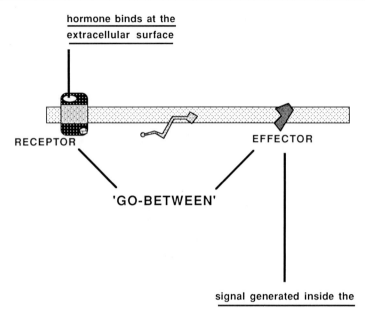

Figure 12.1 G-proteins act as 'go-betweens'. Many receptors cannot of themselves generate an intracellular signal. Rather they depend upon modifying the activity of an effector protein which transmits a signal into the cell interior. Linkage between the receptor and signal generator (effector) in many instances is achieved through specific guanine nucleotide regulatory proteins (G-proteins) which serve as go-betweens

receptors which can either stimulate or inhibit its function. The effects of such receptors are mediated by two distinct G-proteins namely G_s, which serves to couple stimulatory receptors to adenylate cyclase, and G_i, which serves to couple inhibitory receptors to adenylate cyclase (Figure 12.2).

These two G-proteins have been purified and shown to possess apparently identical β (35 kDa) and γ (8–10 kDa) subunits but have distinct α subunits of molecular mass 41 kDa for G_i and 45 kDa for G_s. Interaction of these G-proteins, appropriate receptors and adenylate cyclase involves productive collisions between mobile components in the plane of the membrane [4]. Thus, when a receptor interacts with a G-protein, the latter binds GTP and dissociates, releasing an activated GTP-bound α subunit (Figure 12.3). This, in the case of αG_s, interacts with the catalytic unit of adenylate cyclase which it stimulates. The 'turn-off' reaction is achieved by virtue of the ability of αG_s to hydrolyse GTP to GDP, where the GDP-bound αG_s is unable to stimulate adenylate cyclase and thus re-associates with $\beta\gamma$ subunits to reform the holomeric complex of G_s. The reinstated G-protein complex is then again capable of interacting with an appropriate receptor, thus providing a cycle of events which characterize this signal generation and amplification system.

EXPRESSION AND FUNCTIONING OF G-PROTEINS IN VARIOUS STATES

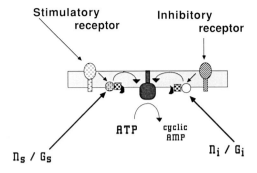

Figure 12.2 Dual control of adenylate cyclase. Regulation of adenylate cyclase is achieved by both stimulatory and inhibitory receptors. These processes are mediated via G_s and G_i respectively, whose heterotrimeric nature is indicated

The G_i-mediated inhibition of adenylate cyclase is achieved in two distinct ways. Firstly, by virtue of the release of (identical) $\beta\gamma$ subunits which serve to inhibit the dissociation of G_s. Secondly, through the action of the GTP-bound αG_i which appears capable of interacting and inhibiting the catalytic unit of adenylate cyclase directly (Figure 12.4).

G_s and G_i are not the only G-proteins known. Transducin, a G-protein involved in coupling rhodopsin to the activation of a cGMP-specific phosphodiesterase, has been well characterized; a G-protein called G_o has been identified and suggested to play a role in regulating Ca^{2+} channels, although its functional role has yet to be unequivocally defined; at least two distinct G-proteins appear able to couple certain receptors to the stimulation of inositol phospholipid metabolism and while these have yet to be identified, one is believed to be capable of being ADP-ribosylated and inactivated by pertussis toxin and the other not (see elsewhere in this book). Molecular, biological and protein purification techniques have also identified a number of guanine nucleotide binding proteins whose function has yet to be defined and circumstantial evidence has indicated that G-proteins may mediate certain forms of exocytosis, intracellular protein transport and certain of insulin's actions.

Over the past few years it has become apparent that there are a number of G_i-like proteins (Figure 12.5). These are proteins which are structurally related to the first protein (G_i), identified as being capable of inhibiting adenylate cyclase and being able to be functionally inactivated by pertussis-toxin-catalysed ADP-ribosylation [1–3]. At least three such species comprise this family, these have been termed G_i1, G_i2 and G_i3 and all have been cloned [5,6]. G_i1, G_i2 and G_i3 are all able to be ADP-ribosylated by pertussis toxin. However, a related species has been identified by molecular cloning [7], although in

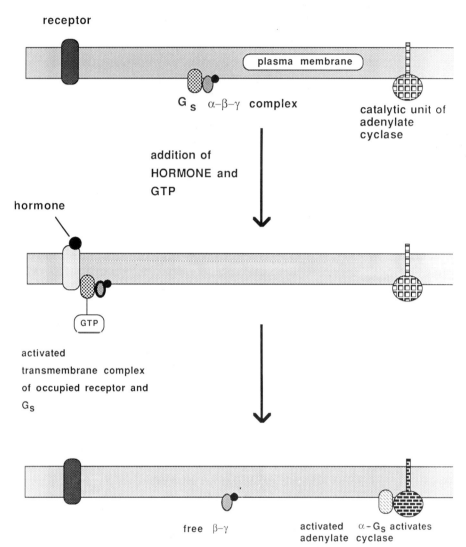

Figure 12.3 Mobile receptor model for activation of adenylate cyclase. Components of this regulatory system are able to undergo lateral diffusion in the bilayer. The receptor, upon binding hormone, interacts with G_s and causes it to bind GTP. The complex then dissociates and the GTP-bound αG_s subsequently activates adenylate cyclase. The 'turn-off' reaction ensues with the hydrolysis of GTP to GDP

this case the cysteine group, which serves as the target in G_i1 to G_i3 (inclusive) appears to be blocked in this species and it cannot be ADP-ribosylated. The functional attributes of these four proteins are still a hotly contested issue. One species (G_i3) is suggested to be responsible for controlling K^+ channel functioning, another (G_i1) is suggested to be the

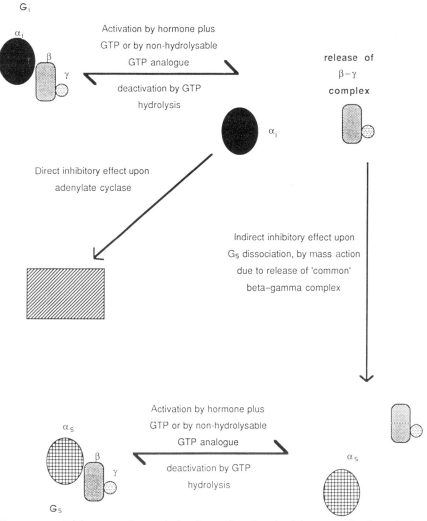

Figure 12.4 Inhibitory regulation of adenylate cyclase. Details of the two modes through which inhibitory regulation of adenylate cyclase can be achieved, i.e. through αG_i interacting with adenylate cyclase and through the common $\beta\gamma$, mass action effect.

pertussis-toxin-sensitive G-protein involved in inositol phospholipid metabolism and the other (G_i2) is suggested to be the 'true' G_i, i.e. capable of inhibiting adenylate cyclase by virtue of its α subunit [8]. However, in HL-60 cells the stimulation of inositol phospholipid metabolism is blocked by pertussis toxin yet no G_i1 is expressed. Thus the 'true' roles of 'G_i' forms are far from established. Indeed this became all the more complicated by experiments which indicated that all G_i 'species' were capable of reconstituting K^+-channel function in a model system [9]. This might be explicable by

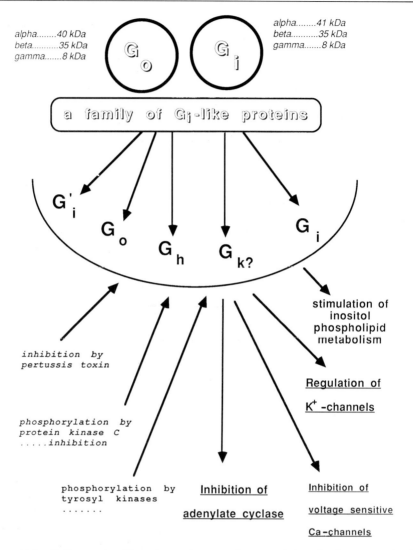

Figure 12.5 G_i-like proteins. This indicates the range of activities believed to be associated with this family of pertussis-toxin-sensitive proteins.

virtue of the considerable homology between the α subunits of these G-protein species which is likely to allow for a degree of promiscuity, especially in the highly artificial reconstitution systems which employ single G-proteins at high concentrations. Perhaps then, greater fidelity occurs in 'natural' systems, although this will require new approaches to address such problems. Thus, the molecular basis of the control of K^+-channel opening has been the subject of fierce debate, which has included the notion that the $\beta\gamma$ complex released from holomeric G_i upon receptor activation may be the species which controls K^+-channel opening [10]. Indeed, this hypothesis has received considerable support from recent experiments [11,12] which indicate that the $\beta(\gamma)$ subunit can cause the indirect rather than direct, opening of K^+ channels. The current hypothesis suggests that the released $\beta\gamma$ subunits serve to activate phospholipase A_2 [13] which then allows either the arachidonic acid produced or the metabolites generated from arachidonate metabolism through the 5-lipoxygenase pathway to induce K^+-channel activation. This role of $\beta\gamma$ subunits appears to add a further and deepening layer of complexity to the G-protein field, where we appear to see signals proliferating subsequent to the activation of both G_s and G_i.

Thus the specificity and role of forms of G_i is far from clear as regards adenylate cyclase inhibition, K^+-channel opening and inositol phospholipid stimulation. Indeed, it may be that all species might inhibit adenylate cyclase, at least to some extent, by virtue of releasing $\beta\gamma$ complexes but that one particular species provides additional inhibiting action by virtue of its interaction of its α subunit with the catalytic moiety of adenylate cyclase.

Detecting G-proteins

G-proteins can be detected either functionally or structurally.

FUNCTIONAL ASSAYS FOR G-PROTEINS

GTPase activity

Interaction of a G-protein with an appropriate occupied receptor causes it to bind GTP and become activated. The ensuing deactivation phase is elicited by virtue of the hydrolysis of GTP to GDP. Thus, G-proteins exhibit a high affinity, ligand-stimulated GTPase activity [1–3]. This provides a means of detecting G-proteins functionally. Indeed, it has proved to be highly useful in human platelets [14,15] and certain cell lines [16], where membranes from those cells exhibit low non-specific GTPase activities (Figure 12.6). However, this assay cannot be used in the majority of instances due to the high non-specific GTPase activities which preclude accurate evaluation of G-protein GTPases.

Binding studies

Alternative functional assays use non-hydrolysable GTP analogues such as p(NH)ppG and GTPγS, to activate G-proteins and stimulate the signal generation system. Also, the interaction of G-proteins with the receptor can often be identified by the ability of non-hydrolysable GTP analogues to alter the affinity of the receptor for its ligand. Conversely, membranes may be incubated with radio-labelled GTP analogues and the effect of receptor occupancy on label binding or dissociation determined. Thus a number

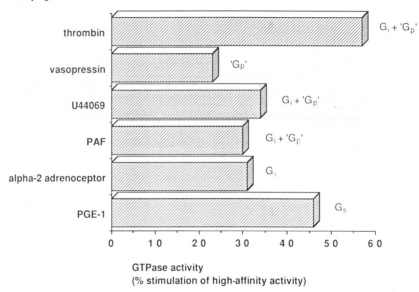

Figure 12.6 Receptor-stimulated GTPase activity in human platelets. This shows ligand stimulating GTPase activity in human platelets. The G-protein believed to be stimulated by the various agonists is indicated. That shown as 'G_p' is the activity of the pertussis-toxin-insensitive G-protein suggested to control the stimulation of inositol phospholipid metabolism (see refs [14,15])

of assays are available to determine the functioning of G-proteins in normal and aberrant states. Such assays, however, cannot be used to quantify G-proteins accurately [1–3].

Coupling to effector systems

G-proteins serve to couple receptors to their appropriate effect or signal generator systems. As the turn-off reaction is the hydrolysis of GTP to GDP, then non-hydrolysable analogues can be used to transform the G-protein into its active 'GTP'-like conformation, hence short-circuiting the need for receptor stimulation. In the case of adenylate cyclase it has been noted by a number of investigators that G_i and G_s have very different affinities for guanine nucleotides. Thus low concentrations of GTPγS can be used to activate selectively the inhibitory G-protein G_i [1–3, 17–19]. From such experiments one can detect the G_i-mediated functional inhibition of adenylate cyclase employing low doses of GTPγS or p(NH)ppG, whereas stimulation, through G_s, is achieved using higher concentrations of these analogues (Figure 12.7).

Such non-hydrolysable analogues have been used to gain information concerning the possible involvement of G-proteins in regulating ion channels exocytosis, insulin action and phosphatidylinositol metabolism (see refs [1–3,20])

The selective activation of G-proteins using different concentrations of guanine nucleotides

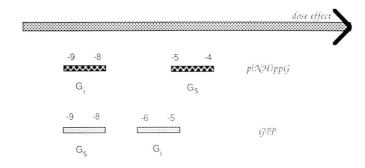

Figure 12.7 Activation of G_i in functional assays. G_i and G_s can be activated selectively by using different concentrations of either GTP or the non-hydrolysable GTP analogue, 5′-guanylyl imidodiphosphate

Reconstitution

Membranes lacking a specific G-protein but with an appropriate receptor and signal generator provide the only way to define unequivocally the role of a G-protein as the addition of an appropriate G-protein should fully reconstitute the system. They also offer a means of quantifying it functionally. Unfortunately, only one such system is available and that is cyc⁻ cells. These are a mutant cell line derived from S49 mouse lymphoma cells, whose membranes possess adenylate cyclase activity and β adrenoceptors but lack αG_s. Thus plasma membranes offer a means of assessing αG_s which can be inserted into these membranes after detergent extraction [1–3].

STRUCTURAL ASSAYS OF G-PROTEINS

The first means offered for quantifying G-proteins was the use of cholera toxin to cause the ADP-ribosylation of G_s and pertussis toxin to cause the ADP-ribosylation of G_i. This, when isolated membranes were used together with [^{32}P]NAD$^+$, led to the radio-labelling of the α subunits of G_s and G_i. It also caused the constitutive activation of G_s and the functional inactivation of G_i [1–3]. Initially toxin-catalysed ADP-ribosylation appeared to offer an attractive means of evaluating the concentration of G-proteins in membranes. However, it was soon realized that such approaches were highly unreliable due to a number of factors. This was because: (a) different membranes contained variable amounts of NADase activity which could rapidly deplete radio-labelled NAD$^+$ with dramatic effects on labelling [21]; (b) it was shown that labelling was markedly different dependent upon the state of association of the G-protein with dissociated G_i not being able to be

ADP-ribosylated by pertussis toxin [22]; (c) a membrane-bound factor, called ARF, was necessary for cholera toxin to ADP-ribosylate G_s, thus changes in the concentration of ARF could alter the ribosylation of G_s [1–3]; and (d) G-proteins other than just αG_s and αG_i, in some cases with similar molecular weights, could be modified by such toxins, thus precluding labelling of particular species with any accuracy [1–6,22–23]. Thus, in discussions below, where data on G-proteins amounts are specified as being determined by toxin-catalysed ADP-ribosylation then a degree of caution should be exercised in interpreting such phenomena until confirmed using immunological reagents.

More recently anti-peptide antisera (Figure 12.8a) have been raised to interact with specific G-protein such as G_s, G_o and the various forms of G_i, which is the subject of review elsewhere in this volume (see Chapter 3). These take advantage of the fact that while G-protein α subunits have a general homology which defines the GTP binding site and the overall domain structure, there are areas of considerable divergence. Such areas presumably account for the differences in specificity between the various G-proteins as regards their interaction with receptor classes and signal generation systems. Thus peptides synthesized against such divergent regions allow for the production of specific antisera. These can be used to quantify G-protein expression using either quantitative western blotting or ELISA techniques as well as providing a means for specific and quantitative immunoprecipitation of G-protein α subunits (see Chapter 3).

A similar approach has been used to identify and quantify mRNA for specific G-proteins, this time using synthetic oligonucleotide probes in northern- and slot-blotting assays (Figure 12.8b) (see Chapter 3).

These reagents allow for the precise measurement of protein and mRNA for G-protein species.

PSEUDOHYPOPARATHYROIDISM

Pseudohypoparathyroidism (PHP) is a genetic disorder associated with a decreased responsiveness of hormones which function through the stimulatory G-protein G_s in order to activate adenylate cyclase [25]. This, was shown by Insel and co-workers [26] to be due to an apparent reduction of functional G_s in patients with type Ia-PHP but not in those with type Ib-PHP.

In these studies the functioning of G_s was assessed by extracting G_s from the platelets of type Ia-PHP patients and reconstituting them into the membranes of a mutant form of S49 mouse lymphoma cells, called cyc$^-$ cells. This mutant cell line expressed adenylate cyclase activity and β-adrenoceptors but does not express G_s. Membranes from these cells thus provide an effective reconstitution assay for G_s. This assay allowed these investigators to show that membranes from type Ia-PHP patients exhibited reduced functional G_s. However, such an assay system did not allow them to determine whether this reduction in activity was due to a reduced amount of G_s in the membranes from platelets of these patients of whether G_s function was attenuated in some manner. While

G_s function was reduced in type Ia-PHP patients, inhibition of platelet adenylate cyclase by the α_2-adrenoceptor was normal, showing that G_i activity was unaltered.

More recently, Levine and co-workers [27] have been able to show that type Ia-PHP patients do in fact appear to exhibit a reduced expression of αG_s as indicated by the reduced concentrations of mRNA for αG_s in fibroblasts from these patients. However, quantification of protein using a specific antibody has yet to be demonstrated. Normal amounts of mRNA from αG_i were found as might have been predicted. These investigators, however, were not able to show a clear correlation between the extent of reduction of mRNA for αG_s and either the extent of reduction of G_s activity in the cells of the clinical phenotype. Thus the relevance of reduced G_s expression to this syndrome and the factors controlling G_s expression are far from clear.

There are two forms of αG_s expressed in cells, one having a molecular mass of 45 kDa and the other, which occurs from alternative RNA splicing, having a molecular mass of 52 kDa. In type Ia-PHP patients, both forms seem to be reduced as determined by quantification of mRNA and cholera-toxin-catalysed ADP-ribosylation (Figure 12.9).

Interestingly, in those type Ia-PHP patients who expressed reduced tissue G_s associated with Albright hereditary osteodystrophy, then both this syndrome and αG_s deficiency appeared to be transmitted together in a dominant, apparently autosomal, pattern. Now, each individual has two αG_s genes. As the pattern of inheritance of αG_s deficiency in these patients is most consistent with an autosomal dominant pattern of inheritance then it would seem that each type Ia-PHP patient has inherited one αG_s allele that can produce functional G_s and one allele that cannot produce αG_s. Certainly this would be consistent with a near 50% reduction in αG_s mRNA being observed in the fibroblasts of these patients. The inability of these investigators to see changed patterns upon restriction mapping of genomic DNA from these subjects, using αG_s probes, suggests that large deletions or rearrangements within the αG_s gene did not occur. Thus the loss of experience of one of these αG_s alleles may result from either a very small deletion or point mutation in a region controlling the expression of αG_s. Certainly an evaluation of control regions for this gene would be most rewarding.

CHRONIC ETHANOL ADMINISTRATION

Acute exposure to ethanol has been shown by a variety of investigators to lead to the stimulation of adenylate cyclase activity [28–31]. This occurs over a range of concentrations where ethanol increases lipid fluidity [28,32–35], then it is possible that this change provides the molecular basis for the stimulatory effect of ethanol. Chronic exposure to ethanol, however, has the opposite effect, i.e., attenuating adenylate cyclase activity, as determined both in cell systems and in lymphocytes and platelets [4,30] from alcoholic subjects. Part of this inhibitory effect may be due to the membrane-lipid-mediated adaptive response to chronic ethanol administration. The leads to increased levels of cholesterol incorporation into membranes, a lipid whose elevation in biological membranes can lead to the attenuation of adenylate cyclase activity [36]. However, it is

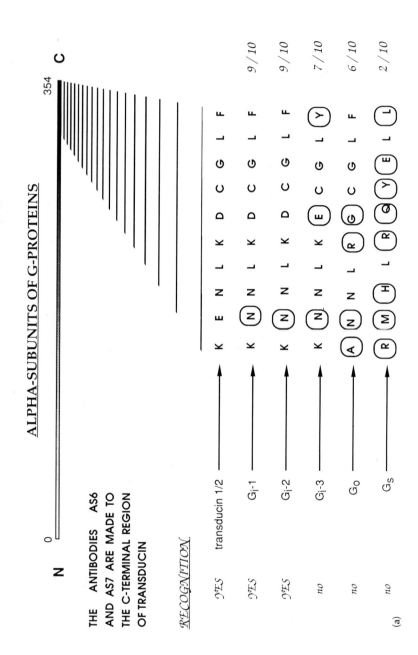

(a)

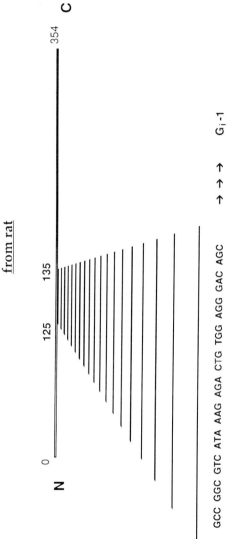

Figure 12.8 Oligonucleotides and anti-peptide antisera for specific G-proteins: (a) How the anti-peptide antisera AS6/AS7 were prepared against the C-terminal decapeptide of transducin (see ref. [58]). (b) How synthetic oligonucleotide probes can be prepared to identify message for the three forms of G_i

in pseudohypoparathyroidism type Ia, there is reduced G_s in all tissues

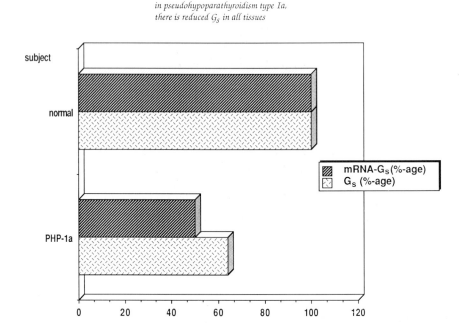

Figure 12.9 G_s in pseudohypoparathyoidism. This figure identifies the change in G_s and mRNA for αG_s in patients with type Ia pseudohypoparathyroidism [26]

clear that such a lipid-mediated effect cannot account for the magnitude of the diminished receptor-coupled adenylate cyclase activity in cells treated chronically with ethanol.

In a recent study [37], however, it has been shown that chronic treatment of a neuroblastoma cell line (NG 108–15) with ethanol led to the reduced expression of αG_s. This was monitored by both reconstitution into cyc$^-$ cells and by using a specific anti-peptide antibody (Figure 12.10). The lesion appeared to be due to decreased synthesis of αG_s rather than enhanced degradation as the mRNA concentrations were also reduced by a comparable amount (Figure 12.10). This reduction in mRNA for αG_s did not appear to reflect a general, non-specific effect on either transcription or mRNA, stability as, at the same time, β-tubulin mRNA was slightly increased in these experiments. Indeed, other studies have shown that chronic ethanol treatment elicited a decrease in mRNA for pro-opiomelanocortin [38] and a large increase in mRNA for the major histocompatibility complex protein [39].

It will be of much interest to define the extent and mechanism(s) whereby ethanol alters the mRNA concentrations of αG_s and other species. Certainly, these changes in mRNA concentrations coupled with its membrane-mediated actions ensure that ethanol exerts complex and pleiotropic effects upon cellular functioning. Importantly, both of these actions are going to have profound effects upon cellular signalling mechanisms.

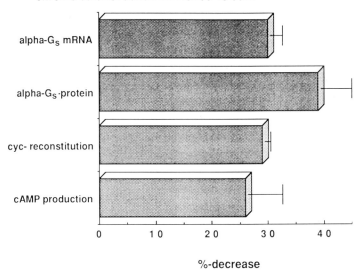

Figure 12.10 Changes in G_s function, protein and mRNA in neuroblastoma cells subjected to chronic ethanol administration. Administration of ethanol to NG 108–15 cells causes a loss of PGE_1-and adenosine-stimulated cAMP accumulation. The basis for this appears to be a reduction in αG_s [36]

GROWTH-HORMONE-SECRETING PITUITARY ADENOMAS

Secretion from somatotrophic cells is known to be controlled by alterations in the intracellular concentration of cAMP. A number of human growth-hormone (GH)-secreting adenomas have been shown [40] to have abnormally high intracellular concentrations of cAMP and to show very high rates of secretion of GH in the absence of stimulation by growth hormone releasing hormone (GHRH). Indeed, in such cells GHRH had little effect on both the intracellular concentration's level of cAMP and on the rate of GH secretion.

When the adenylate cyclase activity of GH-secreting adenomas was studied in isolated membranes, it was shown that this activity was hyper-responsive to stimulation by Mg^{2+}. Now G_s is known to reduce the K_a-value for the activation of adenylate cyclase by Mg^{2+}. Thus, hyper-responsiveness of adenylate cyclase to Mg^{2+} suggests that either G_s or adenylate cyclase, in these tumour cells, has adopted a highly (fully) activated conformation.

These two possibilities have been investigated by a number of very different approaches which indicate that the lesion is due to the presence of a hyper-responsive form of G_s in these cells. One experiment showed that extraction of G_s from adenoma cell membranes and its subsequent incorporation into the membranes of cyc$^-$ cells conferred the same hyper-responsiveness to Mg^{2+} on the adenylate cyclase of cyc$^-$ cells. The second showed that cholera toxin treatment of these adenoma cells, while causing the ADP-ribosylation of the α subunit of G_s, did not elicit activation. This presumably was because G_s was already in an active conformation. However, the cholera toxin experiments did show a reduced level of ADP-ribosylation of αG_s in these cells, which indicated that the activated form of this G-protein was not such a good substrate for the toxin as the native protein. These two experiments clearly differentiate between the two possibilities, showing that a hyper-responsiveness G_s is present. As might be expected, the addition of GHRH effected but a small stimulation of adenylate cyclase activity and with ligands known to activate G_s directly, e.g. Na(Al)F and guanylyl 5'-imidodiphosphate (p(NH)ppG) failed to activate this enzyme, indicating that indeed G_s had adopted a fully activated conformation in these cells.

Such altered G_s was not found in all hypersecreting adenomas, but was observed in a substantial group. It appears that in the 'G_s-activated' group, a mutant form of αG_s is produced which adopts an active conformation. This causes adenylate cyclase to become constitutively activated in these cells, akin to that seen when normal cells are treated with cholera toxin. The molecular basis which gives rise to what is presumably a mutant form of αG_s is likely to be of much interest and should shed light on how the structure of this protein relates to its functioning.

See note added in proof (page 227).

S49 LYMPHOMA CELL LINE

A mutant, cyc⁻, derived from this cell line provided the means for the first unequivocal demonstration of a G-protein, namely G_s. If this were its only contribution to the field then it would deserve mention. However, various investigators have also used mutants from it to make important discoveries in the G-protein field [1–3].

The underlying selection mechanism, which has proved to be so useful, is that elevation of intracellular cAMP levels is cytotoxic to these cells. Thus mutants in cyclic AMP metabolism and in A-kinase can be and have been identified. One particular mutant has been discussed previously, namely the cyc⁻ cell line which fails to express G_s while still expressing β-adrenoceptors and adenylate cyclase. As such, its plasma membranes offer a useful reconstitution system for G_s.

Another mutant cell line called UNC, has also attracted attention [41]. This was identified as having an aberrant G_s protein which although capable of stimulating adenylate cyclase when activated directly, using either Na(Al)F or the non-hydrolysable GTP analogue p(NH)ppG, could not mediate β-adrenoceptor stimulation of this enzyme. This suggested a lesion in the domain of αG_s which interacted with the β-adrenoceptor. Consistent with this was the failure to observe guanine-nucleotide-mediated effects on β-agonist binding to the β-adrenoceptor. Sequence analysis of the gene for αG_s from these cells [42,43] has shown a single point mutation ARG–PRO at position 372, which is at the C-terminal end of this protein. The location of the mutation at this point gives credence to the suggestion that the domain of G-protein α subunits, which interacts with specific receptors, is located at the C-terminal end of the molecule (Figure 12.11).

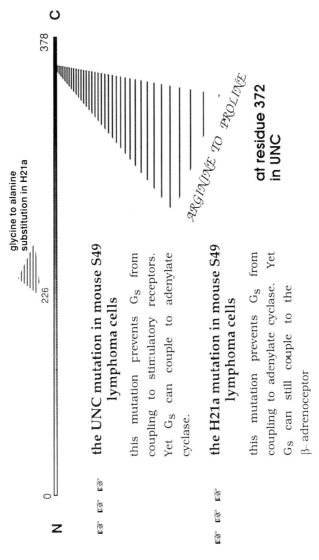

Figure 12.11 Cyc⁻ and H21a: mutations in αG_s. This figure shows the location of point mutations in αG_s found in the UNC and H21a mutations. The numbering of residues is as defined in refs [41,43]

Another mutant form of S49 cells which shows a lesion in G-protein-mediated coupling has been identified. This cell line is called H21a [44]. In this instance, the adenylate cyclase of these cells fails not only to respond to β-adrenoceptor agonist stimulation but also is not activated by ligands which interact with G_s directly, such as Na(Al)F, non-hydrolysable GTP analogues and cholera toxin. Nevertheless, guanine nucleotide analogues are capable of modifying β-agonist binding, indicating that the form of G_s expressed in these cells is capable of coupling to the β-adrenoceptor but is incapable of coupling to adenylate cyclase [44,45]. Sequence analysis of the α subunit of G_s from such a cell line has identified [45] a single point mutation at residue 226 where glycine is replaced by alanine. That the lesion in coupling was due to this defect only was confirmed by expressing the mutant αG_s (H2a) in cyc$^-$ cells, whereby guanine nucleotide sensitivity to β-agonist binding was reconstituted whereas β-agonist stimulation of adenylate cyclase was not.

Interestingly, the region where this mutation lies is highly conserved among the various G-proteins and thus has been considered as an unlikely domain for providing the required specificity to direct αG_s to interest with a specific second messenger system, i.e. adenylate cyclase. It may be, however, that this highly conserved domain is responsible for transducing a conformational change between the GTP binding domain and that domain, postulated to be at the N-terminal region of αG_s, which is responsible for interacting directly with adenylate cyclase. However, one cannot exclude the possibility that the region incorporating residue 226 does not provide part of, or stabilize, the interaction site with adenylate cyclase.

Experiments using tryptic proteolysis to probe for conformational changes induced by guanine nucleotides did show aberrations from the normal, indicating that this single mutation can alter the GTP-induced conformational changes in αG_s [45].

It will be interesting to see if further mutations are found in G-proteins of particular aberrant cells. Certainly, the most obvious instance of this so far is the *ras* gene system which is discussed elsewhere in detail in this book (Chapter 11). Briefly, there is a normal (proto-oncogenic) form of the *ras* gene which is expressed as a 21 kDa protein in many cells. Single point mutations lead to the production of so-called 'activated' oncogenic forms of *ras* which display a reduced GTPase activity compared with the normal protein. Site-specific mutagenesis studies have been performed on this protein in order to identify the domains which are essential for its functioning and for activation (see [46] for example). Such approaches are in progress in various laboratories for probing other G-proteins.

ADRENALECTOMY

Steroid hormones have been shown to alter the ability of a variety of cells and tissue to respond to hormones which act via the regulation of adenylate cyclase [47,48]. This has been noted in both liver [49] and adipose [48] tissue.

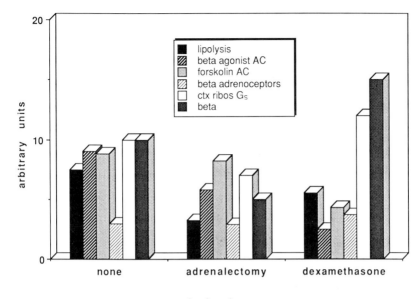

Figure 12.12 Adrenalectomy on adipocyte adenylate cyclase. The effect of adrenalectomy and dexamethasone treatment of adrenalectomized animals on the catechloramine-stimulated adenylate cyclase system from rats is shown [50]. The data show relative activities for lipolysis; β-adrenoceptor agonist-stimulated adenylate cyclase (AC); forskolin-stimulated adenylate cyclase (an index of the amount of catalytic unit); β-adrenoceptor number; cholera toxin (ctx)-mediated ADP ribosylation (ribos) of αG_s and G-protein β subunit (beta)

In the adipocytes of adrenalectomized rats there is an impaired lipolytic response to adrenaline (β-adrenoceptor) stimulation [50]. An analysis of the action of adrenalectomy, upon the adenylate cyclase system of adipocytes from such animals, indicates that glucocorticoids may affect differentially the components of this signal transduction system [50]. Thus, although the number of β-adrenoceptors appears not to be reduced by adrenalectomy and there is no change in the amount of adenylate cyclase, as indicated by forskolin-stimulated adenylate cyclase activity (Figure 12.12), there appears to be a marked reduction in G_s. This, however, was determined by the cholera-toxin-catalysed ADP-ribosylation of αG_s and this needs confirmation by immunological means. However, it is likely that G_s was reduced as a comparable reduction in G-protein β-subunit was also noted by these investigators [50]. Adrenalectomy may thus reduce ligand-stimulated adenylate cyclase activity by attenuating the expression of G_s in adipocytes [50], reticulocytes [51] and hepatocytes [52].

Consistent with this treatment of adrenalectomized animals [53] with dexamethasone enhanced the levels of G_s and thus β-adrenoceptor stimulation of adenylate cyclase in adipocyte membranes (Figure 12.12). However, dexamethasone treatment also appeared to decrease the amount of adenylate cyclase activity, as assessed by forskolin stimulation of the catalytic unit. This observation, with dexamethasone, is in contrast to that found by treating growth-hormone-secreting (GH$_3$) cells with dexamethasone whereupon forskolin-stimulated adenylate cyclase activity was markedly increased [54]. It seems, therefore, that the expression of adenylate cyclase activity may well be affected by glucocorticoids but that the effects are specific to particular cell types. Nevertheless, in GH$_3$ cells, dexamethasone treatment also appeared to enhance the expression of G$_s$ as indicated by an increase in the cholera-toxin-catalysed labelling of αG$_s$ [54].

Analysis of genomic clones for αG$_s$ will thus be of interest in identifying the regulatory regions which presumably determine these effects. Whether defects occur in pathological states remains to be seen. As is discussed below, changes in glucocorticoids and thyroid hormone concentrations are likely to have profound effects upon cellular signalling systems involving G-proteins.

HYPOTHYROIDISM

Hypothyroidism in rats can be induced by treatment with 6-propyl-2-thiouracil. In adipocytes from such hypothyroid animals a number of investigators have noted that receptors which are linked to the inhibition of adenylate cyclase, e.g. α_2-adrenoceptors and A$_2$-adenosine receptors, show an increased responsiveness compared to that noted in control animals [54–56], suggesting that hypothyroidism might lead to an increase in the expression of G$_i$ in adipocytes from these animals. Certainly, in rats this would appear to be the case as hypothyroidism has been shown to lead to an enhanced labelling of the 40 kDa α subunit of G$_i$ [54]. Specific antipeptide antibodies have since been used to provide clear evidence of enhanced expression of αG$_i$ in adipocyte membranes from hypothyroid animals [57]. Indeed, concentrations of αG$_i$ in adipocyte membranes from hypothyroid animals were nearly three-fold greater than controls and nearly six-fold greater than seen in adipocytes from euthyroid animals, where expression of αG$_i$ was reduced relative to the normal animals (Figure 12.13).

Such changes in the amount of αG$_i$ appeared to be paralleled by alterations in β subunit expression. This might indicate some coordination in the expression of these two proteins.

It has yet to be determined whether hypothyroidism elicits similar changes in G$_i$ in other tissues and in other species and whether changes in thyroid status regulate G$_i$ concentrations by increasing expression or decreasing the degradation of αG$_i$ and its β subunit.

DIABETES AND INSULIN-RESISTANT STATES

Treatment of rats with either streptozotocin or alloxan elicits the destruction of the pancreatic β cells which secrete insulin. This induces diabetes, as evidenced by hyperglycaemia and hypoinsulinaemia, and also causes insulin resistance in liver and adipose tissues. Such an effect can, paradoxically, be reversed by insulin therapy.

In hepatocytes [58] and adipocytes [MDH, D. Strassheim, unpublished] from such diabetic rats there appears to be a total loss of functional G_i [58]. This has been assessed by the failure of either p(NH)ppG or GTP to inhibit adenylate activity in membranes derived from diabetic, but not from normal, animals (Figure 12.14). Intriguingly, however, inhibitory coupling of receptors to adenylate cyclase is not abolished in either intact hepatocytes or adipocytes from such diabetic animals. Thus the lesion, which we can show is due to phosphorylation of $\alpha G_i 2$, blocks guanine nucleotide but not receptor-mediated inhibition.

Loss of GTP-activated G_i has a corollary: namely, that it leads to a sensitization of adenylate cyclase to the action of stimulatory hormones. Thus the stimulatory effect of glucagon is enhanced in hepatocyte membranes and the stimulatory β-adrenergic function of adrenaline is enhanced in adipocyte membranes (Figure 12.15). Such observations parallel the effect of treating these cells with pertussis toxin, in order to inactivate G_i, where again the functioning of adenylate cyclase stimulatory hormones is enhanced. As might then be predicted, pertussis toxin treatment of hepatocytes from diabetic animals failed to modify adenylate cyclase activity.

The factor which leads to loss of G_i function remains to be established. Certainly, the fact that two very different compounds, namely alloxan and streptozotocin, elicited this effect suggests that they did not exert any direct action. Indeed, this was confirmed by

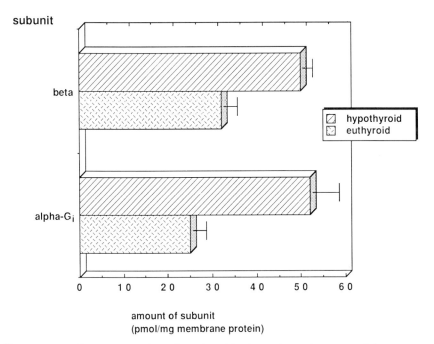

Figure 12.13 G_i expression in hypothyroidism. This figure shows the very much larger amount of αG_i in adipocyte membranes from hypothyroid animals [49]

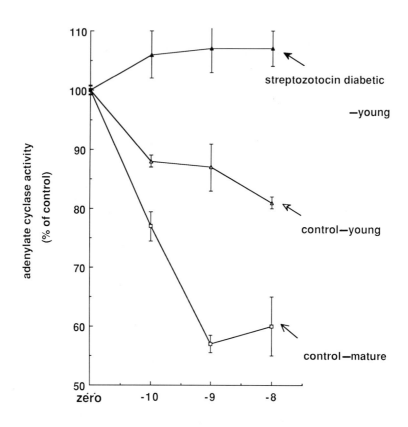

Figure 12.14 Functional G_i in membranes from hepatocytes of normal, young and diabetic rats [50]. In this instance G_i was selectively activated using guanylyl 5′-imidophosphate, in order to inhibit adenylate cyclase activity

the fact that insulin therapy of streptozotocin-diabetic animals restored G_i function.

The loss of G_i function in hepatocytes and adipocytes appears not to be a phenomenon that is either associated with, or occurs as a consequence of, insulin resistance. This is suggested by the observations that loss of G_i function can also be seen in obese Zucker rats and in ob^+/ob^+ mice [59]. Now Zucker rats appear to exhibit a single gene defect which gives rise to obese animals among normal lean littermates as is found similarly for ob^+/ob^+ mice. Insulin resistance, but not diabetes, characterizes these obese animals in the mature state, whereupon loss of functional G_i is apparent in both the hepatocytes and adipocytes from these animals.

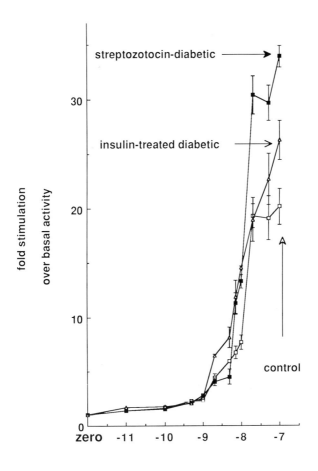

Figure 12.15 Enhanced glucagon stimulation of adenylate cyclase in hepatocyte membranes from diabetic rats. This shows the dose-dependent stimulation of hepatocyte membrane adenylate cyclase activity by glucagon. Membranes were obtained from normal animals, those made diabetic with streptozotocin and diabetic animals undergoing insulin therapy [58]

The apparent loss of G_i function in these various systems could be to: (a) reduced or loss of expression of G_i; (b) expression of an inactive form of G_i; or (c) inactivation of G_i by a post-translational modification (phosphorylation).

Using either pertussis toxin to elicit the NAD^+-dependent ADP-ribosylation of the α subunit of G_i or specific anti-peptide antibodies directly towards αG_i it has been shown that levels of both G_i2 and G_i3 are severely reduced in hepatocyte membranes from

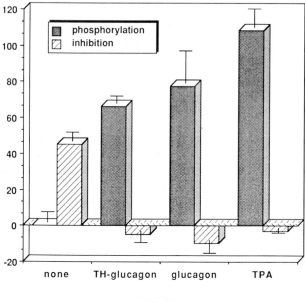

Figure 12.16 Phosphorylation of αG_i in hepatocyte membranes. Intact hepatocytes were treated with either glucagon or TH-glucagon—an analogue that does not activate adenylate cyclase but does stimulate inositol phospholipid metabolism—or the phorbol ester TPA. The functional activity of G_i was assessed as was the incorporation of ^{32}P into its α subunit [39]

diabetic animals. This could explain the failure to observe any function. The residual G_i present in membranes from diabetic animals is thus in an inactive state. Indeed, using adipocytes from streptozotocin-diabetic animals and in both hepatocytes and adipocytes from obese Zucker rats there was no change in G_i expression despite there being a loss of functional G_i. This suggests that G_i is also inactive in these membranes.

A possible explanation for the inactivation of G_i is that it could be phosphorylated. It is known that purified C-kinase can cause the phosphorylation of pure G_i [60] and that treatment of cells with phorbol esters, which activate C-kinase, also inactivates G_i [60]. Recently, we have shown [61] that treatment of hepatocytes with either phorbol esters or hormones which lead to C-kinase activation, causes the phosphorylation and inactivation of G_i (Figure 12.16) and the phosphorylation of αG_i in intact platelets via C-kinase activation has also been noted [62]. It is possible, therefore, that the aberrant activation of C-kinase in cells from diabetic rats and from obese Zucker rats could result in the phosphorylation of αG_i and concomitant loss of function. Indeed, we have been able to show that whereas G_i can be phosphorylated by C-kinase activation in hepatocytes from lean Zucker rats it cannot be phosphorylated in hepatocytes from obese Zucker rats. This suggests that G_i is already phoshorylated in the hepatocytes from the obese animals, thus offering a molecular explanation for the defect seen in the insulin-resistant animals. However, it remains to be seen as to what causes the purported aberrant activation of C-kinase in hepatocytes and adipocytes from these animals. It could be that increased levels of a hormone which stimulates inositol phospholipid metabolism serves to enhance the turnover of this lipid and thus raise the intracellular concentrations of diacylglycerol. Indeed, recently it has been shown [63] that the plasma levels of vasopressin are markedly increased in genetic and streptozotocin-induced diabetes (Figure 12.17). This might account of the increased diacylglycerol levels seen in the myocardium [64] of streptozotocin-diabetic rats. Such changes might be expected to enhance C-kinase activity leading to the inactivation of G_i by phosphorylation, attenuation of the functions of the insulin receptor [65] and to account for the observed down-regulation of β-adrenoceptors [66].

Insulin resistance and loss of functional G_i have also been elicited in vitro using isolated intact adipocytes [67]. This was achieved by incubating adipocytes for up to 72 h with A_1-adenosine receptor agonists, which induced a 50% loss of αG_i, as assessed by pertussis-toxin-catalysed ADP-ribosylation; a 70% loss of cell surface A_1-adenosine

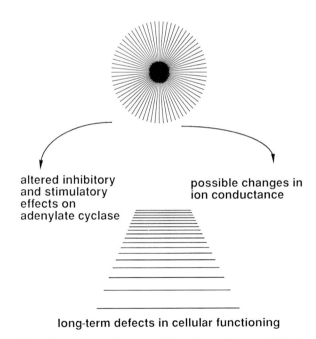

Figure 12.17 Is insulin resistance characterized by a loss of functional G_i? It is suggested that insulin resistance is elicited by enhanced stimulation of the inositol phosholipid pathway which may occur through elevated plasma vasopressin concentrations [63]. This will increase tissue diacylglycerol concentrations [64] which will activate C-kinase. The net effect will be the phosphorylation and inactivation of G_i [61] and the phosphorylation and attenuation of activity of the insulin receptor by C-kinase [65]

receptors and a 60% reduction in the ability of insulin to stimulate glucose transport (in the presence of adenosine deaminase).

There is thus considerable evidence that indicates that loss of functional G_i, whether achieved by reduced expression or inactivation, is associated with various insulin-resistant systems. Certainly there is tantalizing evidence to suggest that the insulin receptor can interact with G_1. Initially this came from observations that treatment of cells with pertussis toxin could attenuate the ability of insulin to inhibit adenylate cyclase activity [68] and to activate a specific cAMP phosphodiesterase in both hepatocytes [69] and adipocytes [70]. However, the former process is not now believed to be G_i-mediated [19]. Furthermore, purified insulin receptor preparations can phosphorylate G_i while it is in its GDP-bound, holomeric state but not when it is activated and dissociated by GTP analogues [71,72]. This suggests a specificity associated with this response. Recent evidence has also shown that insulin can modify the ability of pertussis toxin to catalyse the ADP-ribosylation of αG_i in cell membranes and to modify G_i function. However, such events do not appear to be mediated by insulin catalysing the phosphorylation of G_i.

Begin-Heick and co-workers [73–75] have performed extensive and systematic studies of adenylate cyclase activity in adipocytes and hepatocytes from lean and obese (ob/ob) mice. This system originally gained interest from a large number of investigators due to the fact that adenylate cyclase was resistant to stimulation by isoprenaline in adipocytes from obese animals [76–79]. This appears to be an effect that is specific for isoprenaline as it is not apparent for either glucagon or ACTH, which also effect stimulation of adenylate cyclase. Indeed, ACTH stimulation of adenylate cyclase is more pronounced in membranes from obese animals than in those from lean animals. Thus it is tempting to conclude that the reduced stimulatory action of the β-adrenoceptor is due to its aberrant function rather than to any particular defect in the G-protein system.

Intriguingly, however, Begin-Heick and co-workers [73] have noted recently that G_i function in membrane from both adipocytes and hepatocytes of obese animals is aberrant. Thus no evidence of G_i function was apparent using either p(NH)ppG to inhibit forskolin-stimulated adenylate cyclase or high concentrations of GTP to inhibit isoprenaline-stimulated adenylate cyclase. Consistent with this, treatment of animals with pertussis toxin, to elicit the ADP-ribosylation of G_i enhanced the stimulatory action of hormones in membranes from lean animals but did not affect the system in obese animals. This is despite similar labelling of a 40 kDa protein in membranes from both animals. This animal model appears then to resemble Zucker rats, where G_i is apparently present in the membranes from obese animals but is in an inactive state.

Begin-Heick [80] has suggested that the failure to observe functional G_i in obese states may be due to an enhanced expression of β subunits of this G-protein. We have, however, been able to show that β subunit expression is identical in lean and obese Zucker rats. Our preferred explanation for the lesions is therefore an aberrant αG_i molecule. This is unlikely to result from a mutation in αG_i as the gene for this protein does not map with the ob gene. Rather, we consider the lesion is a post-translational one, namely the phosphorylation and inactivation of G_i. This is likely to be elicited by C-kinase, which can phosphorylate and inactivate G_i. Indeed, C-kinase can phosphorylate and attenuate the functioning of the insulin receptor which could account for insulin resistance [65]. Thus we suggest that aberrant C-kinase activity may be a key lesion in these obese animals.

In the experimental models of diabetes, the enhanced stimulatory effect of hormones is evident. This effect seems clearly due to G_i inactivation as it can be mimicked by treating cells from control animals with pertussis toxin to inactivate G_i, whereas this toxin has no such effect on cells from diabetic animals [58,59]. However, fasting has been shown also to enhance the stimulatory effect of isoprenaline in adipocytes. In this instance, no alteration in G_i responsiveness appears to occur and no change in αG_i labelling is mediated via pertussis toxin [81]. Enhancement of GTP-stimulated adenylate cyclase activity is also apparent in adipocytes from such animals, suggesting that the basis for this effect increase G_s activity. Whether this is due to an increased amount of G_s remains to be determined.

HEART FAILURE

Changes in the intracellular concentration of cAMP are believed to play a fundamental role in the contractile response of cardiac tissue. Thus β-adrenergic stimulation is seen to promote contraction and selective inhibitors of cAMP-specific phosphodiesterases find therapeutic use in treating congestive heart failure. A preliminary study [82] has been performed to evaluate whether changes in G_i occur in heart failure, using pertussis-toxin-promoted labelling of αG_i to quantitate this G-protein. Interestingly, it was found that patients (three) with idiopathic dilated cardiomyopathy showed more intense labelling of αG_i than did non-failing hearts. This accompanied a dramatic decrease in isoprenaline-mediated contraction in the diseased hearts (Figure 12.18). In contrast, a single patient with end-stage heart failure due to myocarditis showed no change in either G_i labelling or in isoprenaline-induced contraction, compared to the control subjects. It is possible then that in idiopathic dilated cardiomyopathy the evaluated levels of G_i could serve to diminish the action of isoprenaline, although functional studies using pertussis toxin to inactivate this protein are needed to confirm such a conclusion. An increase in G_i does not, however, appear to characterize all cases of heart failure as clearly on change occurred in myocarditis. Furthermore, these authors noted in two patients with ischaemic heart disease there was an apparent 40% reduction in G_i present, as assessed by pertussis toxin labelling.

Clearly these are provocative preliminary results which deserve further attention in order both to corroborate and extend such findings. In view of the fact that in animal models of insulin resistance one sees lowered G_i levels in fat and liver it will be of interest to see if this also occurs in heart and if this has a physiological/pathological action. Such studies also need to be extended by using specific antisera and oligonucleotide probes to identify the changes in expression of particular forms of G_i in cardiac tissue. This should prove a particularly interesting avenue of study.

Reduced β-adrenergic responsiveness appears to be characteristic of the myocardium

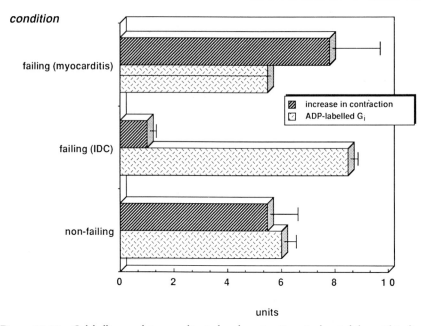

Figure 12.18 G_i-labelling and isoprenaline-induced contractions in heart failure. This figure shows the force of the isoprenaline-induced contractions (mN) in hearts from non-failing and failing (myocarditis and idiopathic dilated cardiomyopathy—IDC) patients together with the pertussis-toxin-catalysed ADP-ribosylation of αG_i in membranes from such hearts (counts × 1000) from ref. [82]

in congestive heart failure [83,84] and certainly enhanced G_i function may offer one contributing factor. Nevertheless, it has been shown that reduced numbers of β-adrenergic receptors are also found in tissue obtained from failing hearts [85,86]. Whether this is sufficient to account for the loss of responsiveness is unclear. Indeed, adenylate cyclase in a number of tissues has been shown to be activated via a collision coupling mechanism [4], which buffers the system considerably against changes in receptor concentration. Thus it has been noted [4] that when hepatocyte adenylate cyclase is stimulated by glucagon, over 75% of glucagon receptors have to be lost before any significant reduction in the ability of glucagon to activate adenylate cyclase occurs. Similar results were also found for the β-adrenergic receptor stimulation of adenylate cyclase in turkey erythrocytes [87], except that the figure was closer to 90%. In the adenylate cyclase system, the key feature which controls the kinetics of activation and amplification is the G-protein G_s. Thus attenuation of its functioning by changes in G_i concentration might well be very important.

Of course, the concentration and properties of G_1 itself are crucial in this system and, in this regard a provocative study [88] has indicated that there may be a dramatic reduction in the concentrations of G_s found in congestive heart failure. Again, toxins were used to assess α subunit concentrations. In this instance cholera toxin labelling of αG_s was reduced by a huge 80% in the lymphocytes of patients with congestive heart failure, compared to normal controls. These patients also exhibited a 40% reduction in lymphocyte β-adrenergic receptor number. Interestingly, however, treatment of these patients with angiotensin-converting enzyme inhibitors, either captopril or lisinopril, led to clinical improvement, a 60% increasing in the number of β-adrenoceptors and a doubling of cholera toxin labelling of αG_s. This is a very provocative study that may shed light on the molecular lesions in signal transduction systems that underly changes in congestive heart failure. However, in order for this to be realized we will have to await not only analyses performed on myocardial tissue but also using immunological probes for G-protein subunits.

GROWTH, DIFFERENTIATION AND AGEING

Little work has been done in this area. However, an interesting study which points to the potential value of such studies is seen with the examination of 3T3-L1 cell differentiation. This is a fibroblast cell line that can be induced to differentiate to adipocytes in the presence of the cAMP phosphodiesterase to inhibit isobutylmethylxanthine and the glucocorticoid, dexamethasone [89]. Such differentiation is accompanied, among many other changes, by the increased ability of lipolytic agents such as the β-adrenoceptor agonist isoprenaline, to stimulate adenylate cyclase activity. Part of this effect may be due to an increased expression of G_s, although this has been surmised only by the enhanced cholera-toxin-catalysed ADP-ribosylation of αG_s. However, using specific antibodies, the major effect is suggested to be due to a decreased expression of G_i, occurring upon differentiation [90]. Interestingly, this change is accompanied by an increase in the expression of G_o.

In contrast to this system, transformation of a fibroblast cell line by over-expression of the normal, N-*ras* proto-oncogene did not lead to any alteration in the expression of either G_s or G_i. This was assessed using both specific antisera and by determining the patterns of ADP-ribosylation of membrane proteins using either pertussis or cholera toxins. Such a study shows that a transformed phenotype can be achieved without any changes in the expression of these major G-proteins [91]. Nevertheless, transformation of cells by either mutant *ras* oncogenes or over-expression of *ras* proto-oncogenes does lead to attenuation of receptor-stimulated adenylate cyclase activity [92]. In a detailed study of transformation induced by over-expression of the N-*ras* proto-oncogene [93], it has been shown that different effects occurred dependent upon whether cells had been transformed recently or for a prolonged period of time. Thus, upon transformation a loss of β-adrenoceptor-stimulated adenylate cyclase activity occurred, partly as a result of receptor down-regulation and partly due to uncoupling of the receptor from the adenylate cyclase system. It is tempting to attribute this to the activation of C-kinase, caused by the increased concentrations of diacylglycerol seen upon *ras* expression, because C-kinase action has been shown to inactivate the β-adrenoceptor [94] and to uncouple the glucagon receptor from stimulation of adenylate cyclase in hepatocytes [95,96]. Chronic transformation with *ras* led also to a lowered activity of adenylate cyclase itself [93]. This was suggested to be due to decreased expression of adenylate cyclase.

Such experiments add detail to a considerable literature which implies that elevation in the concentrations of cAMP can attenuate cellular growth. The recently acquired tools for investigating G-protein expression and functioning may provide insight into whether and how such proteins are affected in the wide variety of transformed states during differentiation.

Age-related changes in the functioning of hormone signalling systems have been noted in a number of systems. However, the underlying molecular details and mechanisms are far from clear. In rat adipocytes, it has been noted that ageing leads to a reduced sensitivity to catecholamines [97] and hypersensitivity to the anti-lipolytic actions of adenosine [98]. Such effects do not appear to be attributable to any change in receptor concentration, which appears similar in both mature and aged animals [97,99,100]. This might indicate that a change in the functioning of the G-protein system might provide an underlying cause. Indeed, a recent study [101], using toxin labelling to assess αG_s and αG_i, appears to have identified major changes in G_i and G_s. If the conclusions from these studies are correct then there is a three-fold increase in the amount of αG_i. This may well account for the increased action of adenosine, through A_1-receptors. However, in order for this to be established we need to know, using immunological probes, which forms of G_i are altered, as adipocytes express all forms of G_i [102]. Also it requires to be established as to which form(s) of G_i actually mediate the inhibitory effect on adenylate cyclase and couple to A_1-adenosine receptors.

This study also implied that a small increase in αG_s also occurred during ageing. Such an observation would not be consistent with the reduced stimulatory effect of β-adrenergic agonists upon adenylate cyclase. However, the reduced β-adrenoceptor action might be attributable either to increase in G_i or to some direct modification of the β-adrenoceptors which leads to their desensitization.

Changes in G-protein expression and function may thus offer a means of determining alterations in the flow of information through signal transduction systems during the ageing process. This is likely to form an intriguing and profitable area of endeavour in the future.

CONCLUSIONS

The availability of highly specific immunological probes for various G-proteins as well as synthetic oligonucleotide probes that are specific for their mRNA should allow for the unambiguous quantification of forms of G-proteins in various tissues during growth and in development as well as in various pathological states. Specific antisera should prove useful in identifying modifications of G-proteins (e.g. phosphorylation) which lead to changes in their activity and for which there is sufficient evidence implicating a role in disease states (e.g. insulin resistance). This information is likely to provide useful indications of the types and importance of particular second messenger systems to such processes. These studies will thus highlight changes that occur in aberrant systems and in pathological states.

NOTE ADDED IN PROOF

Recently, the αG_s genes from a number of GH-secreting human pituitary tumours with constitutively activated adenylate cyclase have been cloned and sequenced [103]. Screening for mutations was done by reverse transcribing the RNA from these cells, amplifying the coding region of the αG_s complementary DNA using the polymerase chain reaction technique and then subcloning the amplified DNA into an appropriate vector in order to sequence the entire coding region from multiple clones. These experiments identified two specific types of mutations in αG_s, which led to its constitutive activation. One mutation occurred at position 227 where glutamine was replaced by arginine (Q227R). This led to the formation of a G_s protein which could constitutively activate adenylate cyclase and had a reduced GTPase activity. Indeed, glutamine 227 is located at a position which is analogous to that of glutamine 61 in $p21^{ras}$ and lies in the GTP-binding domain. The cognate mutation Q61L, in $p21^{ras}$, leads to the production of an 'activated', oncogenic form of $p21^{ras}$ which also has a reduced GTPase activity. Thus G-proteins appear to share a 'common' GTP-binding domain, where cognate mutations lead to similar functional effects.

The second type of mutation occurred at position 201, where the arginine residue was mutated to either cystine (R201C) or histidine (R201H). These mutations are not within the GTP-binding domain and, indeed, there is no cognate domain to this in $p21^{ras}$. Rather, mutation is of the arginine residue which is believed to function as the target for ADP-ribosylation by cholera toxin. Indeed then, we find that either cholera toxin-mediated ADP-ribosylation or these point mutations at position 201 leads to the constitutive activation of this mutant G-protein and to a much diminished GTPase activity. Current ideas indicate that this arginine residue is crucial in maintaining a high GTPase activity for αG_s and its replacement either by another amino acid or its ADP-ribosylation by cholera toxin abolishes this effect and leads to constitutive activation.

These fascinating studies highlight the fact that functionally and structurally distinct domains of αG_s can be identified. Moreover, that precise modifications of this molecule can lead to profound functional changes. It may also be the case that αG_s can be an oncoprotein in those cell types where the increased synthesis of cyclic AMP leads to cellular proliferation, as is the case in pituitary somatotrophs. In this regard one could speculate that if an endogenous enzyme exists which is able to mimic cholera toxin's action in causing the ADP-ribosylation of αG_s in such cells, then such a species might also serve as an oncoprotein.

ACKNOWLEDGEMENTS

Work in the author's laboratory was supported by grants from the MRC, BDA, CRC, AFRC, Lipha Pharmaceuticals and California Metabolic Research Foundation.

REFERENCES

1. Gilman, A. G. (1987) *Ann. Rev. Biochem.* **56**, 615–649.
2. Houslay, M. D. (1984) *Trends Biochem. Sci.*, **9**, 39–40.
3. Birnbaumer, L., Codina, J., Mattera, R., Cerione, R. A., Hildebrandt, J. D., Sunyer, T., Rojas, F. Caron, M. J., Lefkowitz, R. J., and Iyengar, R. (1985) *Mol. Aspects Cell. Reg.*, **4**, 131–182.
4. Houslay, M. D., Dipple, I., and Elliott, K. R. F. (1980) *Biochem. J.*, **186**, 649–658.
5. Jones, D. T., and Reed, R. R. (1987) *J. Biol. Chem.*, **262**, 14241–14249.
6. Beals, C. R., Wilson, C. B., and Perlmutter, R. M. (1987) *Proc. Natl. Acad. Sci. USA*, **84**, 7886–7890.
7. Fong, H. K. W., Yoshimoto, K. K., Eversole-Cire, P., and Simon, M. I. (1988) *Proc. Natl. Acad. Sci. USA*, **85**, 3066–3070.
8. Codina, J., Olate, J., Abramowitz, J., Mattera, R., Cook, R. G., and Birnbaumer, L. (1988) *J. Biol. Chem.*, **263**, 6746–6750.
9. Yatani, A., Mattera, R., Codina, J., Graf, R., Okabe, K., Padrell, E., Iyengar, R., Brown, A. M., and Birnbaumer, L. (1988) *Nature (London)*, **336**, 680–682.
10. Logothetis, D. E., Kurachi, Y., Galper, J., Neer, E. J., and Clapham, D. E. (1987) *Nature (London)*, **325**, 321–326.
11. Kurachi, Y., Ito, H., Sugimoto, T., Shimizu, T., Miki, I., and Ui, M. (1989) *Nature (London)*, **337**, 555–557.
12. Kim, D., Lewis, D. L., Graziada, L., Neer, E. J., Bar-Sagi, D., and Clapham, D. E. (1989) *Nature (London)*, **337**, 557–560.
13. Jelsema, C. L., and Axelrod, J. (1987) *Proc. Natl. Acad. Sci. USA*, **84**, 3625–3627.
14. Houslay, M. D., Bojanic, D., Gawler, D., O'Hagan, S., and Wilson, A. (1986) *Biochem. J.*, **238**, 109–113.
15. Houslay, M. D., Bojanic, D., and Wilson, A. (1986) *Biochem. J.*, **234**, 737–740.
16. McKenzie, F. R., Kelly, E. C. H., Unson, C. G., Spiegel, A. M., and Milligan, G. (1988) *Biochem. J.*, **249**, 653–659.
17. Hildebrandt, J. D., Hanoune, J., and Birnbaumer, L. (1982) *J. Biol. Chem.*, **257**, 14723–14725.
18. Heyworth, C. M., Hanski, E., and Houslay, M. D. (1984) *Biochem. J.*, **222**, 189–194.
19. Gawler, D. J., Milligan, G., and Houslay, M. D. (1988) *Biochem. J.*, **249**, 537–542.
20. Houslay, M. D., and Heyworth, C. M. (1983) *Trends Biochem. Sci.*, **8**, 449–452.
21. Milligan, G., Streaty, R. A., Gierschik, P., Spiegel, A. M., and Klee, W. A. (1987) *J. Biol. Chem.*, **262**, 8626–8630.
22. Tsai, S.-C., Adamik, R., Kanaho, Y., Hewlett, E. L., and Moss, J. (1984) *J. Biol. Chem.*, **259**, 15320–15323.
23. Heyworth, C. M., Whetton, A. D., Wong, S., Martin, B. R., and Houslay, M. D. (1985) *Biochem. J.*, **228**, 593–603.
24. Milligan, G., and Klee, W. (1985) *J. Biol. Chem.*, **260**, 2057–2063.
25. Bourne, H. R., Farfel, Z., and Brickman, A. S. (1981) *Adv. Cyclic Nucleotide Res.*, **14**, 43–50.
26. Motulsky, H. J., Hughes, R. J., Brickman, A. S., Farfel, Z., Bourne, H. R., and Insel, P. A. (1982) *Proc. Natl. Acad. Sci. USA*, **79**, 4193–4197.
27. Levine, M. A., Ahn, T. G., Klupt, S. F., Kaufman, K. D., Smallwood, P. M., Bourne, H. R., Sullivan, K. A., and Van Dop, C. (1988) *Proc. Natl. Acad. Sci. USA*, **85**, 617–621.
28. Whetton, A. D., Needham, L., Dodd, N. J. F., Heyworth, C. M., and Houslay, M. D. (1983) *Biochem. Pharmacol.*, **32**, 1601–1608.
29. Gordon, A. S., Collier, K., and Diamond, I. (1986) *Proc. Natl. Acad. Sci. USA*, **83**, 2105–2118.
30. Luthin, G. R., and Tabakoff, B. (1984) *J. Pharmac. Exp. Therap.*, **228**, 579–587.
31. Bauche, F., Bourdeaux-Jambert, A. M., Guidicelli, Y., and Nordmann, R. (1987) *FEBS Lett.*, **219**, 296–300.
32. Dipple, I., and Houslay, M. D. (1978) *Biochem. J.*, **174**, 179–190.

33. Houslay, M. D., Dipple, I., and Gordon, L. M. (1981) *Biochem. J.*, **967**, 675–681.
34. Houslay, M. D., Gordon, L. M. (1983) *Curr. Topics Mem. Transport*, **18**, 179–231.
35. Diamond, I., Wrubel, B., Estrin, W., and Gordon, A. S. (1987) *Proc. Natl. Acad. Sci. USA*, **84**, 1413–1416.
36. Whetton, A. D., Gordon, L. M., and Houslay, M. D. (1983) *Biochem. J.*, **210**, 437–449.
37. Mochly-Rosen, D., Chang, F. H., Cheever, L., Kim, M., Diamond, I., and Gordon, A. S. (1988) *Nature (London)*, **333**, 848–850.
38. Dove, J. R., Eiden, B. F., Karanian, J. W., and Eskay, R. L. (1986) *Endocrinology*, **110**, 280–286.
39. Parent, L. J., Ehrlich, R., Matis, L., and Singer, D. S. (1987) *FASEB J.*, **1**, 469–473.
40. Vallar, L., Spada, A., and Giannattasio, G. (1987) *Nature (London)*, **330**, 566–568.
41. Haga, T., Ross, E. M., Anderson, H. J., and Gilman, A. G. (1977) *Proc. Natl. Acad. Sci. USA*, **74**, 2016–2020.
42. Rall, T., and Harris, B. A. (1987) *FEBS Lett.*, **224**, 365–371.
43. Sullivan, K. A., Miller, R. T., Masters, S. B., Beiderman, B., Heideman, W., and Bourne, H. R. (1987) *Nature (London)*, **330**, 758–760.
44. Bourne, H. R., Beiderman, B., Steinberg, F., and Brothers, V. M. (1982) *Mol. Pharmacol.*, **22**, 204–210.
45. Miller, R. T., Masters, S. M., Sullivan, K. A., Beiderman, B., and Bourne, H. R. (1988) *Nature (London)*, **334**, 712–715.
46. Hancock, J. F., Marshall, C. J., McKay, I. A., Gardner, S., Houslay, M. D., Hall, A., and Wakelam, M. J. O (1988) *Oncogene*, **3**, 187–193.
47. Davies, A. O., and Lefkowitz, R. J. (1984) *Ann. Rev. Physiol.*, **46**, 119–163.
48. Allen, D. O., and Beck, R. R. (1972) *Endocrinology*, **91**, 504–510.
49. Exton, J. H., Friedman, N., Hee-Aik, A. W., Brinmax, P., Corbin, J. D., and Park, C. R. (1972) *J. Biol. Chem.*, **247**, 3579–3588.
50. Ros, M., Northup, J. K., and Malbon, C. C. (1989) *Biochem. J.*, **257**, 737–744.
51. Stiles, G. L., Stadel, J. M., Delaen, A., and Lefkowitz, R. J. (1981) *J. Clin. Invest.*, **68**, 1450–1455.
52. Garcia-Sainz, J. A., Huerta-Baena, M. E., and Malbon, C. C. (1989) *Amer. J. Physiol.* (in press).
53. Malbon, C. C., Rapiejko, P. J., and Mangano, T. J. (1985) *J. Biol. Chem.*, **260**, 2558–2564.
54. Chang, F. H., and Bourne, H. R. (1987) *Endocrinology*, **121**, 1711–1715.
55. Mills, I., Garcia-Sainz, J. A., and Fain, J. N. (1986) *Biochim. Biophys. Acta*, **876**, 619–630.
56. Saggerson, E. D. (1986) *Biochem. J.*, **238**, 387–394.
57. Milligan, G., Spiegel, A. M., Unson, C. G., and Saggerson, E. D. (1987) *Biochem. J.*, **247**, 223–227.
58. Gawler, D., Milligan, G., Spiegel, A. M., Unson, C. G., and Houslay, M. D. (1987) *Nature (London)*, **327**, 229–232.
59. Houslay, M. D., Gawler, D. J., Milligan, G., and Wilson, A. (1989) *Cell. Signalling*, **1**, 9–22.
60. Katada, T., Gilman, A. G., Watanabe, Y., Bauer, S., and Jakobs, K. H. (1985) *Eur. J. Biochem.*, **151**, 431–437.
61. Pyne, N. J., Murphy, G. J., Milligan, G., and Houslay, M. D. (1989) *FEBS Lett.*, **243**, 77–82.
62. Crouch, M. F., and Lapetina, E. G. (1988) *J. Biol. Chem.*, **263**, 3363–3371.
63. Brooks, D. P., Nutting, D. F., Crofton, J. T., and Shore, L. (1989) *Diabetes*, **38**, 54–57.
64. Okumura, K., Akiyama, N., Hasimoto, H., Ogacia, K., and Satake, T. (1988) *Diabetes*, **37**, 1168–1172.
65. Houslay, M. D., and Siddle, K. (1989) *Brit. Med. Bull.*, **45**, 264–284.
66. Nishio, Y., Kashiwagi, A., Kida, Y., Kodama, M., Abe, N., Saeki, Y., and Shigeta, Y. (1988) *Diabetes*, **37**, 1181–1187.
67. Green, A. (1987) *J. Biol. Chem.*, **262**, 15702–15707.
68. Heyworth, C. M., and Houslay, M. D. (1983) *Biochem. J.*, **214**, 547–552.

69. Heyworth, C. M., Grey, A.-M., Wilson, S. R., Hanski, E., and Houslay, M. D. (1986) *Biochem. J.*, **235**, 145–149.
70. Elks, M. L., Manganiello, V. C., and Vaughan, M. (1983) *J. Biol. Chem.*, **258**, 8582–8587.
71. O'Brien, R. M., Houslay, M. D., Milligan, G., and Siddle, K. (1987) *FEBS Lett.*, **212**, 281–288.
72. Krupinski, J., Rajaram, R., Lakonishok, M., Benovic, J. L., and Cerione, R. A. (1988) *J. Biol. Chem.*, **263**, 12333–12341.
73. Begin-Heick, N. (1985) *J. Biol. Chem.*, **260**, 6187–6193.
74. Begin-Heick, N. (1987) *Mol. Cell. Endocrinol.*, **53**, 1–8.
75. Begin-Heick, N., and Welsh, J. (1988) *Mol. Cell. Endocrinol.*, **59**, 187–194.
76. Dehaye, J. P., Winand, J., Christophe, J. (1978) *Diabetologia*, **13**, 553–561.
77. Shepherd, R. E., Malbon, C. C., Smith, C. J., and Fain, J. N. (1977) *J. Biol. Chem.*, **252**, 7243–7248.
78. Begin-Heick, N. (1980) *Can. J. Biochem.*, **58**, 1033–1038.
79. French, R. R., and York, D. A. (1984) *Diabetologia*, **26**, 466–472.
80. Begin-Heick, N., and Coleman, D. L. (1988) *Mol. Cell. Endocrinol.*, **59**, 171–178.
81. Lacasa, D., Agli, B., and Giudicelli, Y. (1986) *FEBS Lett.*, **202**, 260–266.
82. Neumann, J., Schmitz, W., Scholz, H., Von Meyerinck, L., Doring, V., and Kalmer, P. (1988) *Lancet*, **22**, 936–937.
83. Higgin, C. B., Vatner, S. F., Eckberg, D. L., and Braunwald, E. (1972) *J. Clin. Invest.*, **51**, 715–724.
84. Ginsberg, R., Bristow, M. R., Billingham M. E., Stinson, E. B., Schroeder, J. S., and Harrison, D. C. (1983) *Amer. Heart J.*, **106**, 535–540.
85. Bristow, M. R., Kantrowitz, N. E., Ginsberg, R., and Fowler, M. B. (1985) *J. Mol. Cell Cardiol.*, **17** (suppl.2), 41.
86. Fowler, M. B., Laser, J. A., Hopkins, G. L., Minobe, W., and Bristow, M. R. (1986) *Circulation*, **74**, 1290–1302.
87. Tolkovsky, A. M., and Levitzki, A. (1978) *Biochemistry*, **17**, 3795–3810.
88. Horn, E. M., Corwin, S. J., Steinberg, S. F., Chow, Y. K., Neuberg, G. W., Cannon, P. J., Powers, E. R., and Bilezikian, J. P. (1988) *Circulation*, **78**, 1373–1379.
89. Lai, E., Rosen, O. M., and Rubin, C. S. (1981) *J. Biol. Chem.*, **256**, 12866–12874.
90. Gierschik, P., Morrow, B., Milligan, G., Rubin, C., and Spiegel, A. (1986) *FEBS Lett.*, **199**, 103–106.
91. Milligan, G., Davies, S. A., Houslay, M. D., and Wakelam, M. J. O. (1989) *Oncogene*, **4**, 101–105.
92. Colletta, G., Corda, D., Schettini, G., Cirafaci, A. M., Kohn, C. D., and Consiglio, E. (1988) *FEBS Lett.*, **228**, 37–41.
93. Davies, S-A., Housley, M. D., and Wakelam, M. J. O. (1989) *Biochim. Biophys. Acta*, **1013**, 173–179.
94. Sibley, D. R., and Lefkowitz, R. J. (1985) *Nature (London)*, **317**, 124–129.
95. Heyworth, C. M., and Houslay, M. D. (1983) *Biochem. J.*, **214**, 93–98.
96. Murphy, G. J., Hruby, V. J., Trivedi, D., Wakelam, M. J. O., and Houslay, M. D. (1987) *Biochem. J.*, **243**, 39–46.
97. Guidicelli, Y., and Pecquery, R. (1978) *Eur. J. Biochem.*, **90**, 413–419.
98. Hoffman, B. B., Chang, H., Farabakhsh, Z., and Reaven, G. (1984) *J. Clin. Invest.*, **74**, 1750–1755.
99. Dax, E. M., Partilla, J. S., and Gregerman, R. I. (1981) *J. Lipid Res.*, **22**, 934–943.
100. Hoffman, B. B., Chang, H., Farabakhsh, Z., and Reaven, G. (1984) *J. Clin. Invest.*, **247**, E772–E777.
101. Green, A., and Johnson, J. L (1989) *Biochem. J.*, **258**, 607–610.
102. Mitchell, F. M., Griffiths, S. L., Saggerson, E. D., Houslay, M. D., Knowler, J. T., and Milligan, G. (1989) *Biochem. J.*, **262**, 403–408.
103. Landis, C. A., Masters, S. B., Spada, A., Pace, A. M., Bourne, H. R., and Vallar, L. (1989) *Nature (London)*, **340**, 692–696.

INDEX

adenylate cyclase, 1
  activation mechanisms, 4, 10
  inhibition mechanisms, 6, 10
  coupling to G-proteins, 2
    in yeast, 180, 182
  coupling to receptors, 2
  dual control, 7, 8
  olfaction, 117
adrenoceptors
  coupling of beta adrenoceptors, 2, 4
ageing, 226
antibodies
  antipeptide, 34, 85, 89
  beta-subunit, 33
  functional probes, 40, 85
  G_i forms, 21
arrestin, 18

botulinum toxin, 27

Ca^{2+} channels, 90
collision coupling, 4
cholera toxin, 2, 31, 55
  GTPase action, 75
classification of G-proteins, 48
covalent modification
  myristoylation, 39
  palmitoylation, 176
  phosphorylation, 40, 220
  probes for transducin action, 101, 103

desensitization
  adenylate cyclase, 4
  inositol phospholipid response, 153
disease processes, 196
  adrenalectomy, 214
  alcoholism, 207
  heart failure, 223
  hyperthyroidism, 42, 216
  insulin-resistance, 42, 216
  oncogenes, 179, 225
  pituitary adenoma, 211
  pseudohypoparathyroidism, 206

EF-Tu, 58

G-protein
  alpha-subunits, 2, 9, 11

alternative splicing, 52
antibodies, 33, 48, 86, 92, 97
beta-gamma in transducin action, 108
beta-subunits, 2, 9, 11, 59
  association–dissociation, 8, 9
  genes, 59
blockade of Cl^- channel action, 165
Ca^{2+} channel interaction, 90
cellular localization, 37
chimaeric species, 55, 58
classification, 48
conserved sequences, 52, 100, 108, 174
evolution, 56
distribution, 48
gamma-subunits, 2, 9, 11, 59
  genes, 60
gene structure, 54
  alternative splicing, 52
genetics, 212
  S49, 212
  UNC, 212
  cyc^-, 84, 212
  H21a, 57, 212
GTP hydrolysis, 4, 68, 97, 203
guanine nucleotide binding site, 49, 57, 100, 105, 174
Mg^{2+} regulation, 70
neutrophil form, 68
oligonucleotide probes, 206
postranslational processing, 176
purification, 16
  G_o, 20
purification, 24, 177
quantification using antibodies, 36, 205, 217
structural models, 57, 58, 59, 101, 174
yeast, 61
  RAS, 179
G_i
  coupling to adenylate cyclase, 2, 200, 205
  coupling to ion channels, 90, 144
  coupling to receptors, 2, 200
  GTPase cycle, 67
  interaction with G_s, 8, 11, 199
  subtypes, 49, 86
  subunit dissociation, 7, 9, 11, 200
G_i-family, 21, 49, 86
  antibodies, 21, 86, 208

G_i-family (cont.)
  erythrocyte forms, 22
  oligonucleotide probes, 208
  sequences, 50
G_o, 20
  isoelectric point, 22
  ADP-ribosylation, 22
  purification, 20
G_s
  activating mutations, 57
  alternative splicing, 53
  chimaeras, 58
  coupling to adenylate cyclase, 2, 200, 205
  coupling to beta adrenoceptors, 12
  coupling to ion channels, 137
  coupling to receptors, 2, 200
  effector binding region, 58
  GTPase cycle, 4, 67
  interaction with G_i, 8, 11, 199
  receptor interaction site, 58
  subunit dissociation, 7, 9, 11
GTPase cycle, 4, 41, 67, 96, 201, 203
  basal activity, 76
  cholera toxin action, 75
  GAP, 184
  Mg^{2+}, 70
  pertussis toxin action, 75
  ras, 178, 184

inositol phospholipid response
  cells, 152
  olfaction, 121
  pertussis toxin action, 156
ion channels, 125
  Ca^{2+} channels, 90, 92, 129, 136
  cAMP, 119, 126
  cGMP, 118, 119, 127
  Cl^-, 165
  K^+ channels, 126, 133
    agonist activation, 135
    phosphorylation, 128
    phospholipase A2, 131

kinetics
  adenylate cyclase activation, 4, 10
  G-proteins, 68
  transducin, 96

low molecular weight species, 47, 173, 185
  activating mutations, 182
  arf, 185
  EF-Tu, 51
  interaction with growth factor receptors, 181
  post-translational processing, 176
  purification, 24, 177
  rap, 185
  ras, 164, 214, 225
    crystal structure, 58
    GAP, 168, 184
    palmitoylation, 176
    GTPase, 178, 184, 189
    diacyl glycerol generation, 183
    adenylate cyclase, 180
    phospholipase C, 183
  RAS, 179
  rho, 185
  structure, 174

myristic acid, 39

olfaction, 49, 113

palmitoylation, 176
pertussis toxin, 2, 31, 49
  GTPase action, 75
  ribosylation of alpha-43, 22
  ribosylation of G_i/G_o species, 22
phosphodiesterases
  cyclic GMP activated, 18, 95, 106
phospholipase A_2, 131
phospholipase C, 67, 121, 151, 161, 183

receptors
  agonist affinity and G-protein coupling, 78, 203
  Na ions, 79
  olfaction, 116
reconstruction
  G-proteins and phospholipase C, 165
  receptors and G_i, 86
  receptors and G_s, 86
rhodopsin, 18, 95
ROS membranes, 18

transducin, 17, 95
  antibodies, 97
  beta-gamma action, 108
  cone, 107
  phosphodiesterases
    cyclic GMP activated, 18, 95, 106
  rod, 107
  structure, 99